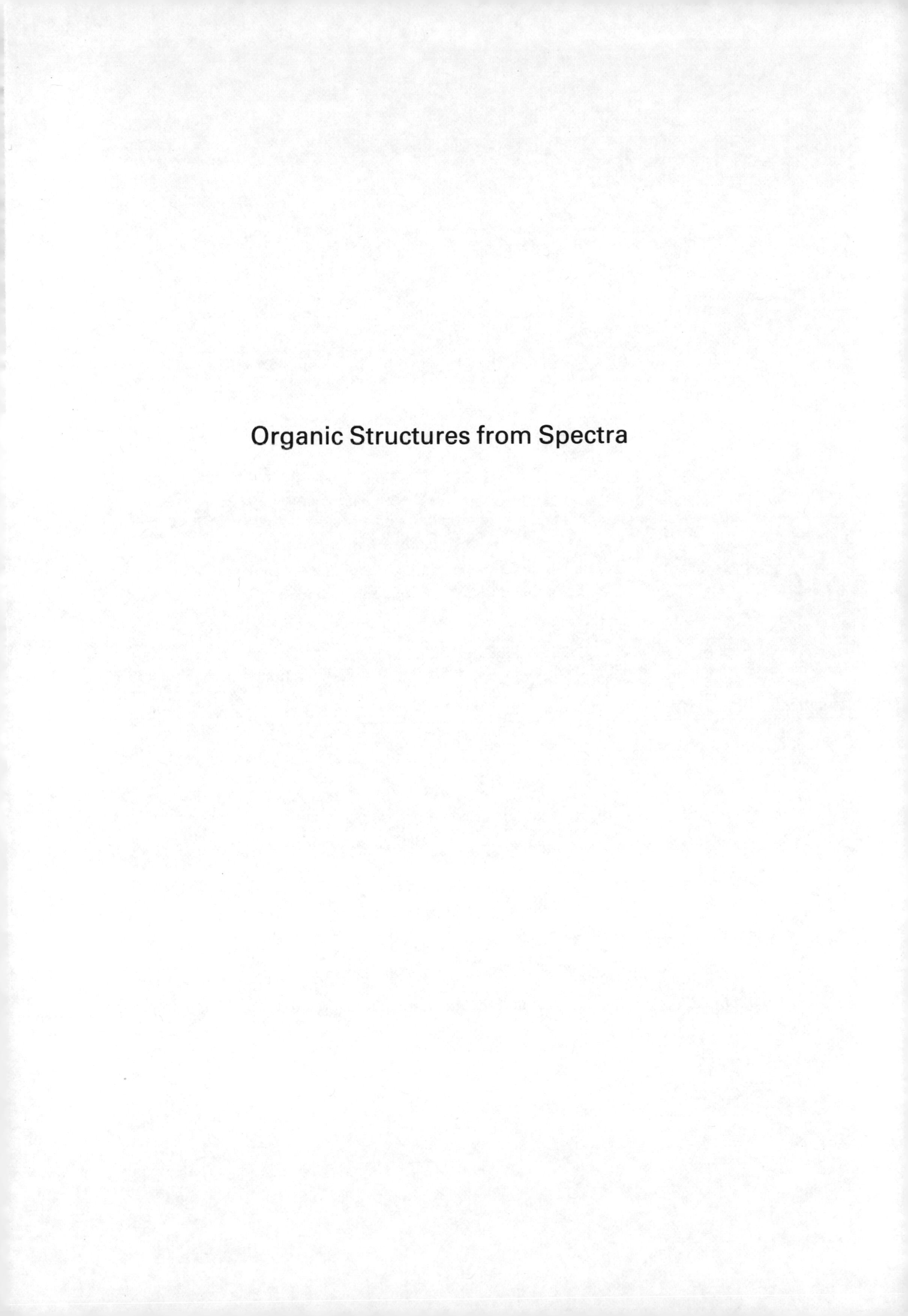

Organic Structures from Spectra

Organic Structures from Spectra

S Sternhell
Professor of Chemistry (Organic), University of Sydney, Australia
J R Kalman
Department of Chemistry, N.S.W. Institute of Technology, Sydney, Australia

JOHN WILEY & SONS LTD
Chichester · New York · Brisbane · Toronto · Singapore

Reprinted with corrections April 1987

Library of Congress Cataloging in Publication Data:

Sternhell, S.
Organic structures from spectra.
Includes index.
1. Spectrum analysis. 2. Chemistry, Organic.
I. Kalman, J. R. II. Title.
QD272.S6S74 1986 547.3'0858 84-22061
ISBN 0 471 90644 1 (cloth)
ISBN 0 471 90647 6 (paper)

British Library Cataloguing in Publication Data:

Sternhell, S.
Organic structures from spectra.
1. Chemistry, Organic 2. Chemistry, Analytic—Qualitative 3. Spectrum analysis
I. Title II. Kalman, J. R.
547.3'46 QD272.S6

ISBN 0 471 90644 1 (cloth)
ISBN 0 471 90647 6 (paper)

Printed in Great Britain

List of Contents

ORGANIC STRUCTURES FROM SPECTRA

PREFACE

The derivation of structural information from spectroscopic data is now an integral part of Organic Chemistry courses at all Universities. At the undergraduate level the principal aim of such courses is to teach students to solve simple structural problems efficiently by using combinations of the major techniques (UV, IR, NMR and MS), and over 18 years we have evolved a course at the University of Sydney, which achieves this aim quickly and painlessly. The text is tailored specifically to the needs and philosophy of this course. As we believe our approach to be successful, we hope that it may be of use in other institutions.

Our course is taught at the beginning of third year, at which stage our students have completed an elementary course of Organic Chemistry (23 lectures) in first year and a mechanistically oriented intermediate course (36 or 45 lectures) in second year. They have also been exposed in their Physical Chemistry courses to elementary spectroscopic theory, but are in general unable to relate it to the material presented in this course.

The course consists of 9 lectures outlining the theory, instrumentation and structure-spectra correlations of the major spectroscopic techniques. The text of this book corresponds to the material presented in the 9 lectures which is given in conjunction with a text book, at present the Fourth Edition of "Spectrometric Identification of Organic Compounds" by R.M. Silverstein, G.L. Bassler and T.C. Morrill (Wiley, New York, 1981), used as a reference and a source of data. Clearly, the treatment is both elementary and condensed and, not surprisingly, the students have great difficulties in solving even the simplest problems at this stage. The lectures are followed by a series of 8, 2-hour problem seminars with 4 problems being presented per seminar. At the conclusion of the course the great majority of the class is quite proficient at solving problems at the level of difficulty shown in this collection and, moreover, has achieved a satisfactory level of understanding of the elements of all methods used. Clearly, the real teaching is done during the problem seminars, which are organized in a manner modelled on that used at the E.T.H. Zurich.

The class (60 - 100 students, attendance is compulsory) is seated in a large lecture theatre in alternate rows and the problems for the day are distributed. The students are permitted to work either individually or in groups and may use any written or printed aids they desire. Staff (generally 4 or 5) wander around giving help and tuition as needed, the empty alternate rows of seats making it possible to speak to each student individually. When an important general point needs to be made, the staff member in charge gives a very brief (up to 5 minutes) exposition at the board. There is a one hour examination consisting of 3 problems and, as mentioned above, the results are in general very satisfactory.

Our philosophy can be summarised as follows: (1) Theoretical exposition must be kept to a minimum consistent with gaining of an understanding of the parts of the technique actually used in the solution of the problems. Our experience indicates that both mathematical background and advanced techniques merely confuse the average student. (2) The learning of data must be kept to a minimum. We believe that it is more important to learn to use a restricted range of data well rather than to achieve a nodding acquaintance with more extensive sets. (3) Emphasis is placed on the concept of the "structural element" and the logic needed to produce a structure out of the structural elements.

Finally, we conclude that the best way of learning to obtain "structures from spectra" is to practise on simple problems. This book was produced principally to assemble a collection of problems we consider satisfactory for our purposes.

Bona fide instructors may obtain a list of solutions by writing to the publishers.

We wish to thank Dr. Leslie D. Field for a critical reading of the manuscript and Dr. Ian Brown for the mass spectra. The professional and technical staff in the Department of Organic Chemistry, University of Sydney, helped us in many ways and thanks are also due to the graduate students who supplied us with samples of compounds used in the problems.

S. Sternhell

J.R. Kalman

July 1984

A. INTRODUCTION

1. GENERAL PRINCIPLES OF ABSORPTION SPECTROSCOPY

The principles summarized below are most obviously applicable to UV and IR spectroscopy and are simply extended to cover NMR spectroscopy. Mass Spectrometry is somewhat different.

Spectroscopy is the study of the quantized interaction of energy (typically electromagnetic energy) with matter. We deal with molecular spectroscopy (bound atoms).

A typical spectrum (Fig. A.1) is a plot of absorption or emission (here absorption) of energy (radiation) against its wave length (λ) or frequency (ν).

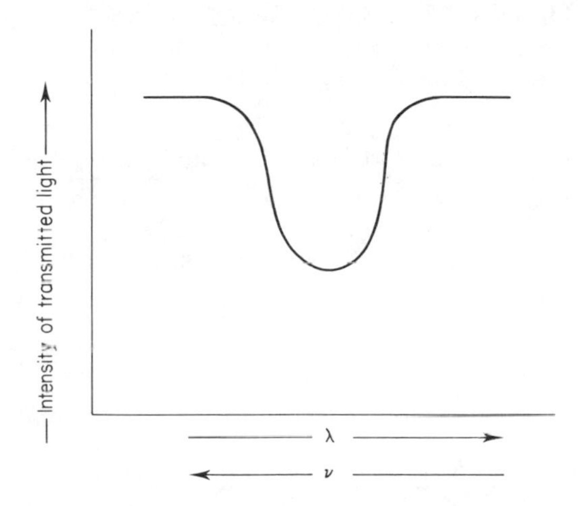

FIG. A.1 : SCHEMATIC ABSORPTION SPECTRUM

Clearly, any absorption band can therefore be primarily characterized by two parameters: the wave length at which maximum absorption occurs and the intensity of absorption at this wave length compared to base-line absorption.

For any spectroscopic transition between energy states (e.g. E_1 and E_2 in Fig. A.2), the change in energy (ΔE) is given by:

$$\Delta E = h\nu$$

where: h is the Planck's constant and ν is the frequency of the electro-magnetic energy absorbed.

Therefore: $\nu \propto \Delta E$

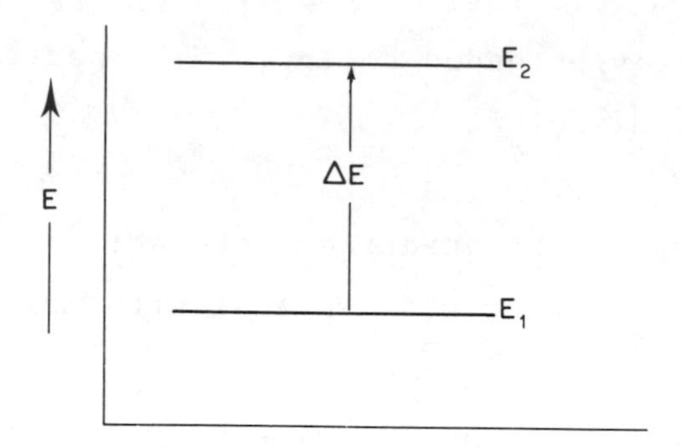

FIG. A.2 : DEFINITION OF A SPECTROSCOPIC TRANSITION

It follows that the x-axis in Fig A.1 is also an energy scale, since

$$\nu\lambda = c \quad \text{(constant speed of light)}$$

and $$\lambda = \frac{c}{\nu}$$

or $$\lambda \propto \frac{1}{\Delta E}$$

A spectrum consists of distinct bands because the absorption, or emission, of energy is quantized. Clearly the energy gap of a transition is a molecular property and is therefore characteristic of molecular structure. The y axis measures the intensity of the band which is proportional to the number of molecules observed (Beer-Lambert Law) and to the probability of the transition, which is also a molecular property.

Both the frequency and the probability of a transition can give structural information.

2. CHROMOPHORE

In general, any spectral feature, i.e. a band or group of bands, is due not to the whole molecule, but to an identifiable part of the molecule, which we shall loosely call a *chromophore*.

A chromophore may correspond to a functional group (e.g. a hydroxyl group or a carbonyl bond) but it may equally well correspond to a single atom within a molecule or to a group of atoms (e.g. a methyl group) which is not normally associated with chemical functionality.

The detection of a chromophore permits us to deduce the presence of a *structural fragment* or a *structural element* in the molecule.

The fact that it is the chromophores and not the molecules as a whole that give rise to spectral features is very fortunate. Otherwise spectroscopy would only permit us to identify known compounds by direct comparison of their spectra with authentic samples. This "fingerprint" technique is often useful, but direct determination of molecular structure is far more powerful, if more difficult.

3. CONNECTIVITY

Even if it is possible to identify enough structural elements in a molecule to account for the molecular formula, one may not be able to deduce the structural formula from this information. We could demonstrate, e.g., that a substance of molecular formula C_3H_5OCl contains the structural fragments

$$-CH_3$$
$$-Cl$$
$$>C=O$$
$$-CH_2-$$

which leaves us with two possible structures

$$CH_3-\underset{\underset{O}{\|}}{C}-CH_2-Cl \quad (1)$$

and

$$CH_3-CH_2-\underset{\underset{O}{\|}}{C}-Cl \quad (2)$$

In other words, not only the presence but the juxtaposition of structural elements must be determined. Fortunately, spectroscopy often gives valuable information concerning *connectivity* of structural elements and in the above example it would be very easy to determine whether we have a ketonic carbonyl group (as in 1) or an acid chloride (as in 2). In addition, we could independently determine whether the methyl and methylene groups are separate (as in 1) or contiguous (as in 2).

4. SENSITIVITY

Sensitivity is generally taken to signify the limits of detectability of a chromophore. However, some methods (e.g. ^{1}H NMR) detect all chromophores accessible to them with equal sensitivity while in other techniques (e.g. UV) the range of sensitivity towards different chromophores spans many orders of magnitude.

In terms of overall sensitivity, i.e. the amount of sample required, it is generally observed that: MS > UV > IR > ^{1}H NMR > ^{13}C NMR, but considerations of relative sensitivity toward different chromophores may be more important.

5. PRACTICAL CONSIDERATIONS

The "big 5" spectroscopic methods (MS, UV IR, ^{1}H NMR and ^{13}C NMR) have become established as the principal tools for the determination of the structures of organic compounds, because between them they detect a wide variety of structural elements.

The instrumentation and skills involved in the use of all five methods are now widely spread, but the ease of obtaining and interpreting the data from each method under real laboratory conditions varies greatly: thus, in very general terms:

(i) While the cost of each type of instrumentation varies greatly (NMR instruments cost between $20,000 and $500,000), as an overall guide, MS and NMR instruments are much more costly than UV and IR spectrometers. With increasing cost goes increasing difficulty in maintenance, thus compounding the total nuisance.

(ii) In routine operation the ease of usage of most UV, IR and NMR instruments is comparable (easy), but some advanced NMR spectrometers are very difficult to operate. Mass spectrometers are very seldom operated by non-specialists.

(iii) The *scope* of each method can be defined as the amount of useful information obtainable and is a function not only of the total amount of information but of interpretability. The scope varies from problem to problem and each method has its aficionados, but the overall utility undoubtedly decreases in the order

$$\text{NMR} > \text{MS} > \text{IR} > \text{UV}$$

with the combined utility of ^{1}H and ^{13}C NMR clearly outstanding.

(iv) The theoretical background needed for each method varies with the nature of the experiment, but the minimum overall amount of theory needed decreases in the order

$$\text{NMR} \gg \text{MS} > \text{UV} \sim \text{IR}$$

B. ULTRAVIOLET (UV) SPECTROSCOPY

1. BASIC INSTRUMENTATION

Basic instrumentation for both UV and IR spectroscopy consists of an energy *source*, a *sample cell*, a *dispersing device* (prism or grating) and a *detector*, arranged schematically in Fig. B.1.

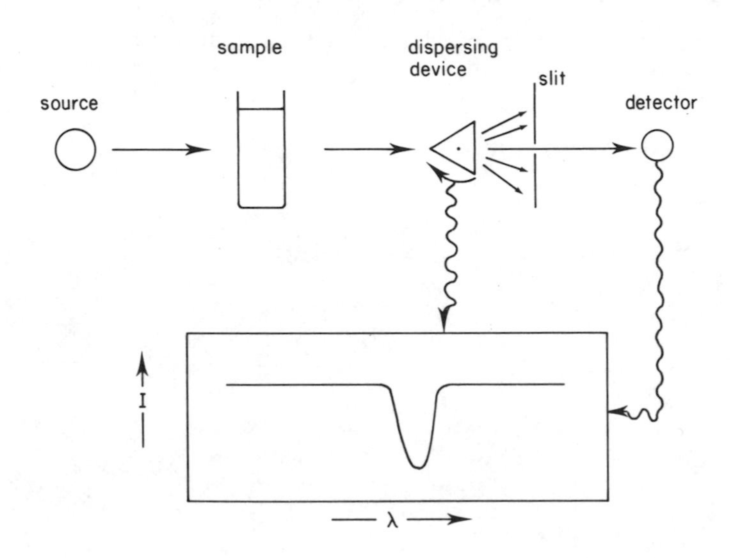

FIG. B.1 : SCHEMATIC DIAGRAM OF A UV OR IR SPECTROMETER

The drive of the dispersing device is synchronized with the x-axis of the recorder so that the latter indicates the wave length of radiation reaching the detector via the slit. The signal from the detector is transmitted to the y-axis of the recorder indicating how much radiation is absorbed by the sample at any particular wave length. In practice *double-beam* instruments are used where the absorption of a *reference cell*, containing only solvent, is subtracted from the absorption of the sample cell. Such instruments also cancel out absorption of the atmosphere in the optical path.

Clearly, the energy source, the dispersing device and the detector must be appropriate for the range of wave length scanned. The solvent (or more generally the medium) and the sample cell must be as transparent as possible in this region because, even with double-beam instruments, the amount of energy available for selective absorption by the sample must be maximized. Quartz cells and ethanol, hexane, water or dioxan are usually chosen for UV measurements.

2. NATURE OF ULTRAVIOLET SPECTROSCOPY

The term "UV spectroscopy" generally refers to *electronic transitions* occurring in the region of the electromagnetic spectrum (200-380 nm) accessible to standard UV spectrometers.

Electronic transitions are also responsible for absorption in the visible region (380-800 nm) which is easily accessible instrumentally but of less importance in the solution of structural problems, because most organic compounds are colourless. An extensive region at wave lengths shorter than $\approx$ 200 nm ("vacuum ultraviolet") also corresponds to electronic transitions, but is not readily accessible with standard instruments, where reliable measurements can usually be obtained only at wavelengths longer than 210 nm.

UV spectra used for determination of structures are invariably obtained in solution.

3. QUANTITATIVE ASPECTS

The y-axis of a UV spectrum may be calibrated in terms of the intensity of transmitted light (i.e. percentage of transmission or absorption) as is shown in Fig. B.2, or it may be calibrated on a logarithmic scale i.e. in terms of *absorptivity* (A) defined in Fig. B.2.

Absorptivity is proportional to concentration and path length (Beer-Lambert Law). For purposes of characterization of a chromophore, intensity of absorption is expressed in terms of *molar absorptivity* (ϵ) given by

$$\epsilon = \frac{M\ A}{C\ l}$$

where M is the molecular weight, C the concentration (in grams per litre) and l the path length through the sample in centimeteres.

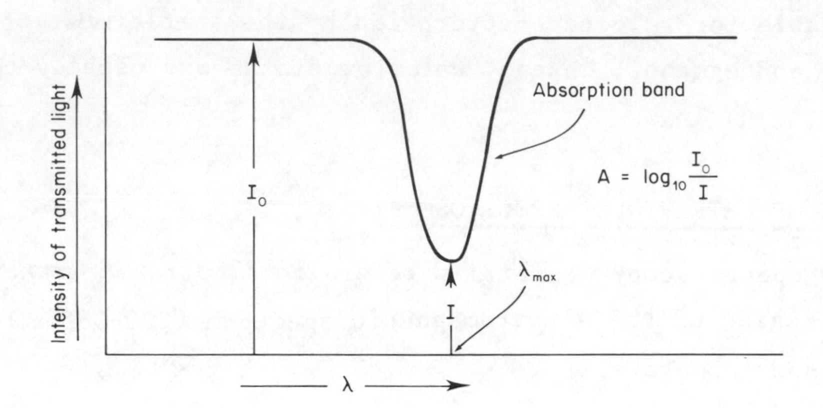

FIG. B.2 : DEFINITION OF ABSORPTIVITY (A)

UV absorption bands (Fig. B.2) are thus characterized by the wave length of the absorption maximum (λ_{max}) and ϵ.

The values of ϵ associated with commonly encountered chromophores vary between 10 and 10^5. The presence of strongly absorbing impurities may therefore lead to erroneous conclusions.

4. CLASSIFICATION OF UV ABSORPTION BANDS

UV absorption bands have fine structure due to the presence of vibrational sub-levels, but this is rarely observed in solution due to collisional broadening. As the transitions are associated with changes of electron orbitals, they are often described in terms of the orbitals involved, e.g.

$$\sigma \rightarrow \sigma^*$$
$$\pi \rightarrow \pi^*$$
$$n \rightarrow \pi^*$$
$$n \rightarrow \sigma^*$$

where n denotes a non-bonding orbital, the asterisk denotes an antibonding orbital and σ and π have the usual significance.

Another method of classification uses the symbols:

B (for benzenoid)
E (for ethylenic)
R (for radical-like)
K (for conjugated - German "konjugierte")

A molecule may give rise to more than one band in its UV spectrum, either because it contains more than one chromophore or because more than one transition of a single chromophore is observed. Thus the ultraviolet spectrum of acetophenone in ethanol contains 3 easily observed bands:

	λ_{max}	ϵ	Assignment	
Ph-CO-CH_3	244	12,600	$\pi \rightarrow \pi^*$	K
	280	1,600	$\pi \rightarrow \pi^*$	B
	317	60	$n \rightarrow \pi^*$	R

However, UV spectra typically contain fewer features (bands) than IR, MS and NMR spectra and therefore have a lower information content.

5. SPECIAL TERMS

Auxochromes (auxilliary chromophores) are groups which have little UV absorption by themselves, but which often have significant effects on the absorption (both λ_{max} and ϵ) of a chromophore to which they are attached. Generally, auxochromes are atoms with one or more lone pairs e.g. -OH, -OR, -NR_2, -halogen.

If a structural change, such as the attachment of an auxochrome, leads to the absorption maximum being shifted to a longer wave length, the phenomenon is termed a *bathochromic shift*. A shift towards shorter wave length is called a *hypsochromic shift*.

6. IMPORTANT CHROMOPHORES

Most of the reliable and useful data is due to relatively strongly absorbing chromophores ($\epsilon > 200$) which are mainly indicative of conjugated or aromatic systems. Examples listed below encompass most of the commonly encountered effects.

(i) Dienes and Polyenes

Extention of conjugation in a carbon chain is always associated with a pronounced shift towards longer wave length, and usually towards greater intensity. Thus:

	λ_{max}(nm)	ϵ
$CH_2=CH_2$	165	10,000
$CH_2=CH-CH=CH_2$	217	20,000
$CH_2=CH-CH=CH-CH=CH_2$ *trans*	263	53,000
$CH_3-(CH=CH)_5-CH_3$ *trans*	341	126,000

Configuration and substitution also influence absorption by the diene chromophore. Thus:

λ_{max} = 214 nm ϵ = 16,000

λ_{max} = 253 nm ϵ = 8,000

(ii) Carbonyl compounds

All carbonyl derivatives exhibit weak ($\epsilon < 100$) absorption between 200 and 300 nm, which is of marginal utility in the determination of structure. However, conjugated carbonyl derivatives always exhibit strong absorption. Thus, in ethanol solution:

	λ_{max}(nm)	ϵ
$CH_2=CH-CHO$	207 328	12,000 20
$CH_3-CH=CH-C(=O)-CH_3$ *trans*	221 312	12,000 40
$(CH_3)_2C=CH-C(=O)-CH_3$	238 316	12,000 60

Empirical rules (Woodward's Rules) of good predictive value are available for the prediction of the positions of the absorption maxima in conjugated olefins and conjugated carbonyl compounds.

(iii) Benzene Derivatives

Benzene derivatives exhibit medium to strong absorption in the UV region.

These bands usually have characteristic fine structure, which is of diagnostic value and both the position and the intensity of the absorption maxima are strongly influenced by substituents. Examples listed below include weak auxochromes ($-CH_3$, -Cl, -OMe), groups which increase conjugation ($-CH=CH_2$, $-\overset{O}{\overset{\|}{C}}-R$, $-NO_2$) and auxochromes whose absorption is pH dependent ($-NH_2$ and -OH).

Compound	λ_{max} (nm)	ϵ
Benzene	184	60,000
	204	7,900
	256	200
Toluene	208	8,000
	261	300
Chlorobenzene	216	8,000
	265	240
Anisole	220	8,000
	272	1,500
Styrene	244	12,000
	282	450
Acetophenone	244	12,600
	280	1,600
Nitrobenzene	251	9,000
	280	1,000
	330	130
Aniline	230	8,000
	281	1,500
Anilinium ion	203	8,000
	254	160
Phenol	211	6,300
	270	1,500
Phenolate ion	235	9,500
	287	2,500

The striking changes in the ultraviolet spectra accompanying protonation of aniline and phenolate ion are exactly those expected in view of loss or reduction of overlap between lone pairs and the π-system of benzene, which may be expressed in the usual Valence Bond terms as e.g.:

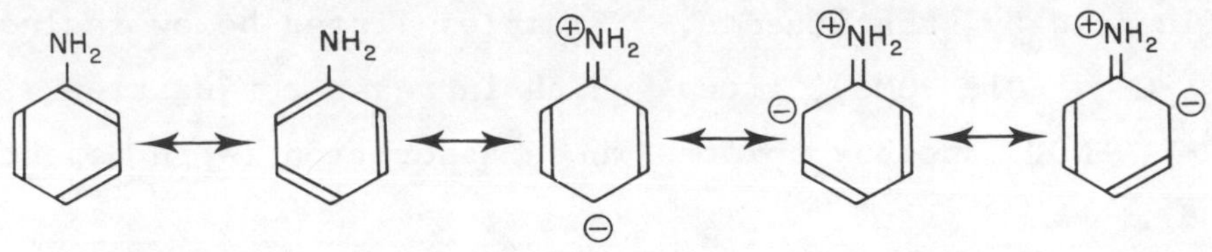

The overlap of lone pairs of auxochromes may also be reduced by steric effects and be reflected in the ultraviolet spectrum. Thus:

λ_{max} = 251 nm ϵ = 15,000

λ_{max} = 248 nm ϵ = 6,360

7. EFFECT OF SOLVENT

Solvent polarity may affect the absorption characteristics, in particular λ_{max}, since the polarity of a molecule usually changes when an electron is moved from one orbital to another. Solvent effects of up to 20 nm may be observed with carbonyl compounds. Thus the $n \rightarrow \pi^*$ absorption of acetone occurs at 279 nm in n-hexane, 270 nm in ethanol, and at 265 nm in water.

C. INFRARED (IR) SPECTROSCOPY

1. ABSORPTION RANGE AND NATURE OF TRANSITIONS RESPONSIBLE FOR ABSORPTION

Infrared absorption spectra are calibrated in wave lengths expressed in micrometres:

$$1\,\mu m = 10^{-6} m$$

or in frequency-related *wave numbers* (cm^{-1}) which are reciprocals of wave lengths. Thus:

$$\text{Wave number } (\bar{\nu} \text{ in } cm^{-1}) = \frac{1}{\text{wave length in } \mu m} \times 10^4$$

The range accessible for standard instrumentation is usually:

$$\bar{\nu} = 4000 \quad \text{to} \quad 666 \quad cm^{-1}$$

or

$$\lambda = 2.5 \quad \text{to} \quad 15 \quad \mu m$$

Infrared absorption intensities are rarely described quantitatively, except for the general classifications of s(strong), m(medium) or w(weak).

The transitions responsible for IR bands are due to molecular *vibrations* i.e. to periodic motions involving stretching or bending of bonds. Polar bonds are associated with strong IR absorption while symmetrical bonds may not absorb at all.

Clearly the vibrational frequency, i.e. the position of the IR bands in the spectrum, will depend on the nature of the bond. Thus shorter and stronger bonds will have their stretching vibrations at the higher energy end (shorter wave length) of the IR spectrum than the longer and weaker bonds. Similarly, bonds to lighter atoms (e.g. hydrogen), vibrate at higher energy than bonds to heavier atoms.

IR bands have rotational sub-structure, but this is normally resolved only in spectra taken in the gas phase.

2. EXPERIMENTAL ASPECTS

The basic layout of IR spectrometers is the same as for UV spectrometers (Fig. B.1), except that all components must now match the different energy range of electromagnetic radiation.

Very few substances are transparent over the whole of the IR range: sodium and potassium chlorides and bromides are the most common. Thus cells used for obtaining IR spectra in solution have NaCl windows and liquids can be examined as films on NaCl plates. Solution spectra are generally obtained in chloroform or carbon tetrachloride but this leads to loss of information at long wave length where there is considerable absorption of energy by the solvent. Alternatively, organic solids may be examined as mulls (fine suspensions) in heavy oils (which absorb in certain regions) or as dispersions in KBr or KCl discs.

To a first approximation, the absorption frequencies due to the important IR chromophores are the same in solid and liquid states.

3. GENERAL FEATURES OF IR SPECTRA

As molecules generally consist of large assemblages of bonds and as each bond may have several IR-active *vibrational modes*, IR spectra have many features.

The characteristic frequencies of many such vibrations are strongly influenced by small changes in molecular structure, thus making it difficult to deduce the presence of structural fragments from IR data.

The above effects make IR spectra very useful for identifying compounds by direct comparison with spectra from authentic samples (*"fingerprinting"*), but limit their usefulness in deducing structures from spectroscopic data. Fortunately, some groups of atoms form chromophores which are readily recognized from IR spectra.

An IR chromophore will tend to be useful for the determination of structure if it meets some of the following criteria:

(i) The chromophore should not absorb in the most crowded region of the spectrum (600-1400 cm^{-1}) where strong overlapping stretching absorptions from C-X single bonds (X = O, N, S, P and halogens) make assignment difficult.

(ii) The chromophores should be strongly absorbing to avoid confusion with weak harmonics. However, in otherwise empty regions e.g. 1800-2500 cm^{-1}, even weak absorptions can be assigned with confidence.

(iii) The absorption frequency must be structure dependent in an interpretable manner. This is particularly true of the very important bands due to the C=O stretching vibrations, which generally occur between 1630 and 1850 cm^{-1}.

4. IMPORTANT CHROMOPHORES

(i) -O-H Stretch

Not hydrogen-bonded ("free") 3600 cm^{-1}

Hydrogen-bonded 3100-3200 cm^{-1}

This shift is clearly related to the weakening of the O-H bond as a consequence of the hydrogen atom participating in hydrogen bonding.

(ii) Carbonyl groups always give rise to strong absorption between 1630 and 1850 cm^{-1} due to C=O stretching vibrations. Moreover various types of carbonyl compounds (ketones, esters etc.) are associated with well defined regions of absorption, the most important of which are listed below:

	$\bar{\nu}$ cm^{-1}
Ketones	1700 - 1725
Aldehydes	1720 - 1740
Aryl or α,β-unsaturated aldehydes or ketones	1660 - 1715
Cyclopentanones	1740 - 1750
Cyclobutanones	1760 - 1780
Carboxylic acids	1700 - 1725
Esters	1735 - 1750
Aryl or α,β-unsaturated esters	1715 - 1730
δ-lactones	1735 - 1750

γ-lactones	1760 - 1780
Amides	1630 - 1690
Acid chlorides	1785 - 1815
Acid anhydrides (two bands)	1740 - 1850

Even though the ranges for individual types often overlap, it may be possible to make a definite decision from information derived from other regions of the IR spectrum. Thus esters also exhibit strong C-O stretching absorption between 1200 and 1300 cm^{-1} while carboxylic acids exhibit O-H stretching absorption generally near 3000 cm^{-1}.

The characteristic shift toward lower frequency associated with the introduction of α,β-unsaturation can be rationalized by considering the Valence Bond description of an enone:

$$\underset{\underline{a}}{>C=C<\;\; >C=O} \longleftrightarrow \underset{\underline{b}}{>C=C<\;\; >C^{\oplus}-O^{\ominus}} \longleftrightarrow \underset{\underline{c}}{{}^{\oplus}>C-C<\;\; =C-O^{\ominus}}$$

The additional structure c, which cannot be considered for an unconjugated carbonyl derivative, implies that the carbonyl band in an enone has more single bond character and is therefore weaker.

The involvement of a carbonyl group in hydrogen bonding reduces the frequency of the carbonyl stretching vibration by about 10 cm^{-1}. This can be rationalized in a manner anologous to that proposed above for free and H-bonded O-H vibrations.

(iii) Chromophores Absorbing in the Region Between 1900 and 2600 cm^{-1}

The absorptions listed below often yield useful information because, though usually of only weak or medium intensity, they occur in regions largely devoid of absorption by commonly occurring chromophores.

-C≡C-	2100 - 2300 cm^{-1}
-C≡N	~ 2250 cm^{-1}
-N=C=O	~ 2270 cm^{-1}
-N=C=S	~ 2150 cm^{-1} (broad)
>C=C=C<	~ 1950 cm^{-1}

(iv) Miscellaneous polar functional groups can often be identified with the aid of IR data. This is particularly useful for groups not containing magnetic nuclei and thus not identifiable by NMR spectroscopy.

Some of the most common of these groups are listed below:

		cm^{-1}
-N-H		3300 - 3500
>C=N-		1480 - 1690
$-\overset{+}{N}(=O)O^-$	and	1500 - 1650 (s) 1250 - 1400 (m)
>S=O		1010 - 1070 (s)
O=S=O	and	1300 - 1350 (s) 1100 - 1150 (s)
$-SO_2-N<$ $-SO_2-O-$	and	1140 - 1180 (s) 1300 - 1370 (s)
-C-F		1000 - 1400 (s)
-C-Cl		580 - 780 (s)
-C-Br		560 - 800 (s)
-C-I		500 - 600 (s)

While the essentially symmetrical carbon-carbon double bonds have only weak infrared absorption, the more polar carbon-carbon double bonds in enol ethers and enones may absorb strongly between 1600 and 1700 cm^{-1}. The N-H bonding absorption in amides and sulphonamides also occurs in this region.

D. MASS SPECTROMETRY

Mass spectrometry is based on a single principle: it is possible to determine the mass of an ion in the gas phase. The mass spectrum thus consists of a plot of masses of ions against their relative abundance.

Strictly speaking, it is only possible to obtain the mass/charge ratio (m/e), but as multicharged ions are very much less abundant than those with a single electronic charge (e = 1), m/e is for all practical purposes equal to the mass of the ion, m.

The principal experimental problems thus consist of volatilizing the substance, which implies high vacuum, and of ionizing the neutral molecules.

1. IONIZATION PROCESSES

The most common method involves *electron impact* and there are two general courses of events following a collision of a molecule M with an electron e. By far the most probable event involves electron ejection which yields an odd-electron positively charged *cation radical* $[M]^{+\cdot}$ of the same mass as the initial molecule M [equation (i)].

$$M + e \longrightarrow [M]^{+\cdot} + 2e \qquad \text{(i)}$$

This cation radical is known as the *molecular ion* and its mass gives a direct measure of the molecular weight of a substance. An alternative, far less probable process also takes place and it involves the capture of an electron to give a negative *anion radical*, $[M]^{-\cdot}$

$$M + e \longrightarrow [M]^{-\cdot} \qquad \text{(ii)}$$

Electron impact mass spectrometers are generally set up to detect only positive ions, but negative-ion mass spectrometry is also possible.

The most important ionization method, besides electron impact, is *chemical ionization* where an intermediate substance (generally methane) is introduced at a higher concentration than that of the substance being investigated. The carrier gas is ionized by electron impact and the substance is then ionized by collisions with these ions.

All subsequent discussion is limited to positive-ion electron-impact mass spectrometry.

The energy of the electron responsible for the ionization process [equation (i)] can be varied. It must be sufficient to knock out an electron and this threshold, typically about 10-12 eV, is known as the *appearance potential*. In practice much higher energies (~70 eV) are used and this large excess energy (1 eV = 95 kJ mol^{-1}) causes further *fragmentation* of the molecular ion.

The two important types of fragmentation are:

$$[M]^{+\cdot} \longrightarrow A^{+} \text{ (even electron cation) } + B^{\cdot} \text{ (radical)}$$

or

$$[M]^{+\cdot} \longrightarrow C^{+\cdot} \text{ (cation radical) } + D \text{ (molecule)}$$

As only species bearing a positive charge will be detected, the mass spectrum will show signals due not only to $[M]^{+\cdot}$ but also due to A^{+}, $C^{+\cdot}$ and to fragment ions resulting from subsequent fragmentation of A^{+} and $C^{+\cdot}$.

As any species may fragment in a variety of ways, the typical mass spectrum consists of many signals.

2. INSTRUMENTATION

In the most common type of mass spectrometer (Fig. D.1) the positively charged ions of mass m and charge e (generally e = 1) are subjected to an accelerating voltage V and passed through a magnetic field H which causes them to be deflected into a circular path of radius r. These quantities are connected by the relationship given in equation (iii)

$$\frac{m}{e} = \frac{H^2 r^2}{2V} \qquad \text{(iii)}$$

The values of H and V are known, that of r is determined experimentally and e is assumed to be unity thus permitting us to determine the mass m.

In practice the magnetic field is scanned so that streams of ions of different mass pass sequentially through the slit before striking the detecting system (ion collector).

The whole system (Fig. D.1) is under high vacuum (10^{-6} Torr) to permit the volatilization of the sample and requires a complex inlet system.

The magnetic scan is synchronized with the x-axis of a recorder and calibrated to appear as *mass number* (strictly m/e). The amplified current from the ion collector gives the relative abundance of ions on the y-axis.

The signals are usually pre-processed by a computer which assigns a relative abundance of 100% to the strongest peak (*base peak*). All mass spectra in this collection of examples are of this type.

3. MASS SPECTRAL DATA

(i) Mass of ions can be easily determined to the nearest unit value. Thus the position of $[M]^{+\cdot}$ gives a direct measure of molecular weight. By use of *double focussing*, a mass of any ion can be determined to approximately $\pm$ 0.0001 of a mass unit. E.g. one can distinguish between CO and N_2:

	Low Resolution	High Resolution
CO	28 (12 + 16)	27.9949
N_2	28 (14 + 14)	28.0062

By simple extension of this example it can be shown that by examining a mass spectrum at sufficient resolution one can obtain the composition of each ion. Most importantly the *composition of* $[M]^{+\cdot}$ *gives the molecular formula of the compound.*

Because most elements occur as mixtures of isotopes, nearly all ion peaks are accompanied by satellite peaks. It is particularly easy to locate peaks at higher masses than the parent ion $M^{+\cdot}$, because they cannot be confused with peaks due to fragment ions.

In principle it is possible to calculate the molecular formula of a compound from the exact ratios of abundances of $[M]^{+\cdot}$, $[M+1]^{+\cdot}$, $[M+2]^{+\cdot}$ etc. because the contribution of various heavier isotopes differs for various elements. In practice this is rarely used but the presence of those elements (see Table D.1) which contain significant proportions ($\geq$ 1%) of minor isotopes can often be demonstrated by inspection.

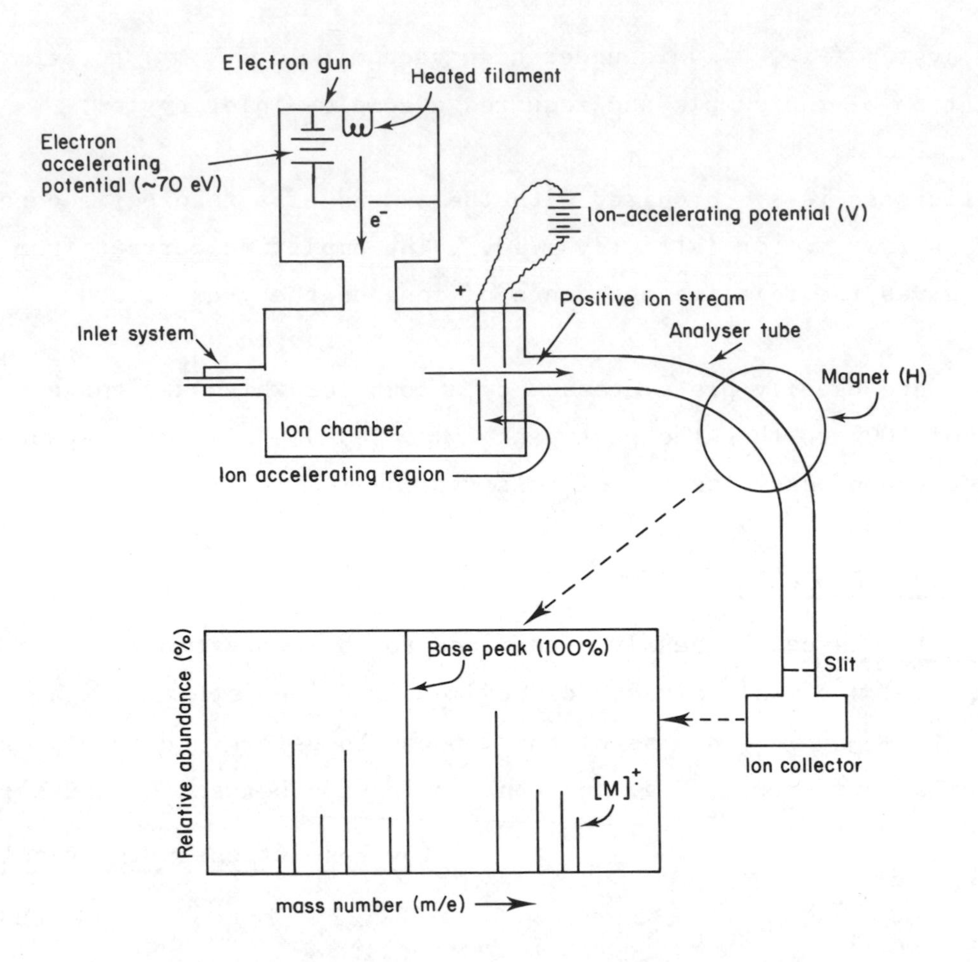

FIG. D.1 : SCHEMATIC DIAGRAM OF AN ELECTRON-IMPACT MASS SPECTROMETER

(ii) Fragmentation Pattern

The mass spectrum (Fig. D.1) consists, in addition to the molecular ion peak, of a number of peaks at lower mass number (Section D.1). The principles determining the mode of fragmentation are reasonably well-understood, (Section D.5 and D.6) and it is possible to derive structural information from the fragmentation pattern in several ways. Firstly, the appearance of prominent peaks at certain mass numbers is empirically correlatable with certain structural elements (Table D.2), e.g. a prominent peak at m/e = 43 is a strong indication for the presence of a $CH_3-\underset{\overset{\|}{O}}{C}-$ fragment.

TABLE D.1

MASSES OF SELECTED ISOTOPES

Isotope	Relative Abundance	Mass
^{12}C	100	12.0000
^{13}C	1.08	13.00336
^{32}S	100	31.9721
^{33}S	0.78	32.9715
^{34}S	4.40	33.9679
^{35}Cl	100	34.9689
^{37}Cl	32.5	36.9659
^{79}Br	100	78.9183
^{81}Br	98	80.9163

Clearly, this type of information is analogous to the interpretation of IR spectra.

In the case of the mass spectrum, however, information can also be obtained from *differences* between the masses of two peaks. Thus a prominent fragment ion occurring 15 mass numbers below the molecular ion suggests strongly that a loss of a methyl group has occurred and therefore that a methyl group was present in the substance examined.

Secondly, the knowledge of the principles governing the mode of fragmentation of ions makes it possible to confirm the structure assigned to a compound and, quite often, to determine the juxtaposition of structural fragments and thus to distinguish between isomeric substances.

For example, the mass spectrum of benzyl methyl ketone, $Ph\text{-}CH_2\text{-}\underset{O}{\overset{}{\underset{\|}{C}}}\text{-}CH_3$ contains a strong peak at m/e = 91 due to the stable ion $Ph\text{-}CH_2^{\oplus}$, but this ion is absent in the mass spectrum of the isomeric propiophenone $Ph\text{-}CO\text{-}CH_2CH_3$ where the structural elements Ph- and $-CH_2-$ are separated. Instead, a prominent peak occurs at m/e = 105 due to the presence of the structural element $Ph-\underset{O}{\underset{\|}{C}}-$.

TABLE D.2

MASSES OF COMMONLY OCCURRING FRAGMENT IONS

29	$-\overset{O}{\overset{\|}{C}}-H$, $-CH_2CH_3$	57	C_4H_9, $-\underset{O}{\underset{\|}{C}}-CH_2CH_3$
30	NO	60	$CH_2=\overset{OH}{\overset{\mid}{C}}-OH$
31	$-CH_2OH$	65	C_5H_5
39	C_3H_3	77	$-C_6H_5$
41	$-CH_2-CH=CH_2$	91	$-CH_2-C_6H_5$
43	$-C_3H_7$, $-\overset{O}{\overset{\|}{C}}-CH_3$	92	$C_5H_4N-CH_2$
45	$-\overset{O}{\overset{\|}{C}}-OH$	105	$-\overset{O}{\overset{\|}{C}}-C_6H_5$
46	$-NO_2$	127	I
55	C_4H_7		
56	C_4H_8		

(iii) Meta-stable Peaks

If the fragmentation process

$$a^+ \longrightarrow b^+ \quad + \quad c \quad \text{(neutral)}$$

takes place in the ion-accelerating region of the mass spectrometer (Fig. D.1) ion peaks corresponding to the masses of a^+ and b^+ (m_a and m_b) may be accompanied by a broader peak at mass m*, such that

$$m^* = \frac{m_b^2}{m_a}$$

This often permits positive identification of a particular fragmentation path.

4. REPRESENTATION OF FRAGMENTATION PROCESSES

As fragmentation reactions in a mass spectrometer involve the breaking of bonds, they can be represented by the standard "arrow notation" of organic chemistry. For some purposes a radical cation (e.g. a generalized ion of the molecular ion corresponding to an alkyl-ethyl ether) can be represented without an attempt to localize the missing electron:

$$[M]^{+\cdot} \qquad \text{or} \qquad [H_3C\text{-}CH_2\text{-}O\text{-}R]^{+\cdot}$$

However, to show a fragmentation process it is generally necessary to indicate "from where the electron is missing" even though no information about this exists. Thus in the case of the molecular ion corresponding to an alkyl-ethyl ether, it can be reasonably inferred that the missing electron resided on the oxygen.

This done, the application of standard arrow notation permits us to represent a commonly observed process, viz. the loss of 15 units of mass from this molecular ion:

$$H_3C\text{-}CH_2\text{-}\ddot{O}^{\oplus}\text{-}R \longrightarrow H_3\dot{C} \quad H_2C{=}O^{\oplus}\text{-}R \longleftrightarrow H_2C^{\oplus}\text{-}O\text{-}R$$

5. FACTORS GOVERNING FRAGMENTATION PROCESSES

Three factors dominate the fragmentation processes.

(i) Weak bonds tend to be broken

(ii) Stable fragments (not only ions, but also the accompanying radicals and molecules) tend to be formed.

(iii) Some fragmentation processes depend on the ability of molecules to assume cyclic transition states.

Favourable fragmentation processes naturally occur more often and ions thus formed give rise to strong peaks in the mass spectrum.

6. EXAMPLES OF COMMON TYPE OF FRAGMENTATION PROCESSES

(i) Cleavage of aliphatic carbon skeletons at branch points is favoured as it leads to the formation of more substituted carbonium ions.

E.g.

$$H_3C\text{-}C(CH_3)_2\text{-}CH_2\text{-}CH_2\text{-}CH_3 \xrightarrow{-e} H_3C\overset{+}{\cdot}\,C(CH_3)_2\text{-}CH_2\text{-}CH_2\text{-}CH_3 \longrightarrow H_3C\cdot \quad \oplus C(CH_3)_2\text{-}CH_2\text{-}CH_2\text{-}CH_3$$

(ii) Cleavage tends to occur β to heteroatoms, double bonds and aromatic rings because delocalized carbonium ions result in each case.

(a)

$$R-\ddot{X}-\overset{|}{\underset{|}{C}}-\overset{|}{\underset{|}{C}}- \xrightarrow{-e^-} R-\overset{\oplus\cdot}{X}-\overset{|}{\underset{|}{C}}-\overset{|}{\underset{|}{C}}- \longrightarrow R-\ddot{X}-\overset{\oplus}{C}< \;\longleftrightarrow\; R-\overset{\oplus}{X}=C< \quad \cdot C\lessdot$$

X = O, N, S, halogen

(b)

$$>C=C<\text{-}\overset{|}{\underset{|}{C}}-\overset{|}{\underset{|}{C}}- \xrightarrow{-e^-} >\overset{\oplus}{C}-\dot{C}<\text{-}\overset{|}{\underset{|}{C}}-\overset{|}{\underset{|}{C}}- \longrightarrow >C=C<\text{-}\overset{\oplus}{C}< \;\longleftrightarrow\; >\overset{\oplus}{C}-C=C< \quad \cdot\overset{|}{\underset{|}{C}}-$$

(c)

$$C_6H_5-\overset{|}{\underset{|}{C}}-\overset{|}{\underset{|}{C}}- \xrightarrow{-e^-} C_6H_5-\overset{\oplus}{C}\cdot\overset{|}{\underset{|}{C}}- \longrightarrow (\text{resonance structures of } C_6H_5\overset{\oplus}{C}<,\ \text{positive charge at ortho, para and benzylic positions}) \quad \cdot\overset{|}{\underset{|}{C}}-$$

(iii) Cleavage tends to occur α to carbonyl groups to give stable acylium cations. R may be an alkyl, -OH or -OR group.

$$-\overset{|}{\underset{|}{C}}-\overset{O}{\overset{\|}{C}}-R \xrightarrow{-e^-} -\overset{|}{\underset{|}{C}}-\overset{\cdot\overset{+}{O}\cdot}{\overset{\|}{C}}-R \longrightarrow -\overset{|}{\underset{|}{C}}\cdot \quad \overset{\oplus}{O}\equiv C-R \longleftrightarrow O=\overset{\oplus}{C}-R$$

(iv) Cleavage may also occur α to heteroatoms, e.g. in the case of ethers:

$$-\overset{|}{\underset{|}{C}}-\ddot{\underset{..}{O}}-R \xrightarrow{-e^-} -\overset{|}{\underset{|}{C}}-\overset{\cdot +}{\underset{..}{O}}-R \longrightarrow -\overset{|}{\underset{|}{C}}\oplus \quad \cdot O-R$$

(FREE RADICAL)

(v) Cyclohexene derivatives may undergo a retro Diels-Alder reaction:

$-e^-$

$+\cdot$

OR

(vi) Compounds where the molecular ion can assume the appropriate 6-membered cyclic transition state usually undergo a cyclic fragmentation, known as the McLafferty rearrangement. This involves a transfer of a γ hydrogen atom to an oxygen and is often observed with ketones, acids and esters:

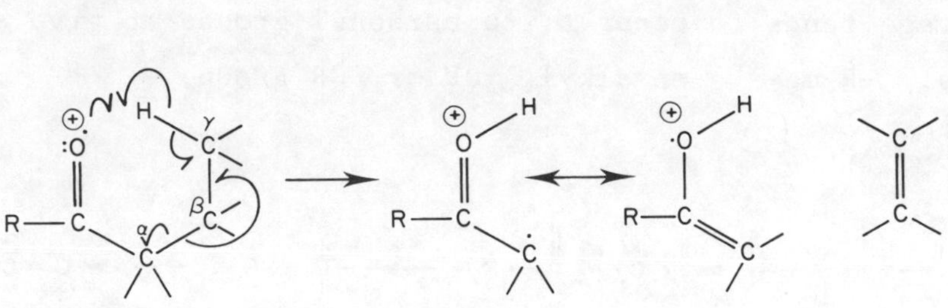

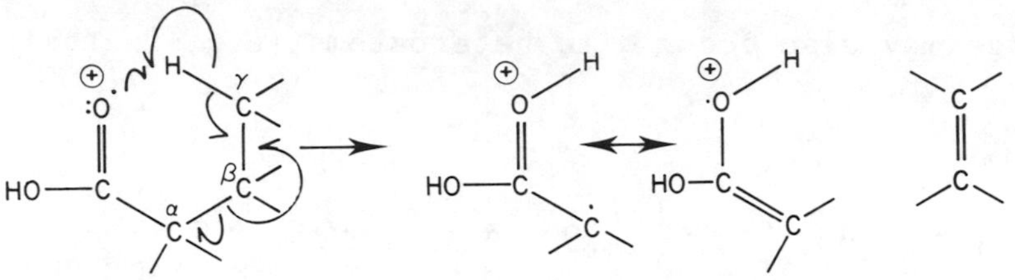

With primary carboxylic acids $R-CH_2-\underset{\underset{O}{\|}}{C}-OH$, this fragmentation leads to a characteristic peak at m/e = 60

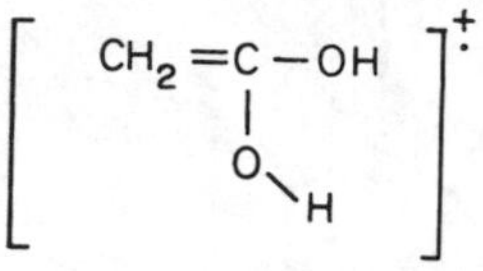

$$\left[\begin{array}{c} CH_2=C-OH \\ | \\ O-H \end{array}\right]^{+\cdot}$$

With carboxylic esters, two types of McLafferty rearrangements may be observed:

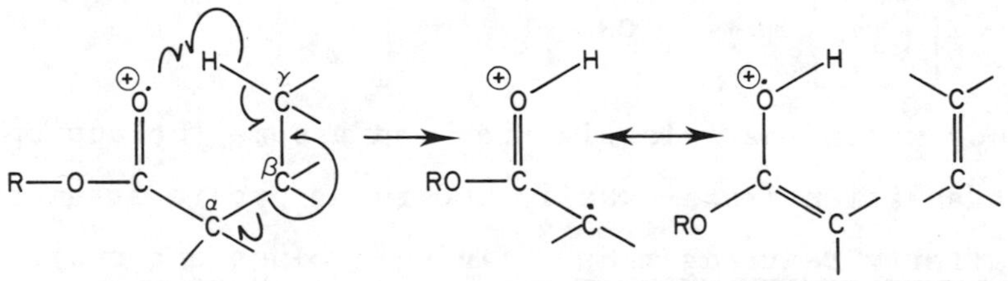

and

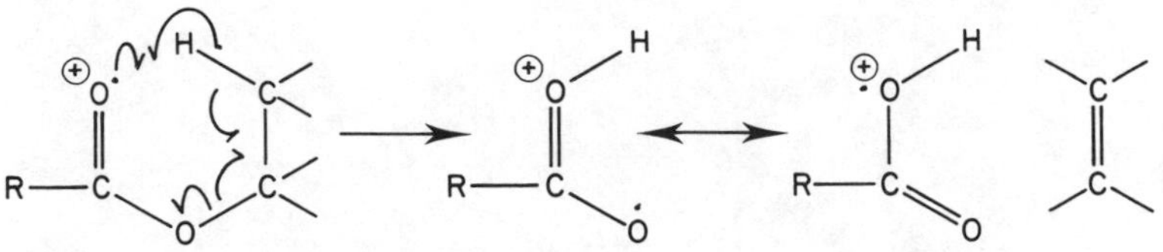

E. NUCLEAR MAGNETIC RESONANCE (NMR) SPECTROSCOPY

1. PRINCIPLES AND INSTRUMENTATION

(i) The Physics of Nuclear Spins

All nuclei have charge by virtue of containing protons and some of them also behave as if they spun. A spinning charge is equivalent to a conductor carrying a current (i) and therefore will be associated with a magnetic field H (Fig. E.1). Such nuclear magnetic dipoles are characterized

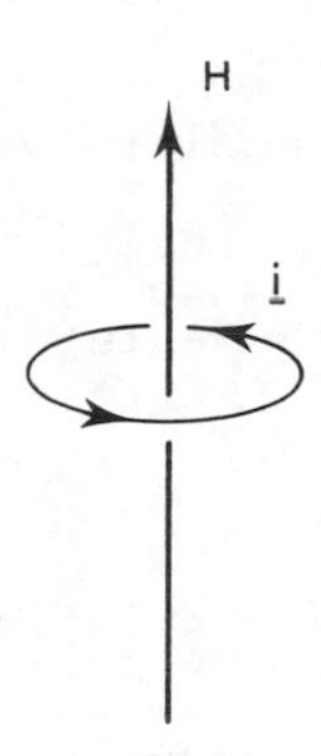

FIG. E.1 : THE RELATIONSHIP BETWEEN THE SPINNING CHARGE AND RESULTING MAGNETIC FIELD

by nuclear magnetic spin quantum numbers which are designated by the letter I and can take up values equal to 0, ½, 1, 3/2 ..... etc.

From the point of view of nuclear magnetism, it is useful to consider three types of nuclei:

Type 1: I = 0, which means that the nuclei do not interact with the applied magnetic field and are therefore not NMR chromophores. The reason for this behaviour is quite straightforward: all nucleons (protons and neutrons) have spins and the spins pair off. Nuclei composed of even numbers of protons and neutrons therefore have no net spin. A corollary of this is that nuclear spin is a property characteristic of certain isotopes rather than of certain atoms. In fact the most prominent examples of nuclei with I = 0 are ^{12}C and ^{16}O, the dominant isotopes of carbon and oxygen, but both of these elements also have magnetic isotopes.

Type 2: $I = \frac{1}{2}$, which means that the nuclei have a non-zero magnetic moment (dipole), but no nuclear electric quadrupole (Q). By chance, the two most important nuclei ^{1}H (ordinary hydrogen) and ^{13}C (the non-radioactive isotope of carbon occurring to the extent of 1.06% at natural abundance) belong to this category as do the two other commonly observed nuclei (^{19}F and ^{31}P). Together, NMR data for ^{1}H and ^{13}C account for well over 90% of all observations in the literature and the discussion and examples in this book all refer to these two nuclei. However, the spectra of all nuclei with $I = \frac{1}{2}$ can be understood easily on the basis of common theory.

Type 3: $I > \frac{1}{2}$, which means that the nuclei have both a magnetic moment and an electric quadrupole. This group includes some common isotopes (e.g. ^{2}H and ^{14}N both have I = 1) but they are difficult to observe and will not be discussed further.

The most important consequence of nuclear spin is that in a uniform magnetic field, a nucleus of spin I may assume 2I + 1 orientations. For nuclei with $I = \frac{1}{2}$, there are clearly just 2 permissible orientations, as $2 \times \frac{1}{2} + 1 = 2$. These two orientations will be of unequal energy (by analogy with the parallel and antiparallel orientations of a bar magnet in a magnetic field) and it is possible to induce a spectroscopic transition (spin-flip) by the absorption of a quantum of electromagnetic energy of the appropriate frequency obeying the usual (page 1) relation to the energy gap involved:

$$\nu = \frac{\Delta E}{h} \qquad \text{(i)}$$

In the case of the nuclear spin flip, however, there is another variable affecting the size of the energy gap between the two spin states, namely the strength of the applied field H_o. It is found that

$$\nu = KH_o \qquad \text{(ii)}$$

where K is a constant characteristic of the nucleus observed. Equation (ii) is known as the Larmor equation and is the fundamental relation in NMR spectroscopy. It can be seen that, unlike with other forms of spectroscopy, the frequency of the absorbed electromagnetic radiation is not an absolute value for any particular transition, but may assume an infinite number of values depending on the strength of the applied magnetic field. Any one of the possible sets of matching values of ν and H_o corresponds to the condition of *resonance* and this is the origin of the term *Resonance* in Nuclear Magnetic Resonance Spectroscopy. Thus for ^1H and ^{13}C *resonance frequencies* corresponding to magnitudes of applied magnetic field (H_o) commonly found in commercial instruments are as follows:-

$\nu\,^1H$(MHz)	$\nu\,^{13}C$(MHz)	H_o(Tesla)
60	15.087	1.4093
80	20.1115	1.8790
90	22.629	2.1139
100	25.144	2.3488
200	50.288	4.6975
400	100.577	9.3950

In common jargon, NMR spectrometers are referred to as "60MHz", "200MHz" or "400MHz" instruments, even if the spectrometer is set to observe a nucleus other than ^1H.

It can be seen that all the frequencies listed correspond to the radio frequency region of the electromagnetic spectrum and inserting these values into equation (i) gives the size of the energy gap involved: 100MHz corresponds to 4×10^{-5}kJmol^{-1}. This is, of course, a negligibly small value on the chemical energy scale. Thus NMR spectroscopy is, for all practical purposes, a ground-state phenomenon.

Any absorption signal observed in a spectroscopic experiment must originate from excess of the population in the lower energy state, the so called *Boltzmann excess* which is equal to $N_\beta - N_\alpha$, where N_β and N_α are the populations in the lower (β) and upper (α) energy states. Now for molar quantities, the general Boltzmann relation (iii) shows that:

$$\frac{N_\beta}{N_\alpha} = e^{\frac{\Delta E}{RT}} \qquad \text{(iii)}$$

Clearly, as the energy gap (ΔE) approaches zero, the right hand side of equation (iii) approaches e^0, i.e., 1 and the Boltzmann excess becomes very small. For the NMR experiment it is typically of the order of 1 in 10^5 which renders the method inherently insensitive. However, inspection of equations (i) and (ii) shows that the energy gap and therefore ultimately the sensitivity, increases with increasing applied magnetic field. This is one of the reasons for the use of high magnetic fields in NMR spectrometers.

Even at the highest fields, the NMR experiment would not be practicable if mechanisms did not exist for the restoration of the Boltzmann equilibrium which is perturbed as the result of the absorption of electromagnetic radiation [equation (i)]. These mechanisms are known by the general term of *relaxation* and are not confined to NMR spectroscopy; however because of the small magnitude of the Boltzmann excess they are more critical in the NMR experiment.

The most important relaxation processes in NMR involve interactions with other nuclear spins which are in the state of random thermal motion. This so called *spin-lattice relaxation* mechanism is a simple exponential decay process characterized by a time constant T_1 for restoration of equilibrium, and can operate efficiently only in liquids, gases and solutions. NMR spectra of solids exhibit entirely different characteristics and will not be discussed any further.

The break-down of the relaxation conditions necessary for the production of sharp NMR signals may be caused by high viscosity of the medium or the presence of paramagnetic impurities. The increase of the power (amplitude) of the electromagnetic radiation may also upset the Boltzmann equilibrium by producing the condition known as *saturation*, whose first symptoms are also line-broadening.

(ii) Acquisition of the NMR spectrum

As the NMR phenomenon is not observable in the absence of an applied magnetic field, a magnet is an essential component of any NMR spectrometer. Such magnets may be permanent (as in many routine instruments), electromagnets, or they may be based on superconducting solenoids cooled by liquid helium (as in modern research instruments). They all must share the following characteristics: (a) The magnetic field must be strong. This is partly due to the sensitivity factors discussed above, but even more importantly it ensures superior dispersion of signals and, in the case of ^{1}H NMR, also very important simplification of the spectrum. These last two factors will be discussed below. (b) The magnetic field must be homogeneous so that all portions of the sample examined experience exactly the same magnetic field. Any inhomogeneity of the magnetic field will result in broadening and distortion of spectral bands and, in extreme cases, in the appearance of spurious multiple sets of signals. These effects follow directly from the nature of the Larmor equation [equation (ii)] and are deliberately induced in the *NMR scanners* used in medicine. However, for the determination of the structure of organic compounds the highest attainable degree of field homogeneity is desirable, because useful information may be lost if the width of the NMR spectral lines exceeds as little as 0.2 Hz. Clearly, 0.2 Hz in, say, 100 MHz implies a homogeneity of 2 in 10^{9}, a very stringent requirement. (c) The magnetic field must be very stable, so that it does not drift during the acquisition of the spectrum, which may take from several seconds to several hours. The temporal stability required is generally beyond the realms of available technology and is attained by *locking* the value of the field and the electromagnetic frequency using electronic feed-back devices.

(a) Continuous Wave (CW) NMR Spectroscopy

Inspection of the Larmor equation [equation (ii)] shows that for any nucleus the condition of resonance may be achieved by keeping the field constant and changing (or sweeping) the frequency or, alternatively, by keeping the frequency constant and sweeping the field. Figure E.2 shows a schematic diagram of the latter arrangement, but it is important to note that the two modes of obtaining the "swept" (or *continuous wave*) NMR spectrum are essentially equivalent. In fact, once the spectrum has been calibrated as shown in section E.2 (i), it is impossible to determine which mode had been used to obtain it.

For protons the intensity of the signal (which may be electronically integrated as shown in Fig. E.2) is directly proportional to the number of nuclei undergoing a spin-flip and proton NMR spectroscopy is therefore a quantitative method. However, for ^{13}C NMR this is no longer generally true. These differences are partly due to different ranges of T_1 for these nuclei and partly to the nuclear Overhauser effect [section E.5 (i)].

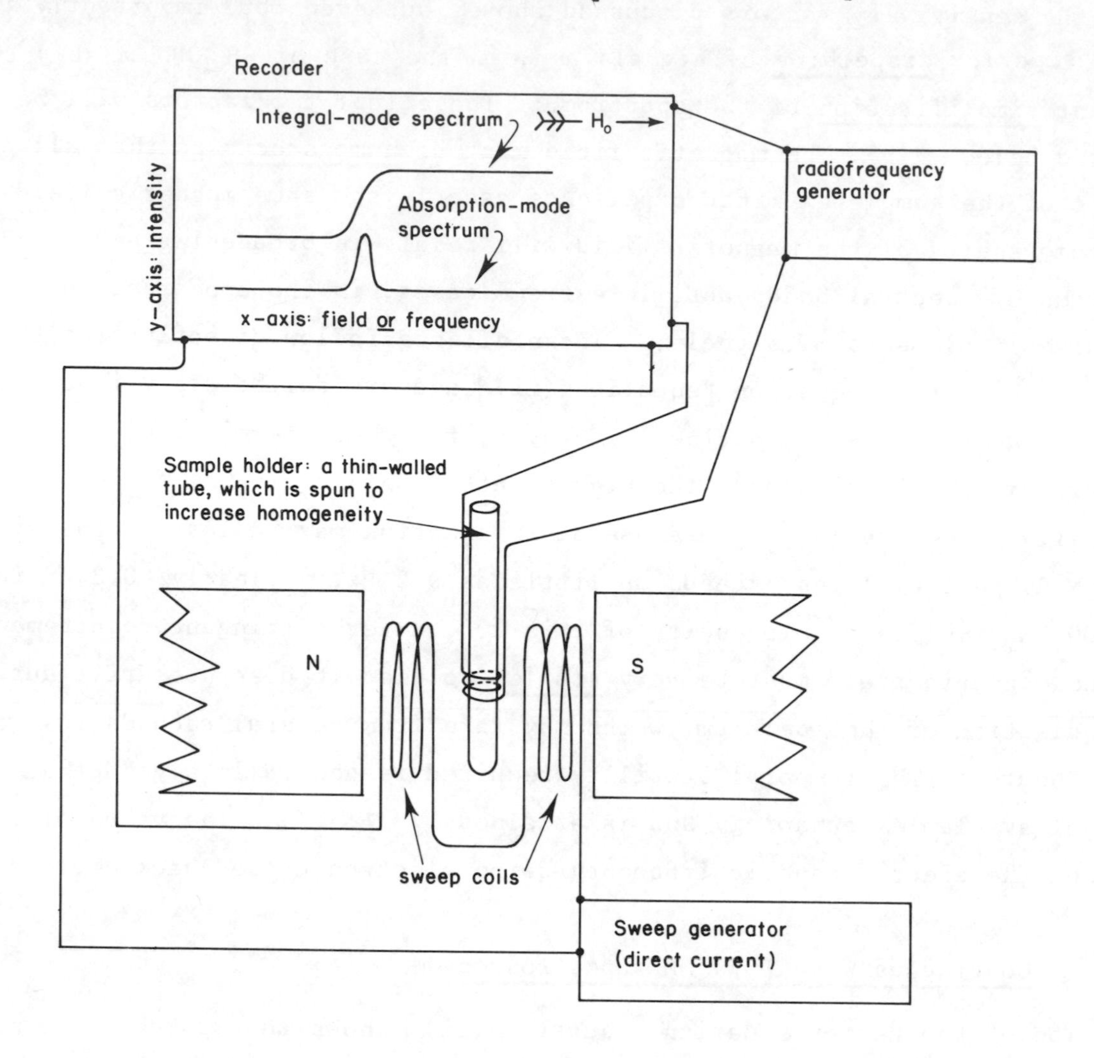

FIG. E.2 : A FIELD-SWEEP CW NMR SPECTROMETER

(b) Fourier-Transform (FT) NMR Spectroscopy

As an alternative to the CW method, an intense pulse of electromagnetic energy can be applied causing all of the resonating nuclei to become magnetized, i.e. flipped to their higher energy α-state. It is possible to sample this magnetization as a function of time and, for a single resonance, one obtains an exponential decay curve (*free induction decay* or FID) shown in Fig. E.3a. This type of spectrum, known as a *time-domain* spectrum can be converted into the ordinary, or *frequency-domain* spectrum (Fig. E.3b) by performing a mathematical operation known as *Fourier transformation*.

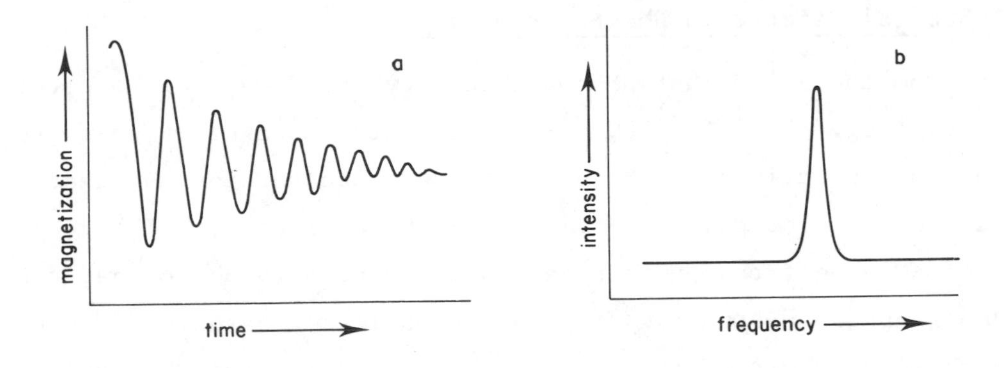

FIG. E.3 : TIME-DOMAIN AND FREQUENCY-DOMAIN NMR SPECTRA

Most NMR spectra consist of a number of signals and their time-domain spectra appear as a superposition of a number of traces of the type shown in Fig. E.3a. Such spectra are quite uninterpretable by inspection, but Fourier transformation converts them into ordinary frequency-domain spectra. The time-scale of the FID experiment is of the order of seconds during which the magnetization may be sampled many thousands of times. This is accomplished by a dedicated minicomputer, which is also used to perform the Fourier transformation.

The principal advantage of FT NMR spectroscopy is a great increase in sensitivity per unit time of experiment: a CW scan generally takes of the order of one hundred times as long as the collection of the equivalent FID. Thus during the time it would have taken to acquire a CW spectrum, the minicomputer can accumulate many FID scans and add them up in its memory. This results in an increase in sensitivity (signal/noise) proportional to the square root of the number of scans. It is the increase in sensitivity brought about by the introduction of FT NMR spectroscopy which has permitted the routine observation of ^{13}C NMR spectra.

In addition, the FID can be manipulated mathematically to enhance sensitivity (e.g. for routine ^{13}C NMR) at the expense of resolution, or resolution (often important for ^{1}H NMR) at the expense of sensitivity. Furthermore, it is possible to devise sequences of pulses which result, after suitable mathematical manipulation, in NMR spectroscopic data which are of great value. However, such methods (e.g. two-dimensional NMR) are beyond the scope of this book.

(iii) Chemical Effects in NMR Spectroscopy

It is clear from the above that NMR spectroscopy could be used to detect certain nuclei (e.g. ^{1}H, ^{13}C, ^{19}F, ^{31}P) and, in the case of ^{1}H, also to estimate them quantitatively. However, the expense and inconvenience of NMR instrumentation would hardly be justified on these grounds. The real usefulness of NMR spectroscopy is based on two secondary phenomena, the *chemical shift* and *spin-spin coupling* and, to a lesser extent, on phenomena related to the *time-scale* of the NMR experiment. Both the chemical shift and spin-spin coupling reflect the chemical environment of the nuclear spins whose spin-flips are observed and can therefore be considered as chemical effects in NMR.

2. THE CHEMICAL SHIFT

The magnitude of the applied magnetic field (H_o) in the Larmor equation [equation (ii)], refers to the field at the nucleus. When the nucleus is contained within a molecule the value of H_o is subtly altered by its molecular environment. Thus the statement on page 31 which implies that in an applied magnetic field of 2.1139 T, protons will resonate at 90 MHz is only an approximation: in fact protons in most molecular environments will resonate over a range of about 1000 Hz at this field. However, a near neighbouring nucleus to ^{1}H, ^{19}F, resonates (again over a range of frequencies) at 84.699 MHz, i.e. 5,331,000 Hz away. It can thus be seen that the range of frequencies exhibited by any nucleus as a result of changing molecular environment (hence the term chemical shift) is really only the fine structure of a single resonance. The ability to resolve this resonance amounts to a practical definition of *high resolution NMR spectroscopy*.

As the chemical shift of a nucleus reflects the molecular structure, it can be used to obtain structural information. Further, as hydrogen and carbon (and therefore ^{1}H and ^{13}C nuclei) are almost universal constituents of organic compounds the amount of structural information available from ^{1}H and ^{13}C NMR spectroscopy greatly exceeds in value the information available from other forms of molecular spectroscopy. In other words, every hydrogen and carbon atom in an organic molecule becomes "a chromophore" (see page 2) and, in the case of hydrogen, it is also possible to estimate the relative amounts of hydrogen in various molecular environments.

(i) The Dimensionless Chemical Shift Scale

Before proceeding to examine the relationships between the chemical shifts and molecular structure, it is essential to establish a chemical shift scale. The Larmor equation [equation (ii)] shows that it is impossible to specify the frequency of a resonance in terms of a single number. All one could say is that a particular nucleus resonates at a particular frequency when subjected to a particular applied magnetic field.

The problem can be solved by expressing the value of the field in terms of the basic frequency used to detect this type of nucleus at this field. Thus using a ^{1}H NMR spectrometer operating at a field of 2.1139 T means that the experimentally obtained frequency could be expressed in dimensionless units by dividing it by 90 MHz (see page 31). However, it is much more convenient to use a relative rather than an absolute frequency scale (thereby avoiding very awkward numbers). For this reason, resonance frequencies are always expressed relative to a standard substance.

The substance chosen for the standard for both ^{1}H and ^{13}C is tetramethylsilane $(CH_3)_4Si$ abbreviated to TMS. It is added to the solution of the substance to be examined and has the following advantages: (i) It is a relatively inert low boiling (b. 26.5^{o}) liquid which can be easily removed after use. (ii) It gives a single signal in both ^{1}H and ^{13}C by virtue of having only one type of both hydrogen and carbon. (iii) The chemical environment of both carbon and hydrogen in TMS is unusual due to the presence of silicon and hence they resonate outside the normal range. Thus the reference signal is unlikely to be superimposed on a signal originating from the substance examined. (iv) The chemical shift of TMS is not affected by complexation because of absence of polar groups etc.

We can thus express the chemical shift in terms of "Hertz from TMS" and, by convention, the absence of sign implies a shift in the direction of lower field (i.e. *downfield*). This can be converted to dimensionless units by dividing by the basic operating frequency of the spectrometer and, to avoid very small numbers, it is multiplied by 10^{6}. The resulting chemical shift is then defined as δ, which is a dimensionless number corresponding to "parts per million" (ppm).

Thus:

$$\text{Chemical shift in ppm } (\delta) = \frac{\text{Chemical shift from TMS in Hz}}{\text{Spectrometer frequency in Hz}} \times 10^6$$

The chemical shift range for protons covers roughly 10 ppm and that for ^{13}C roughly 200 ppm. For protons there is an alternative scale which expresses the chemical shifts in ppm on a τ (tau) scale, so that:

$$\tau = 10 - \delta$$

Graphically the two scales correspond to each other in the following manner:

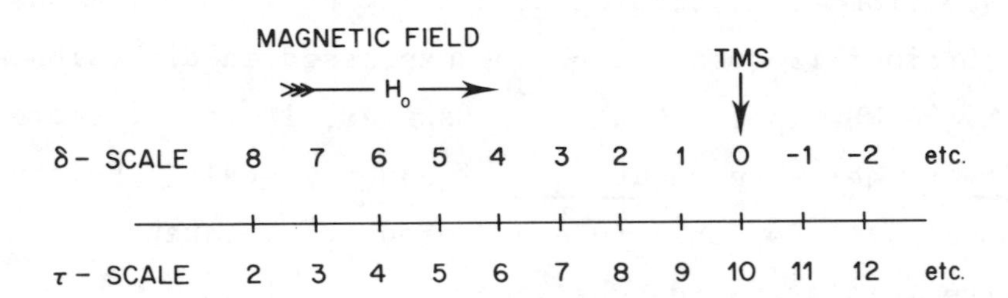

NMR chart paper (particularly that used for ^{1}H NMR) may be calibrated in both scales, as well as in Hz because this gives direct information related to spin-spin coupling (Chapter E.4).

(ii) Factors influencing the chemical shifts of ^{1}H and ^{13}C

Under the influence of the applied magnetic field, the electrons surrounding the magnetic nuclei circulate in a manner analogous to the relationship between conductors carrying a current and a magnetic field. Such circulating electrons create magnetic fields of their own (see e.g., Fig. E.1), which may augment or oppose the applied magnetic field over a volume of space.

In general, the electrons surrounding a magnetic nucleus will tend to generate a magnetic field which will oppose the applied magnetic field in the region of the nucleus. Thus the applied magnetic field needed to bring this nucleus into the condition of resonance will be higher than it would have been in the absence of the circulating electrons and it is said that the electrons *shield the nucleus.* Clearly any effect which alters the density or spatial distribution of the electrons around the nucleus will alter the precise degree of this shielding and be ultimately reflected in the value of the frequency of the electromagnetic radiation needed to bring the nucleus to resonance in a constant applied field. It follows that <u>any</u> structural change should be reflected in the chemical shift : if the molecular environment of two nuclei (say two hydrogens or two carbons) is not precisely (by symmetry) identical, their chemical shifts will not be the same unless by *accidental equivalence.* When two nuclei have identical molecular environments and hence the same chemical shift, they are said to be *chemically equivalent* or *isochronous.*

Both hybridization and changes of electron density due to substituents lead to greater effects at the carbon atom than at the hydrogen atom attached to it. This is illustrated in Table E.1:

TABLE E.1: RELATIVE EFFECTS ON THE CHEMICAL SHIFTS OF ^{13}C AND ^{1}H

Compounds	^{13}C (ppm from TMS)	^{1}H (ppm from TMS)
CH_4	-2.1	0.23
CH_3Cl	23.8	3.05
CH_2Cl_2	52.9	5.33
$CHCl_3$	77.3	7.24
CH_3CH_3	7.3	0.86
$CH_2=CH_2$	122.1	5.25
benzene	128.0	7.26
CH_3 (of CH_3-C(H)=O)	31.2	2.20
H-C=O	200.5	9.80
CH_3 (of CH_3-CH_2-CH_2-Cl)	11.5	1.06
CH_2	26.5	1.81
CH_2	46.7	3.47
Cl		

In fact, the range of chemical shifts of ^{13}C in typical organic compounds spans approximately 200 ppm while the corresponding range in ^{1}H chemical shifts is only approximately 10 ppm.

The last three entries in Table E.1 illustrate the fact that substitutent effects diminish rapidly with the increasing number of bonds separating the observed nucleus from the substituent.

The chemical shift of a nucleus may also be affected by the presence in its vicinity of a *magnetically anisotropic* group. In such a group the circulation of electrons induced by the applied magnetic field of the spectrometer is not cancelled out by the random thermal tumbling experienced by the sample and therefore such groups are surrounded by volumes of space in which the resonating nuclei being observed are *shielded* (resulting in an upfield shift toward smaller δ values) or *deshielded* (resulting in a downfield shift toward larger δ values). Thus aromatic rings shield nuclei placed above and below their planes and deshield nuclei in their planes. The latter effect is the principal cause of the difference in the chemical shifts between the protons in ethylene and benzene (Table E.1). While, in principle, such effects also operate in ^{13}C NMR spectroscopy, the very large range of shifts caused by direct effects (Table E.1) makes these *through space* effects less obvious.

The chemical shift of any magnetic nucleus within a molecule is thus dependent upon all the other groups and therefore is unique for any position in any molecule. For the purposes of solving simple structural problems, such as those presented in this text, it is only necessary to grasp the principles stated above and the broad trends illustrated in Tables E.1, E.2 and E.3.

The ultimately achievable width of a spectral line is between 0.5 and 0.05 Hz (i.e. it is not infinitely small) and the signals may be much wider due to spin-spin coupling (Section E.4). Both of these limitations are in Hz and independent of the operating field of the spectrometer, while the difference between chemical shifts of nuclei (expressed in Hz) <u>is</u> dependent on the operating frequency. It follows that signals due to protons in different molecular environments may overlap in spectra taken at lower fields but be clearly separated in spectra taken at higher fields. It can thus be said that operation at higher magnetic field offers better <u>dispersion</u>, which is a major advantage.

TABLE E.2: CHEMICAL SHIFTS (δ IN PPM) OF ^{1}H IN COMMON ORGANIC ENVIRONMENTS

Tetramethylsilane $(CH_3)_4Si$ 0

Methyl groups attached to sp^3 hybridized carbon atoms 0.8-1.2

Methylene groups attached to sp^3 hybridized carbon atoms .1-1.5

Methine groups attached to sp^3 hybridized carbon atoms ...1.2-1.8

Electron withdrawing substitutents (-OH, -OCOR, -OR, $-NO_2$, halogen) cause a downfield shift of 2-4 ppm when present at C_α and have less than half of this effect when present at C_β.

sp^2 hybridized carbon atoms (carbonyl groups, olefinic fragments, aromatic rings) cause a downfield shift of 1-2 ppm when present at C_α and have less than half of this effect when present at C_β.

Acetylenic protons 2-3

Olefinic protons 5-8

Aromatic and heterocyclic protons 6-9

Aldehydic protons 9-10

<u>Labile</u> protons (-OH, $-NH_2$, -SH etc.) have no characteristic chemical shift ranges. However, such resonances can always be positively identified by *in-situ* exchange with D_2O, which cause them to disappear from the spectrum.

NMR spectra are almost invariably obtained in solution. The solvents of choice: (a) Should have adequate dissolving power. (b) Must not associate strongly with solute molecules as this is likely to produce appreciable effects on chemical shifts. This requirement must sometimes be sacrificed in the quest of adequate solubility. (c) Should be essentially free of interfering signals. Thus for ^{1}H NMR, the best solvents are proton-free. (d) Should preferably contain deuterium, which provides a convenient "locking" signal for stabilization of the magnetic field (see page 33).

The most commonly used solvent is deuterochloroform, $CDCl_3$, which is only weakly associated with most organic substrates, contains no protons and has a deuterium atom.

TABLE E.3: CHEMICAL SHIFTS (δ IN PPM) OF ^{13}C IN COMMON ORGANIC ENVIRONMENTS

Environment	δ (ppm)
Tetramethylsilane $(CH_3)_4Si$	0
Methyl with only H or R at both C_α and C_β	0-30
Methylene with only H or R at both C_α and C_β	20-45
Methine with only H or R at both C_α and C_β	30-60
Quaternary carbon with only H or R at both C_α and C_β	30-50
CH_3-O	50-60
CH_3-N	15-45
C≡C	75-95
C=C	105-145
C-aromatic	110-155
C-heteroaromatic	105-165
-C≡N	115-125
C=O (acids, esters and amides)	155-185
C=O (Ketones and aldehydes)	185-225

In ^{1}H NMR spectroscopy, the invariably present signal due to $CHCl_3$ (isotopic impurity) at δ 7.24 provides a secondary standard. In ^{13}C NMR spectroscopy the signal due to the carbon in $CDCl_3$ is weaker than one would expect due to the absence of the nuclear Overhauser effect (see section E.5) and appears as a triplet due to spin-spin coupling to deuterium (see section E.4), centred at δ 77.3.

3. THE NMR TIME-SCALE

Two magnetic nuclei situated in different molecular environments must give rise to separate signals in the NMR spectrum, say $\Delta\nu$ Hz apart (Fig. E3a). However, if some process is taking place which interchanges the environments of the two nuclei at a rate (k) much faster than $\Delta\nu$ times per second, the two nuclei will give rise to a single signal at an intermediate frequency (Fig. E.3c). When the rates (k) of the exchange process are comparable to $\Delta\nu$, one observes typical *exchange broadened* spectra (Fig. E3b), from which one can, in fact, derive the rate constant of the process.

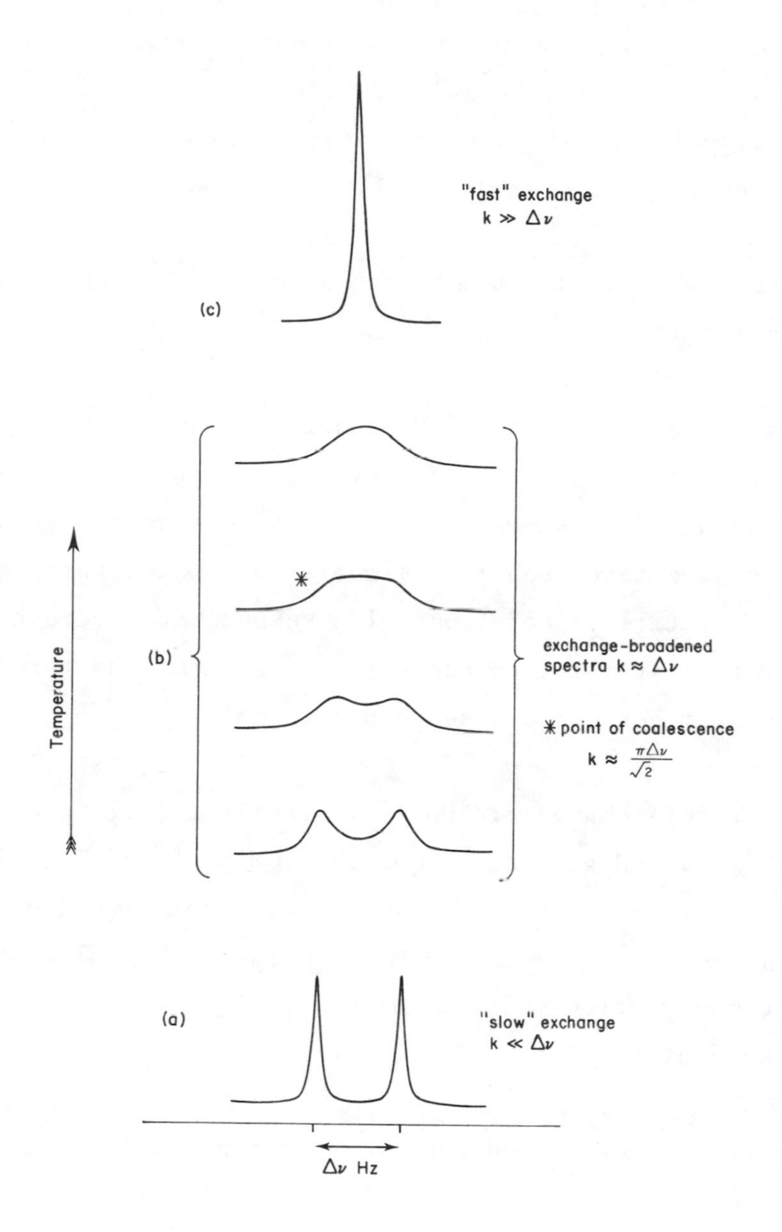

FIG. E.3: NMR SPECTRA OF EXCHANGING NUCLEI

In practice, a compound in which an exchange process operates can give rise to a series of spectra of the type shown in Fig. E.3, as observation at different temperatures alters k.

The averaging effects would apply to any process, but the *NMR time-scale* happens to coincide with the rates of a number of chemical processes which lead to the averaging of the chemical environments of nuclei within molecules. This is so, because the frequency unit Hz corresponds to "one cycle per second" and inverse seconds happen to be within the kinetic range of several common chemical processes. The terms "slow" and "fast" introduced in Fig. E.3, refer to the *NMR time-scale*, which is much slower than the time-scale of other forms of spectroscopy, e.g. UV or IR.

The common processes which give rise to variation of the appearance of NMR spectra with temperature are:

(i) Some conformational processes. Fortunately most of these are so fast on the NMR time-scale that normally only averaged spectra are observed. In particular, consideration of symmetry would require a methyl group attached to a chiral carbon to give three separate signals due to the three hydrogen atoms being in different spatial orientations with respect to the rest of the molecule. However, fast rotation ensures averaging of the three environments so that the three hydrogens give rise to one signal.

(ii) Intermolecular interchange of labile (slightly acidic) protons, such as those in -OH, $-NH_2$ and -SH groups. Thus the -OH protons of a mixture of two different alcohols may give rise to either an averaged signal or to separate signals, depending on the temperature of the sample, the polarity of the solvent and the concentrations of the solutes, all of which influence the rate of exchange of the protons.

(iii) Valence-bond tautomerism and rotation about partial double bonds (e.g. the peptide bond) often occur at rates which give rise to observable phenomena in NMR spectra.

4. SPIN-SPIN COUPLING

(i) The electron-coupled spin-spin interaction

A typical organic molecule contains more than one magnetic nucleus and the energy change associated with the spin flip of the observed nucleus may be affected by the other magnetic nuclei which are present in their various quantized energy states. This causes the appearance of additional fine structure (multiplicity) in many signals and is not only the principal cause of difficulty in interpreting ^{1}H NMR spectra, but also provides very valuable structural information when correctly interpreted.

A nucleus "senses" the energy state of another nucleus via the agency of the electrons in the chemical bonds separating them, and the proper name for the phenomenon known as *spin-spin coupling* is "electron-coupled spin-spin interaction".

Two important observations follow from this: (i) No inter-molecular spin-spin coupling is observed. (ii) Spin-spin coupling falls off (in energy terms) with the number of bonds separating the interacting nuclei, generally becoming unobservable when more than 3 bonds separate the two interacting nuclei.

The magnitude of the spin-spin interaction is the difference in the energy of the nuclear transition caused by the interaction (coupling) with another spin. It is expressed in terms of a coupling constant symbolized as J in units of Hz. The coupling constant is related to, but not always identical with, the splitting of spectral lines. Because J depends only on the number, type and spatial arrangement of the bonds separating the two nuclei, it is independent of the applied magnetic field. Conversely, the magnitude of J or even the mere presence of detectable interaction, constitutes valuable structural information.

The elementary theory of spin-spin multiplicity developed below will be given in terms of proton-proton interactions but applies equally well to all nuclei with the spin quantum number, $I = \frac{1}{2}$. There are certain simplifications associated with ^{13}C NMR spectra which are due to the low natural abundance of ^{13}C and these will be dealt with in section E.4 (v) below.

To obtain structurally useful information one must solve two separate problems. Firstly, one must analyse the spectrum (sections (ii) and (iii)) to obtain the NMR parameters (δ and J) for all the protons and, secondly, one must interpret the values of the coupling constants in terms of structure (section (vi)).

(ii) Definitions

(a) A spin system is a group of coupled protons. Clearly, a spin system cannot extend beyond the bounds of a molecule, but it may not include a whole molecule. Thus *iso*propyl propionate (1) comprises two spin systems, a five-proton system for the propionic acid residue and a seven-proton system for the *iso*propanol residue because the ester group provides a barrier of 5 bonds against effective coupling between the two parts.

$$CH_3CH_2\text{-}\overset{\overset{\displaystyle O}{\|}}{C}\text{-}O\text{-}CH(CH_3)_2$$

(1)

(b) Strongly and weakly coupled spins These terms refer not to the actual magnitude of J, but to the ratio of the separation of chemical shifts expressed in Hz ($\Delta\nu$) to the coupling constant J. For most purposes $\Delta\nu/J$ larger than ~ 3 can be defined as "large" and such spins are said to be *weakly coupled*. When this ratio is smaller than ~ 3, the spins are said to be *strongly coupled*. Two important conclusions follow:

(1) Because here the chemical shift separation ($\Delta\nu$) is expressed in Hz, rather than in the dimensionless δ units, its value will change with the operating frequency of the spectrometer, while the value of J remains constant. It follows that two spins will become progressively more weakly coupled as the spectrometer frequency increases. Now it turns out that weakly coupled spin systems are much more easy to analyse than strongly coupled spin systems and thus operation at higher frequencies (and therefore at higher applied magnetic fields) will yield more interpretable spectra. This has been a very important reason for the development of NMR spectrometers operating at ever higher magnetic fields.

(2) Within a spin system, some pairs of nuclei or groups of nuclei may be strongly coupled and others weakly coupled. Thus a spin-system may be *partially strongly coupled*.

(c) Magnetic equivalence A group of protons is *magnetically equivalent* when they not only have the same chemical shift (chemical equivalence) but also have identical spin-spin coupling to each individual nucleus outside the group.

(d) Conventions used in naming spin systems Consecutive letters of the alphabet (e.g. A, B, C, D, ...) are used to describe groups of protons which are strongly coupled. Subscripts are used to give the number of protons which are magnetically equivalent. Primes are used to denote protons which are chemically equivalent (page 39) but not magnetically equivalent. A break in the alphabet indicates weakly coupled groups. Thus

ABC denotes a strongly coupled 3-spin system

AMX denotes a weakly coupled 3-spin system

ABX denotes a partially strongly coupled 3-spin system

A_3BMXY denotes a 7-spin system in which the three magnetically equivalent A nuclei are strongly coupled to the B nucleus, but weakly coupled to the M, X and Y nuclei. The nucleus X is strongly coupled to the nucleus Y but weakly coupled to all the other nuclei. The nucleus M is weakly coupled to all the other 6 nuclei

AA'XX' is a 4-spin system described by two chemical shift parameters (for the nuclei A and X) but where $J_{AX} \neq J_{AX'}$. A and A' (as well as X and X') are pairs of nuclei which are chemically equivalent but magnetically non-equivalent.

(iii) Analysis of NMR Spectra

The process of deriving the NMR parameters δ and J from a set of multiplets due to a spin system is known as the analysis of the NMR spectrum. In principle, any spectrum arising from a spin system, however complicated, can be analysed by quantum mechanical calculations performed by a computer.

Two separate operations are involved in the commonly employed analytical programmes. Firstly, the programme will generate a spectrum corresponding to any set of NMR parameters (i.e., chemical shifts and spin-spin coupling constants). Clearly, such a simulated spectrum may be compared with the spectrum being analysed and if considered identical, the parameters are taken to correspond to the actual chemical shifts and coupling constants. This operation, while very easy to perform and inexpensive in computation time, is very rarely sufficient by itself.

The second stage involves the comparison of frequencies of all resolvable lines of the experimental spectrum with the frequencies of all lines of a simulated spectrum, which is similar to, but not identical with the experimental spectrum. The programme then performs an iterative operation wherein the input parameters are varied until the best match is obtained. The "best" input parameters are then taken as the actual chemical shifts and coupling constants. It must be realized, however, that the validity of any such analysis finally depends on the match between the experimental spectrum and the computed (simulated) spectrum.

In practice any spin system beyond about 9 nuclei is extremely difficult to analyse unless very much simplified by weak coupling or symmetry.

Fortunately, in a very large number of cases multiplets can be correctly analysed by inspection and direct measurements. Spectra of that type are known as *first order spectra* and they arise from weakly coupled spin systems. At high applied magnetic fields, a large proportion of ^1H NMR spectra are nearly pure first-order and there is a tendency for simple molecules, e.g. those exemplified in the problems in this text, to exhibit first-order spectra even at moderate fields.

The physical basis of the phenomenon of multiplicity in some NMR spectral lines is quite straightforward, as can be shown in the following examples:

Consider a system of two protons (H_A and H_X), and let the two allowed spin states be α (high energy) and β (low energy). Then for upward transitions of the nucleus H_A we can have:

$$H_A\beta \text{ to } H_A\alpha \quad \text{with} \quad H_X \text{ in state } \alpha$$

and

$$H_A\beta \text{ to } H_A\alpha \quad \text{with} \quad H_X \text{ in state } \beta$$

These two transitions are of a different energy depending on the strength of interaction between H_A and H_X and therefore the resonance due to H_A will be split into a doublet.

As the population distribution of H_X between the α and β states is almost completely equal (recall the vanishingly small Boltzmann excess discussed on page 31), the two transitions are of equal probability and hence H_A will give rise to a symmetrical doublet.

Similarly, for a system of three spins, one H_A and <u>two</u> H_X, we can have the following upward transitions for H_A:

$H_A\beta$ to $H_A\alpha$ with one H_X in state α and the other in state β

$H_A\beta$ to $H_A\alpha$ " " " " " β " " " " " α

$H_A\beta$ to $H_A\alpha$ with both H_X nuclei in state α

$H_A\beta$ to $H_A\alpha$ " " " " " " β

The first two transitions are equivalent in energy (degenerate) and hence H_A will give rise to a triplet with the intensity ratios 1:2:1.

However, while this type of operation is the first necessary step in deriving the proper quantum-mechanical basis for the analysis of spin-spin multiplets, it is insufficient to derive even the restricted *first-order rules*, which may be legitimately used to analyse weakly coupled spin-systems. For this reason, the first-order rules are listed empirically. As with the foregoing discussion, the rules will refer to "protons" rather than to "nuclei of spin quantum number = ½", because for all practical purposes the analysis of NMR spectra deals with proton spectra, ^{13}C NMR being somewhat simpler in this respect (see section E.4v).

<u>Rule 1</u>: A group of n magnetically equivalent protons will split a resonance of an interacting group of protons into n+1 lines. For example, the resonance due to the A protons in an A_nX_m system will be split into m+1 lines, while the resonance due to the X protons will be split into n+1 lines. More generally, splitting by n nuclei of spin quantum number I, results in 2nI+1 lines. This simply reduces to n+1 for protons where I = ½.

<u>Rule 2</u>: The spacing (measured in Hz) of the lines in the multiplet will be equal to the coupling constant. In the above example all spacings in both parts of the spectrum will be equal to J_{AX}.

Rule 3: The true chemical shift of each group of interacting protons lies in the centre of the (always symmetrical) multiplet.

Rule 4: The relative intensities of the lines within each multiplet will be in the ratio of the binomial coefficient of $(a+b)^n$. This leads to the following numerical relationships:

Number of protons responsible for splitting	multiplicity	relative ratio of intensities in the multiplet
1	doublet (2)	1:1
2	triplet (3)	1:2:1
3	quartet (4)	1:3:3:1
4	quintet (5)	1:4:6:4:1
5	sextet (6)	1:5:10:10:5:1
6	septet (7)	1:6:15:20:15:6:1
7	octet (8)	1:7:21:35:35:21:7:1
8	nonet (9)	1:8:28:56:70:56:28:8:1

Clearly, in the case of higher multiplets the outside components may be lost in the instrumental noise, e.g. a septet may appear as a quintet. The intensity relationship is the first to be significantly distorted in non-ideal cases, but this does not lead to serious errors in spectral analysis.

Rule 5: When a group of magnetically equivalent protons interacts with more than one group of protons, its resonance will take the form of a multiplet of multiplets. Thus the resonance due to the A protons in a system $A_nM_pX_m$ will have the multiplicity of $(p+1)(m+1)$.

The appropriate coupling constants will control splittings and relative intensities will obey rule 4.

Rule 6: Magnetically equivalent protons do not give rise to splitting, e.g. any system A_n will give rise to a singlet.

Rule 7: Spin systems which contain groups of chemically equivalent protons which are not magnetically equivalent cannot be

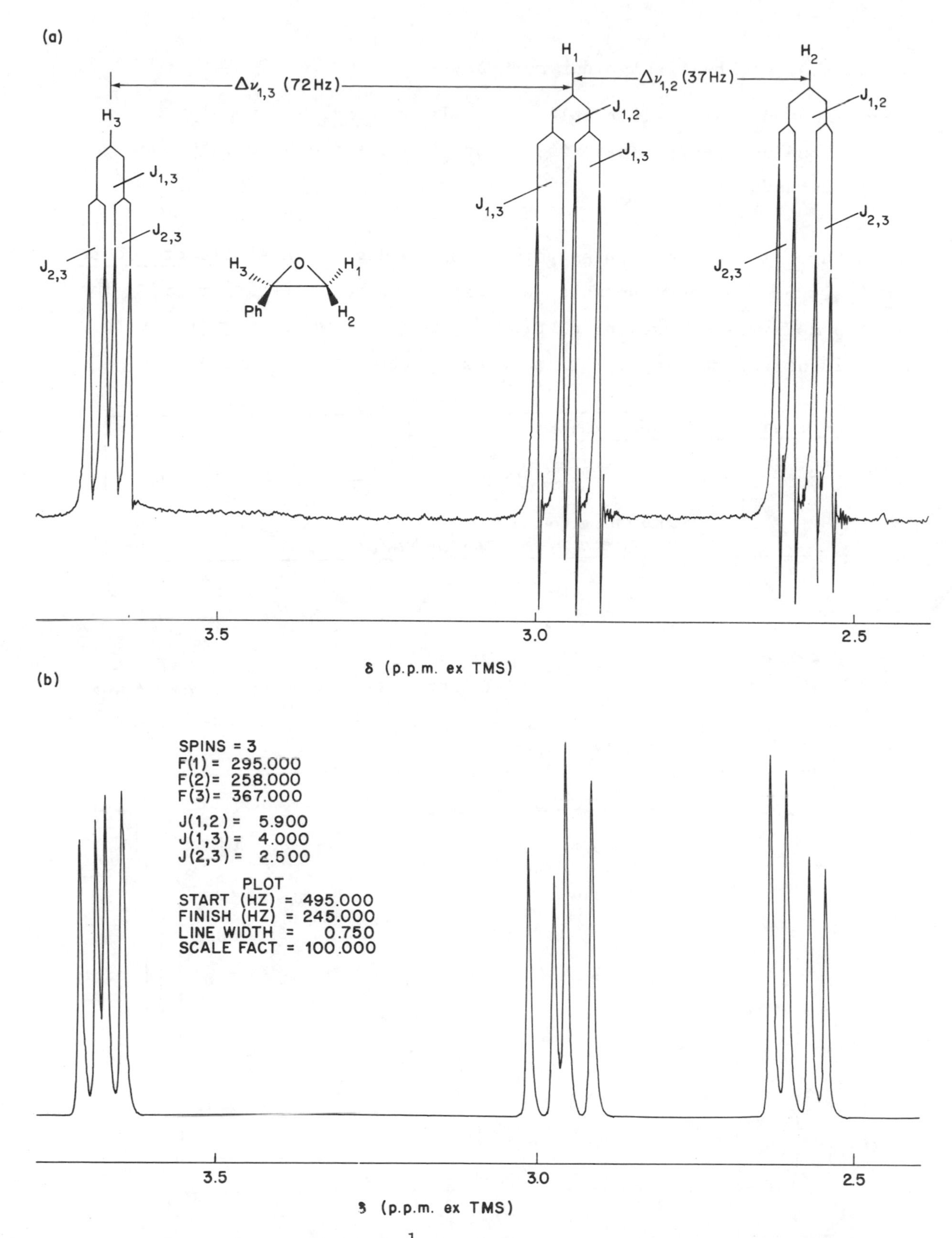

FIG. E.4 : A PORTION OF 100 MHz ^{1}H SPECTRUM OF STYRENE OXIDE IN CARBON TETRACHLORIDE (5%). THE EXPERIMENTAL SPECTRUM TOGETHER WITH THE SPLITTING DIAGRAM IS SHOWN IN FIG. E.4a. THE FIRST-ORDER PARAMETERS DERIVED FROM THIS SPECTRUM ARE LISTED IN FIG. E.4b, TOGETHER WITH THE COMPUTED SPECTRUM CORRESPONDING TO THEM.

analysed by first-order methods.

<u>Rule 8</u>: Not only strongly coupled, but also <u>partially</u> strongly coupled spectra (page 46) cannot be analysed by first-order methods.

The knowledge of the above rules permits us to develop a simple <u>procedure</u> for the analysis of any spectrum which we suspect of being first order. The first step consists of drawing a *splitting diagram*, which permits us to identify identical spacings. This is exemplified in Fig. E.4a.

<u>Check for validity of first-order analysis</u>:

$\frac{\Delta\nu_{AM}}{J_{AM}} = \frac{60}{1.5} = 40$ $\quad \frac{\Delta\nu_{MX}}{J_{MX}} = \frac{63}{3.5} = 18$ $\quad \frac{\Delta\nu_{AX}}{J_{AX}} = \frac{60+63}{6} = \frac{123}{6} = 20.5$

each ratio is greater than 3
first-order analysis is therefore justified

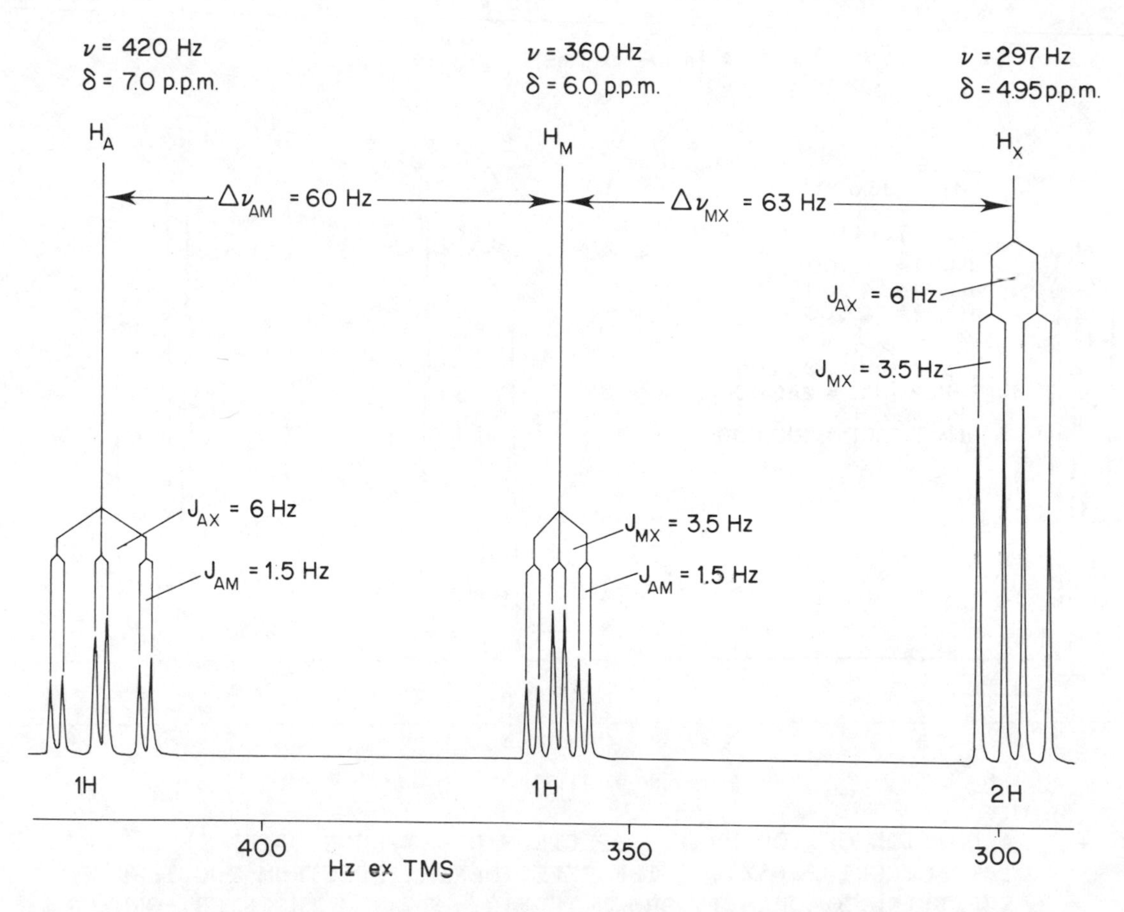

<u>FIG. E.5</u> : 60 MHz ^{1}H NMR SPECTRUM OF A 4-SPIN SYSTEM WITH SPLITTING DIAGRAM AND FIRST-ORDER COUPLING CONSTANTS.

Identification of corresponding splittings allows us to derive first-order coupling constants between relevant protons or groups of protons.

It is immaterial whether the spectral chart is calibrated directly in Hz or in ppm (as in Fig. E.4), because, provided that one knows the operating frequency of the spectrometer, the units can be readily derived from one another by means of the relationship given on page 38.

Having obtained the coupling constants, the chemical shifts can be simply obtained by measuring the centres of each multiplet. After the completion of this procedure, it is desirable to check if the first order assumptions have, in fact, been justified. This is most easily done by obtaining the values of $\Delta\nu$ in Hz for each pair of the interacting protons (or groups of protons) and checking if the ratio of $\Delta\nu/J$ is "large" i.e. >3 in each case. In the example shown in Fig. E.4a, even the most strongly coupled pair of protons (H_1 and H_2) where, by coincidence, both the smallest $\Delta\nu$(37Hz) and the largest J(5.9 Hz) occur, has $\Delta\nu/J = 6.27$, i.e. far in excess of the number 3 (see above). Thus one concludes that the 100 MHz spectrum of the 3 aliphatic protons of styrene oxide (Fig. E.4a) is indeed a first-order spectrum, i.e. an AMX system.

A more sophisticated way of checking the validity of a first-order analysis is to use the parameters derived by this procedure to generate a simulated spectrum with the aid of a computer program (see page 47). It can be seen (Fig. E.4b) that in this case the simulated spectrum is for all practical purposes identical with the experimental spectrum, again proving that the first-order analysis was justified.

Another example of first order analysis is shown in Fig. E.5. Again, consideration of $\Delta\nu$ and J values leads to the conclusion that first-order analysis was justified and that the system is indeed the weakly coupled AMX_2 type. This example is set out in terms of a model answer applicable to a number of problems in the last section of this book.

A spin system comprising just two protons (i.e. an AX or an AB system) is always exceptionally easy to analyse because, independently of the value of the ratio of $\Delta\nu$ /J, it always consists of just four lines with each pair separated by J_{AB}. The only distortion from the first-order pattern consists of the gradual reduction of intensities of the outer lines in favour of the inner lines, a characteristic "sloping towards" the coupling partner. A series of simulated spectra of two-spin systems are shown in Fig. E.6 and illustrate the above point.

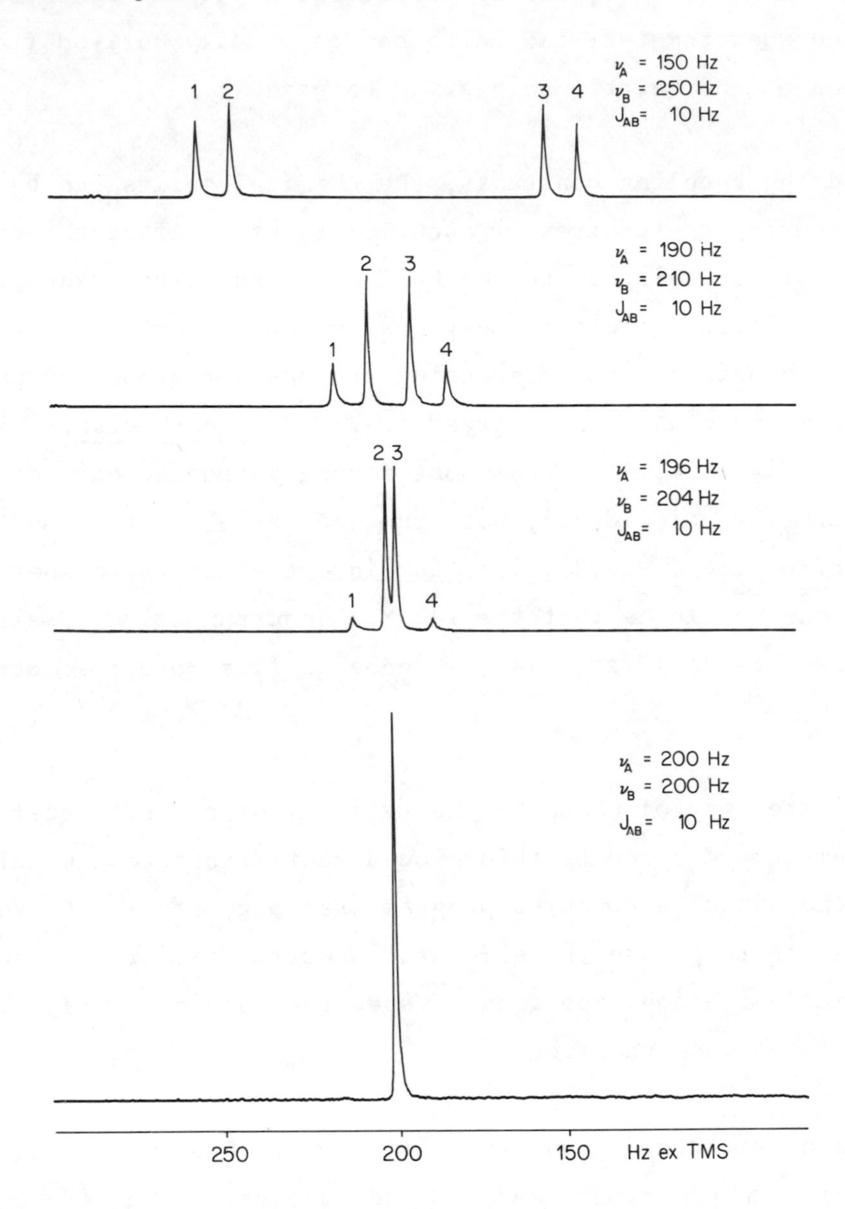

FIG. E.6 : SIMULATED ^{1}H NMR SPECTRA OF TWO-SPIN SYSTEMS

A number of problems in this book are designed for practice of first-order analysis and are of the same type as those given in Figs. E.4 and E.5

One can also gain an understanding of the operation of first-order rules by building up NMR spectra from a list of parameters. In such cases, the reader acts as a simplified version of the non-iterative program referred to on page 47. A number of such problems are also included in this collection.

(iv) Spin decoupling

It is possible to collapse the multiplicity due to spin-spin coupling by introducing an additional radio-frequency into the NMR spectrometer at the exact frequency of a group of coupled nuclei. This causes rapid upwards and downward transitions of the irradiated nuclei. As a consequence, any group of nuclei spin-spin coupled to the nuclei being irradiated will not experience two separate states from the irradiated nuclei and is said to be *decoupled* with consequent loss of multiplicity in the observed resonance.

The capacity of spin-decoupling techniques for the simplification of NMR spectra is obvious. Unfortunately, for technical reasons, it is not possible to decouple strongly coupled nuclei.

(v) Special features of ^{13}C NMR spectra

^{13}C is a rare isotope (1.06% at natural abundance) and this leads to certain consequences in ^{13}C NMR spectroscopy connected with spin-spin coupling phenomena.

The most obvious consequence of the low natural abundance of ^{13}C is the very small probability of a second atom of ^{13}C being present at any particular position in a molecule where one ^{13}C atom is present. It follows that ^{13}C, ^{13}C spin-spin coupling has no consequences on the appearance of ^{13}C NMR spectra unless special techniques are employed. In effect, every ^{13}C resonance observed is due to a particular isotopoisomer present to the extent of 1.06% in the mixture of the isotopoisomers which comprise the sample. This must be contrasted with ^{1}H NMR spectra where all signals arise from the same single isotopoisomer because ^{1}H is, for all practical purposes, present in 100% natural abundance.

Thus complications due to spin-spin coupling in ^{13}C spectra arise solely from coupling to protons. The proton-carbon coupling constants fall into two distinct categories by size: the direct (across one bond) coupling constants, which are between 125 and 250 Hz and the much smaller (generally below 10 Hz) coupling constants between ^{13}C and ^{1}H separated by two or more bonds.

Because the ^{13}C and ^{1}H chemical shift ranges are separated by tens of megahertz at all usable fields, there is no technical difficulty in observing ^{13}C spectra with the simultaneous decoupling of all protons using *broad band* or *noise* decoupling. In such *proton decoupled* ^{13}C spectra the individual resonances for carbon atoms in different environments appear as singlets, making these spectra very simple to interpret. It must be recalled, however, that the relative intensities of these signals do not (see pages 33 and 59) give reliable information about the relative numbers of carbon atoms in different environments.

Another common mode of obtaining ^{13}C NMR spectra consists of applying a strong decoupling signal at a single frequency just outside the range of proton resonances. This is usually referred to as *off resonance decoupling* or SFORD (Single Frequency Off Resonance Decoupling). The effect of this technique on the ^{13}C spectrum is to "shrink" the values of splittings due to all carbon-proton coupling. Because all couplings other than across one bond are relatively small, they simply cease having any effect on the multiplicity in the carbon spectrum. The larger one bond C-H couplings remain, making it possible to distinguish by inspection whether a carbon atom is a part of a methyl group (quartet), methylene group (triplet), a methine (CH) group (doublet) or a quaternary carbon (a singlet, just as in the fully decoupled spectrum).

The SFORD mode is thus a very powerful tool for structure determination and most of the problems in this book show both fully decoupled and SFORD ^{13}C NMR spectra. Unfortunately, the appearance of the quartets, triplets and doublets in SFORD spectra sometimes departs from the ideal form because of the possibility of strong coupling between protons (carbons and protons are always weakly coupled because of the enormous $\Delta\nu_{^{13}C,\,^{1}H}$). It must also be stressed that only the multiplicity of the signals and not the actual spacings within the multiplets is of significance in the SFORD spectra. The spacings are governed principally by the numerical value of the difference between the decoupling frequency and the resonance frequencies of the protons involved and are thus dependent not only on molecular structure but on experimental parameters.

(vi) Correlation of coupling constants with structure

Interproton spin-spin coupling constants are of obvious value in obtaining structural information from spectra, in particular in linking together structural elements. The best known of these relationships are summarized below:

(a) Saturated systems

Geminal coupling (^2J) in (2) is defined as J_{AB}. The typical range is 10-16 Hz,

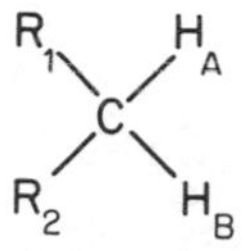

$J_{AB} \sim 10-16$ Hz

(2)

$J_{AB} \sim 6-8$ Hz

(3)

but values between 0 and 22 Hz have been recorded in less usual structures.

Vicinal coupling (^3J) is defined as J_{AB} in structure (3) and can take up values between 0 and 16 Hz depending largely on the dihedral angle Ø between the coupled protons. This dependence takes the form of the so-called Karplus relationship which is approximately stated in equations (i) and (ii):

$$J_{AB} = 10 \cos^2\phi \text{ for values of } \phi \text{ between 0 and } 90^\circ \quad \text{.......... (i)}$$

$$J_{AB} = 15 \cos^2\phi \text{ for values of } \phi \text{ between 90 and } 180^\circ \quad \text{........ (ii)}$$

The Karplus relationship has proved of great value in determining the stereochemistry of organic molecules but must be treated with caution even in its more sophisticated forms, because vicinal coupling constants also depend on the nature of substituents.

In systems assuming average conformation, such as the ethyl group, ^3J generally lies between 6 and 8 Hz.

(b) Unsaturated and aromatic systems

All coupling constants in the olefinic fragment (4) depend upon the nature of substituents, but unless substituents of quite unusual nature are present, the *cis* ($^3J_{AB}$) and *trans* ($^3J_{AC}$) ranges do not overlap thus permitting assignment of geometry. J_{cis} also depends on the ring size in cyclic structures.

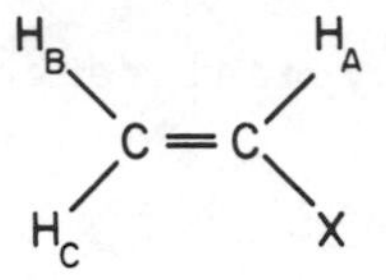

$^3J_{AB}$ (cis) = 6 – 11 Hz

$^3J_{AC}$ (trans) = 12 – 19 Hz

$^2J_{BC}$ (gem) = 0 – 3 Hz

(4)

The magnitude of the long-range allylic coupling, $^4J_{AB}$ in structure (5) is controlled by the dihedral angle between the C-H_A bond and the plane of the double bond in a relationship reminiscent of the Karplus relation.

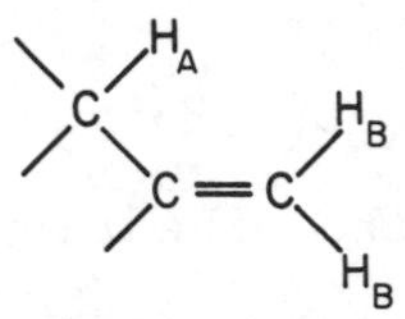

$^4J_{AB}$ = 0-3 Hz

(5)

The *ortho* (^3J), *meta* (^4J) and *para*(^5J) coupling constants in benzene derivatives generally fall in the ranges indicated in structure (6). In practice, spin-spin splitting between *para* related protons is usually not

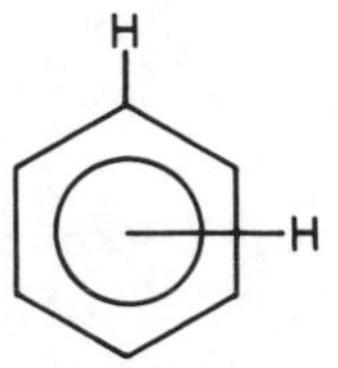

$^3J_{ortho}$ = 6 - 10 Hz

$^4J_{meta}$ = 1-3 Hz

$^5J_{para}$ = 0-1.5 Hz

(6)

resolved. Coupling constants in heterocyclic systems assume similar values except that *ortho* coupling in five-membered rings and next to the heteroatoms is significantly smaller.

5. MISCELLANEOUS TOPICS

This section deals briefly with a number of topics in NMR spectroscopy, which while not directly pertinent to the solution of the problems of the type collected in this book, give some idea of the power of the method in the solution of complex structural problems.

(i) The Nuclear Overhauser Effect (NOE)

Irradiation of one nucleus while observing the resonance of another may result in a change of multiplicity of the signal if the nuclei are spin-spin coupled (section E.4), but may also result in a change in the amplitude of the observed resonance. This phenomenon, known as the *nuclear Overhauser effect* (NOE) is quite independent of spin-spin coupling and has its origin in the relaxation process (see page 32). Typically an enhancement of a few per cent (theoretically up to 50%) is observed between protons which are in close spatial proximity (the effect is inversely proportional to the sixth power of distance), but because it can be measured quite accurately by difference spectroscopy, it is a very powerful means for the determination of structures of organic compounds.

The intensities of ^{13}C resonances may increase up to 200% when directly bonded ^{1}H nuclei are irradiated. This effect is very important in increasing the sensitivity in the proton-decoupled ^{13}C spectra (page 56) but the variability of the proton/carbon NOE contributes to the non-quantitative nature of ^{13}C NMR spectra (page 34).

It must be clearly understood that spin-spin coupling and NOE are different phenomena and that the coupling interaction takes place via the bonds separating the nuclei, while the NOE effect is transmitted through space.

(ii) Aromatic Solvent Induced Shift (ASIS) and Lanthanide Induced Shift (LIS)

If a solvent associates significantly with the solute and in addition has magnetically anisotropic groups (page 40), the position of resonances of the solute is likely to be significantly different from the values obtained in the usual (page 41) solvents. The most useful solvents for the purpose of inducing such *solvent-shifts* are aromatic solvents, in particular perdeuterobenzene, and the effect is named *aromatic solvent induced shift* (ASIS). The numerical values of ASIS are usually of the order of 0.1 - 0.5 ppm and they vary with the molecule studied depending mainly on the geometry of the complexation. The most obvious utility of ASIS is the removal of accidental equivalence (page 39) in ^{1}H NMR spectra.

Much larger effects (up to 10 ppm) may be obtained when the complexing reagent is a lanthanide derivative, typically a tris-β-diketonate of europium or praseodymium. Such *lanthanide induced shifts* (LIS) not only simplify spectra drastically by "expanding" the chemical shift scale, but the values of LIS for different groups within a molecule may be correlated with distances from the complexing functionality, such as a carbonyl or a hydroxyl group.

(iii) Advanced Techniques

Since the advent of time-domain NMR spectroscopy (page 34) a number of advanced techniques have been devised which yield enormously valuable information for the solution of complex structural problems. The principles behind such techniques are well beyond the scope of this book and the spectrometer output may not even have the appearance of conventional NMR spectra such as those shown in the problems in this collection. We shall merely list below the sort of structural information which is available from some of these techniques.

One group of advanced techniques results in correlations: thus it is possible to correlate the chemical shifts of carbons with the chemical shifts of the protons directly coupled to them. This technique is based on spin-decoupling (page 55) and is of obvious utility. It is also possible to correlate in a similar manner which protons are coupled to other protons, even if all the resonances are not resolved, and to obtain NOE (page 59) maps.

Another group of useful techniques are the so-called editing sequences, which permit the assignment of ^{13}C resonances to methyl, methylene, methine and quaternary carbons much more reliably than the SFORD mode (page 56) used in the problems collected in this book.

Possibly the most exciting recent development is a technique which gives the carbon connectivity within the organic molecule.

(iv) The NMR Spectra of "Other Nuclei"

As mentioned above (page 30) ^{1}H and ^{13}C NMR spectroscopy accounts for the overwhelming proportion of all NMR observations. However, there are many isotopes with $I \neq 0$ and they include the common isotopes ^{19}F, ^{31}P and ^{2}H. The NMR spectroscopy of these "other nuclei" has had surprisingly little impact on the solution of structural problems and will not be discussed here. It is however important to be alert for the presence of other magnetic nuclei in the molecule, because they often cause additional, and otherwise inexplicable, multiplicity in ^{1}H and ^{13}C NMR spectra due to spin-spin coupling.

F. PROBLEMS

As mentioned in the preface, the principal purpose of this book is to present a collection of suitable problems.

Problems 1 - 115 are all of the basic "structures from spectra" type, are generally considerably simpler than those found in other collections and are arranged roughly in order of increasing complexity.

No solutions to the problem are given so that they can be used for examination purposes. When we have used these problems for this purpose, we have requested the examinee not only to deduce the structures but also to: "assign NMR spectra as completely as you can, rationalize all numbered peaks in the mass spectrum and account for all significant features of the UV and IR spectra". Clearly, the selection process we have used in the numbering of some peaks in the Mass Spectra is connected with this format of examining: the peaks were chosen to illustrate the common types of fragmentation reactions in the mass spectrometer (pages 25-28).

Bona fide instructors may obtain a list of solutions to problems 1 - 115 by writing to the publishers.

The last group of problems (116 - 131) are of a different type and deal in one way or another with interpretation of simple ^{1}H NMR spin-spin multiplets. To the best of our knowledge, problems of this type are not available in other collections and they are included here because we have found that the interpretation of multiplicity in ^{1}H NMR spectra is the greatest single cause of confusion in the minds of students.

The spectra presented in problems were obtained under conditions stated on individual problem sheets. Mass spectra were obtained on an AEI MS-9 spectrometer. 100 MHz ^{1}H NMR spectra were obtained on a Varian XL-100 or a Varian HA-100 spectrometer. 400 MHz ^{1}H NMR spectra were obtained on a Bruker WM-400 spectrometer and 60 MHz ^{1}H NMR spectra were obtained on a Jeol

FX60-Q spectrometer. ^{13}C NMR spectra were obtained at 15 MHz on a Jeol FX60-Q spectrometer, at 20 MHz on a Varian CFT-20 spectrometer and at 100 MHz on a Bruker WM-400 spectrometer.

Ultraviolet spectra were recorded on a Perkin-Elmer 402 spectrophotometer and Infrared spectra on a Perkin-Elmer 710B spectrometer.

While there is no doubt in our minds that the only way to acquire facility in obtaining "organic structures from spectra" is to practise, some students have found the following general approach to solving structural problems by a combination of spectroscopic methods helpful:

1. Perform all routine operations:
   (a) Determine the molecular weight from the Mass Spectrum.
   (b) Determine relative numbers of protons in different environments from the ^{1}H NMR spectrum.
   (c) Determine the number of carbons in different environments and the number of quaternary carbons, methine carbons, methylene carbons and methyl carbons from the ^{13}C NMR spectrum.
   (d) Examine the problem for any additional data concerning composition and determine the molecular formula if possible.
   (e) Determine the molar absorptivity in UV, if applicable.

2. Examine each spectrum in turn for obvious structural elements
   (a) IR spectrum for carbonyl groups, hydroxyl groups, NH, C≡C, etc.
   (b) Mass spectrum for typical fragments e.g.

   $PhCH_2$-, CH_3CO-, CH_3-, etc.

   (c) UV spectrum for conjugation, aromatic rings etc.
   (d) ^{1}H NMR spectrum for CH_3-, CH_3CH_2-, aromatic protons, $-CH_nX$, exchangeable groups etc.

3. Write down all structural elements you have determined. Note that some are monofunctional (i.e. must be end-groups, such as $-CH_3$) whereas some are bifunctional (e.g. -CO-, $-CH_2-$, -COO-), or trifunctional (e.g. $-\overset{|}{\underset{|}{C}}H$, $\overset{|}{\underset{/\backslash}{N}}$).
   Add up the atomic weights of your structural elements and compare this sum with the molecular weight of the unknown. The difference (if any) may give a clue to the nature of the undetermined structural elements (e.g. an ether oxygen). At this stage elements of symmetry may become apparent.

4. Try to assemble the structural elements. Note that there may be more than one way of fitting them together. Spin-spin coupling data or information about conjugation may enable you to make a definite choice between possibilities.

5. Return to each spectrum in turn and rationalize all major features (especially all major fragments in the mass spectrum and all features of the NMR spectra) in terms of your proposed structure.

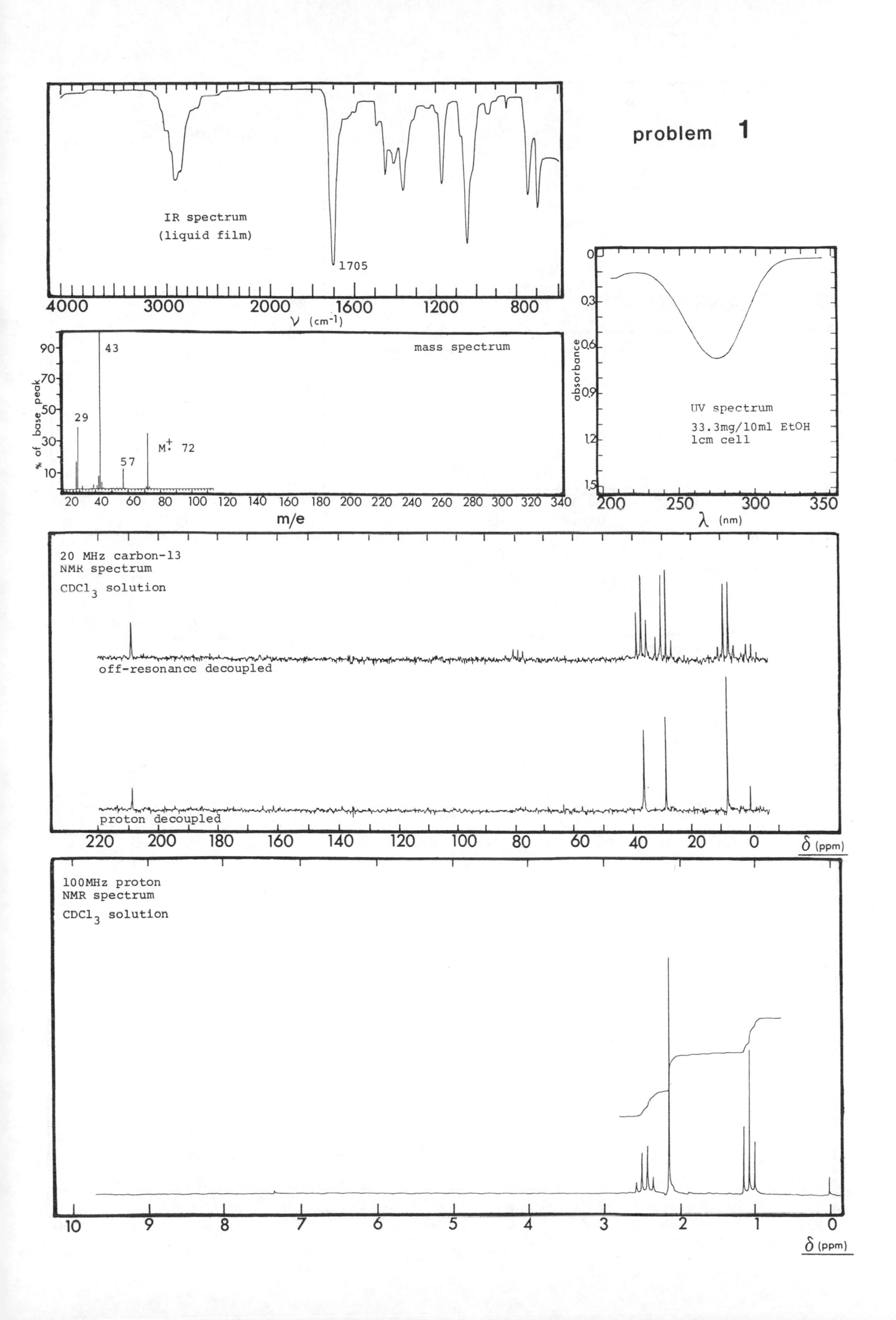
problem 1
IR spectrum
(liquid film)
1705
4000
3000
2000
1600
1200
800
ν (cm⁻¹)
mass spectrum
% of base peak
43
29
57
M⁺˙ 72
m/e
UV spectrum
33.3mg/10ml EtOH
1cm cell
absorbance
λ (nm)
20 MHz carbon-13
NMR spectrum
CDCl3 solution
off-resonance decoupled
proton decoupled
δ (ppm)
100MHz proton
NMR spectrum
CDCl3 solution
δ (ppm)

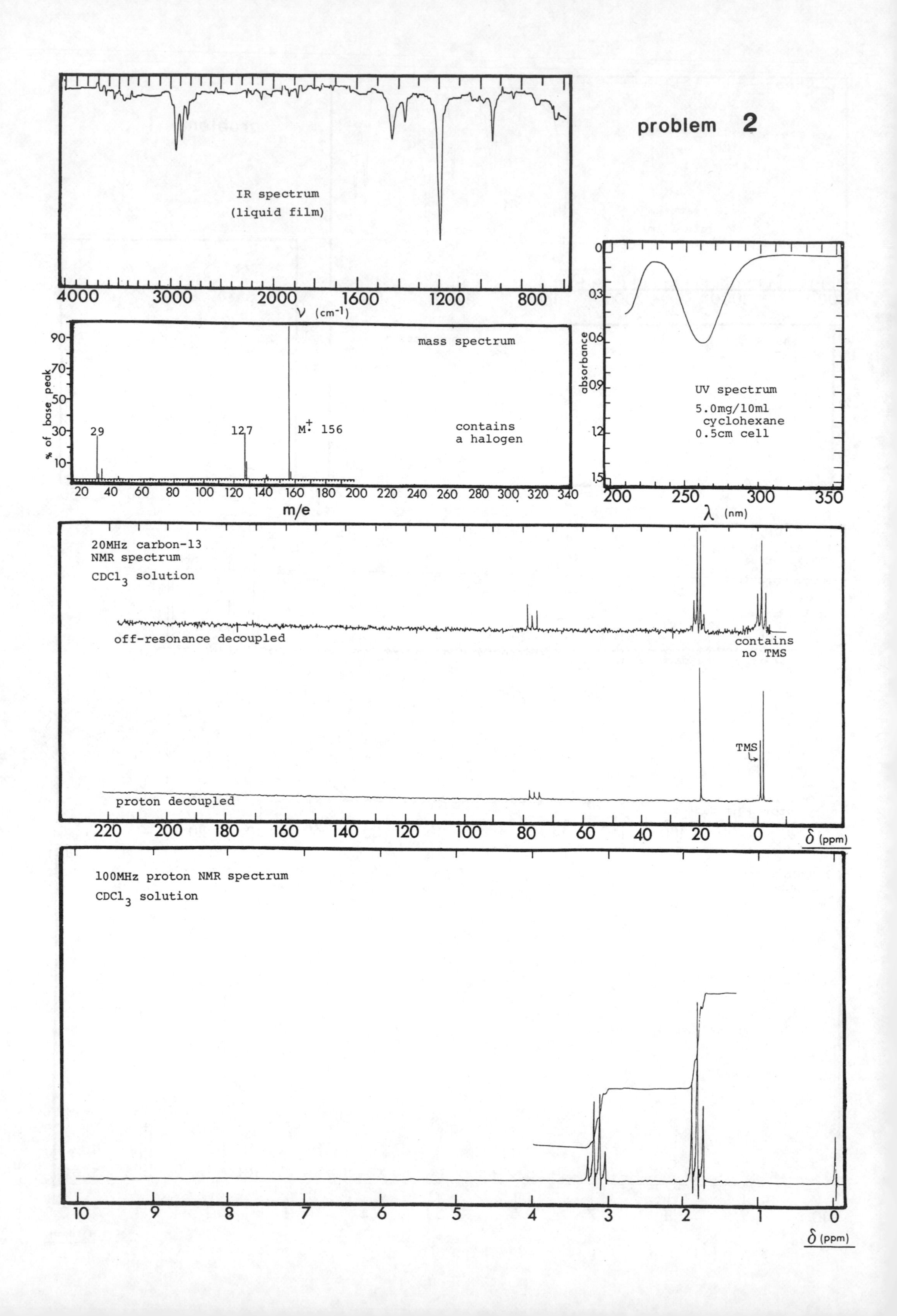

problem 2
IR spectrum
(liquid film)
4000
3000
2000
1600
1200
800
ν (cm-1)
mass spectrum
% of base peak
90
70
50
30
10
29
127
M+· 156
contains
a halogen
20
40
60
80
100
120
140
160
180
200
220
240
260
280
300
320
340
m/e
absorbance
0
0.3
0.6
0.9
1.2
1.5
UV spectrum
5.0mg/10ml
cyclohexane
0.5cm cell
200
250
300
350
λ (nm)
20MHz carbon-13
NMR spectrum
CDCl3 solution
off-resonance decoupled
contains
no TMS
TMS
proton decoupled
220
200
180
160
140
120
100
80
60
40
20
0
δ (ppm)
100MHz proton NMR spectrum
CDCl3 solution
10
9
8
7
6
5
4
3
2
1
0
δ (ppm)

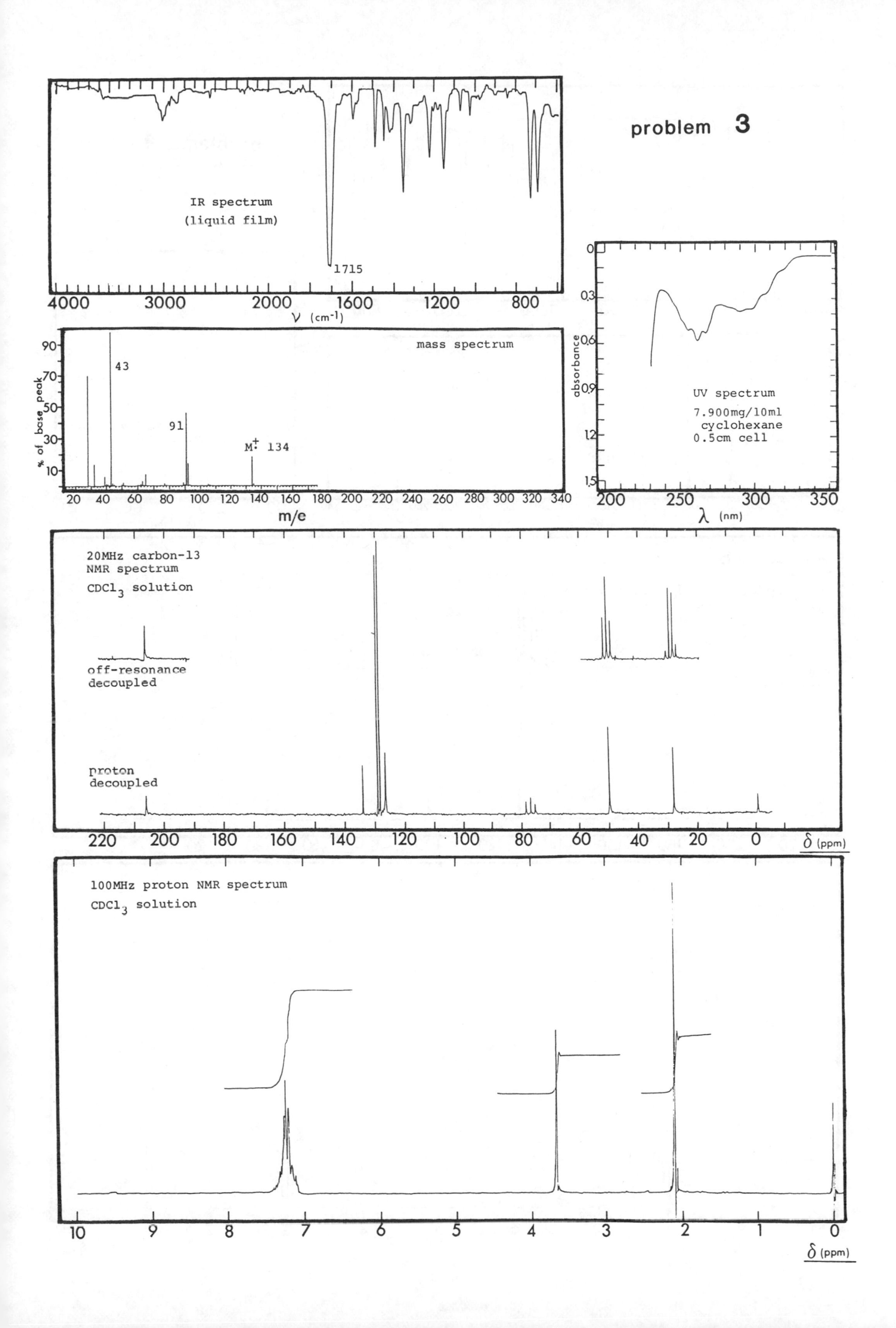

problem 3
IR spectrum
(liquid film)
1715
4000
3000
2000
1600
1200
800
ν (cm-1)
mass spectrum
43
91
M+ 134
% of base peak
m/e
UV spectrum
7.900mg/10ml
cyclohexane
0.5cm cell
absorbance
λ (nm)
20MHz carbon-13
NMR spectrum
CDCl3 solution
off-resonance
decoupled
proton
decoupled
δ (ppm)
100MHz proton NMR spectrum
CDCl3 solution
δ (ppm)

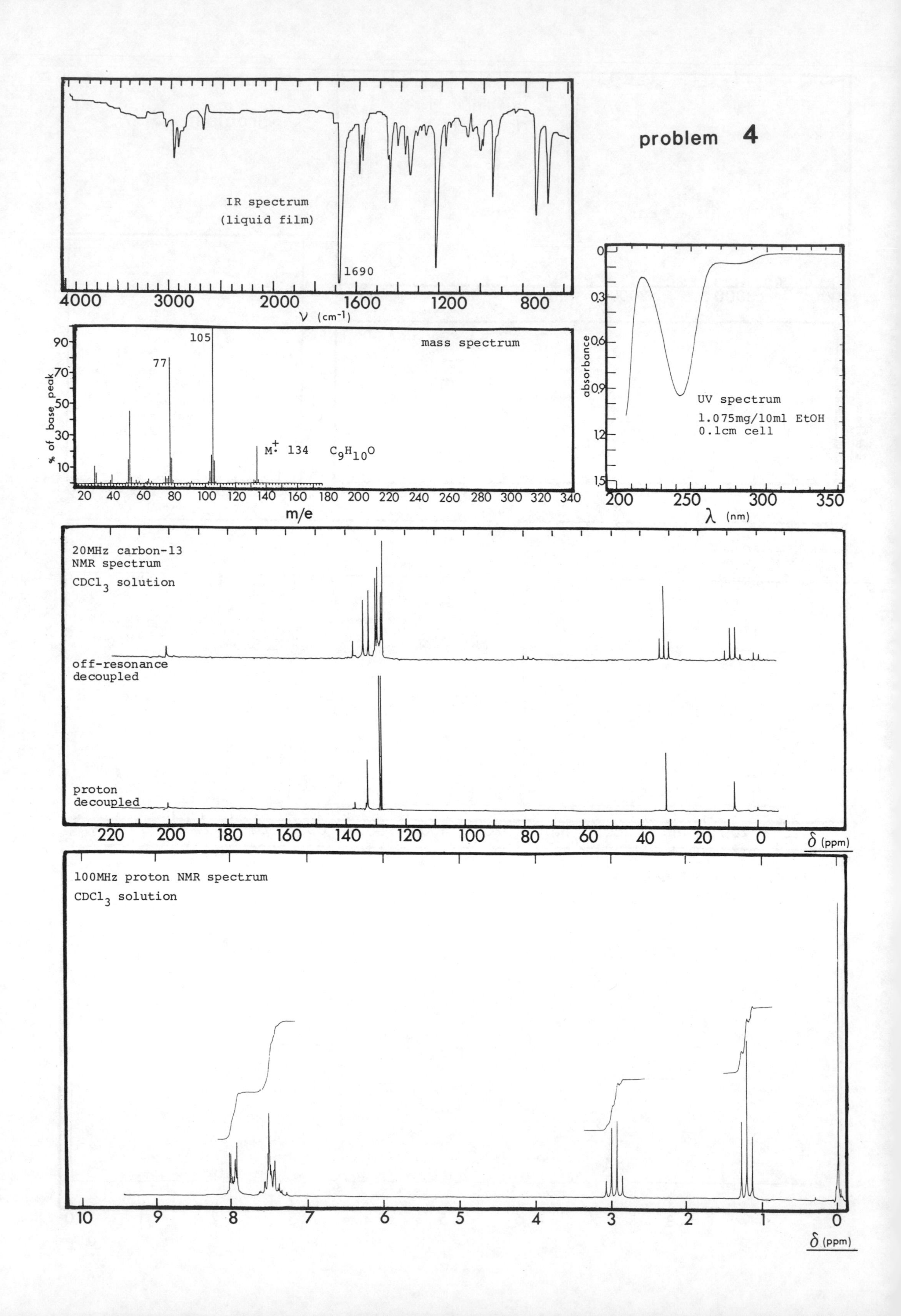
problem 4
IR spectrum
(liquid film)
1690
4000
3000
2000
1600
1200
800
ν (cm⁻¹)
mass spectrum
105
77
M⁺· 134 C9H10O
% of base peak
m/e
UV spectrum
1.075mg/10ml EtOH
0.1cm cell
absorbance
λ (nm)
20MHz carbon-13
NMR spectrum
CDCl3 solution
off-resonance
decoupled
proton
decoupled
δ (ppm)
100MHz proton NMR spectrum
CDCl3 solution
δ (ppm)

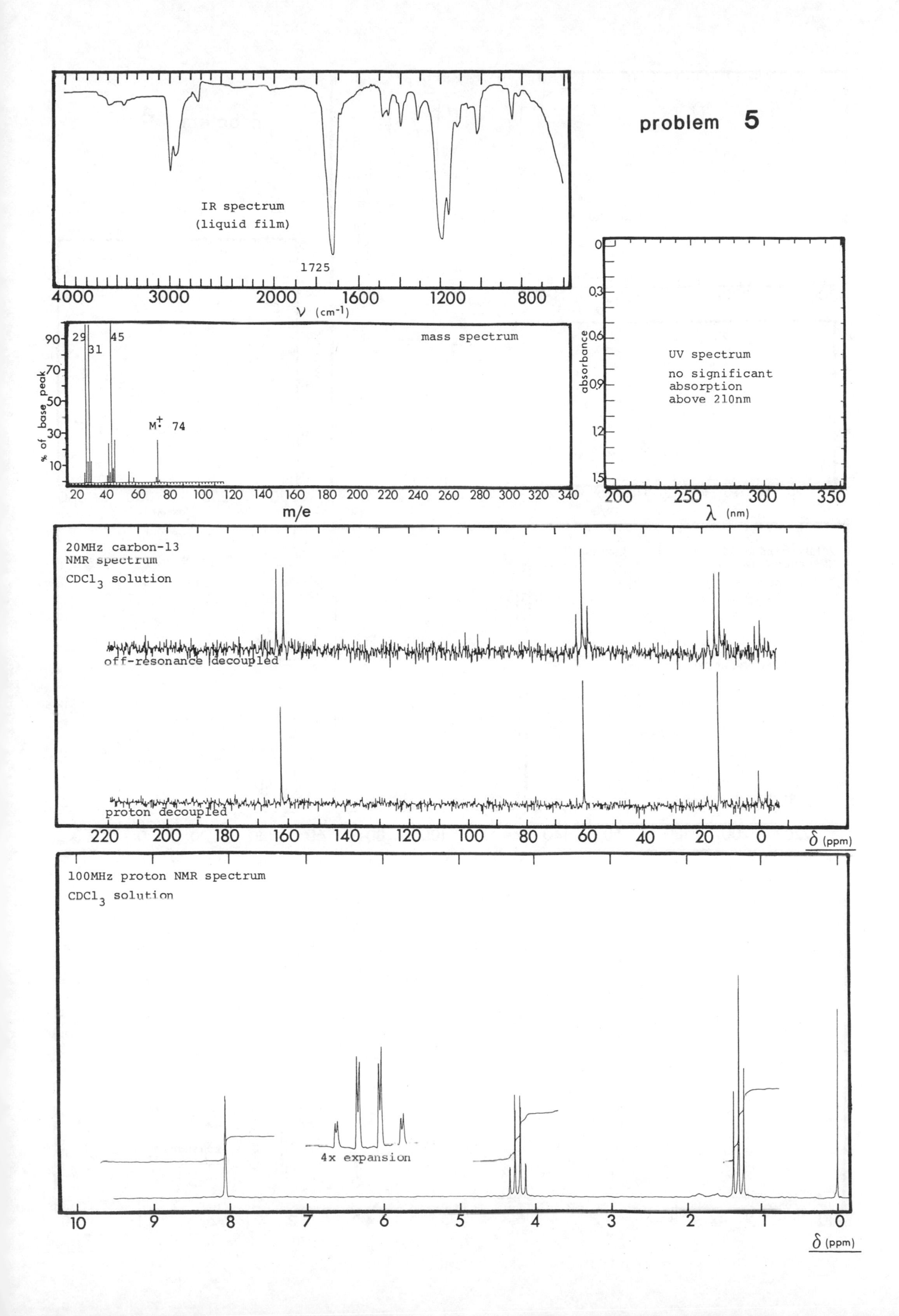

problem 5
IR spectrum
(liquid film)
1725
4000
3000
2000
1600
1200
800
ν (cm-1)
mass spectrum
29
31
45
M+· 74
% of base peak
m/e
UV spectrum
no significant
absorption
above 210nm
absorbance
λ (nm)
20MHz carbon-13
NMR spectrum
CDCl3 solution
off-resonance decoupled
proton decoupled
δ (ppm)
100MHz proton NMR spectrum
CDCl3 solution
4x expansion
δ (ppm)

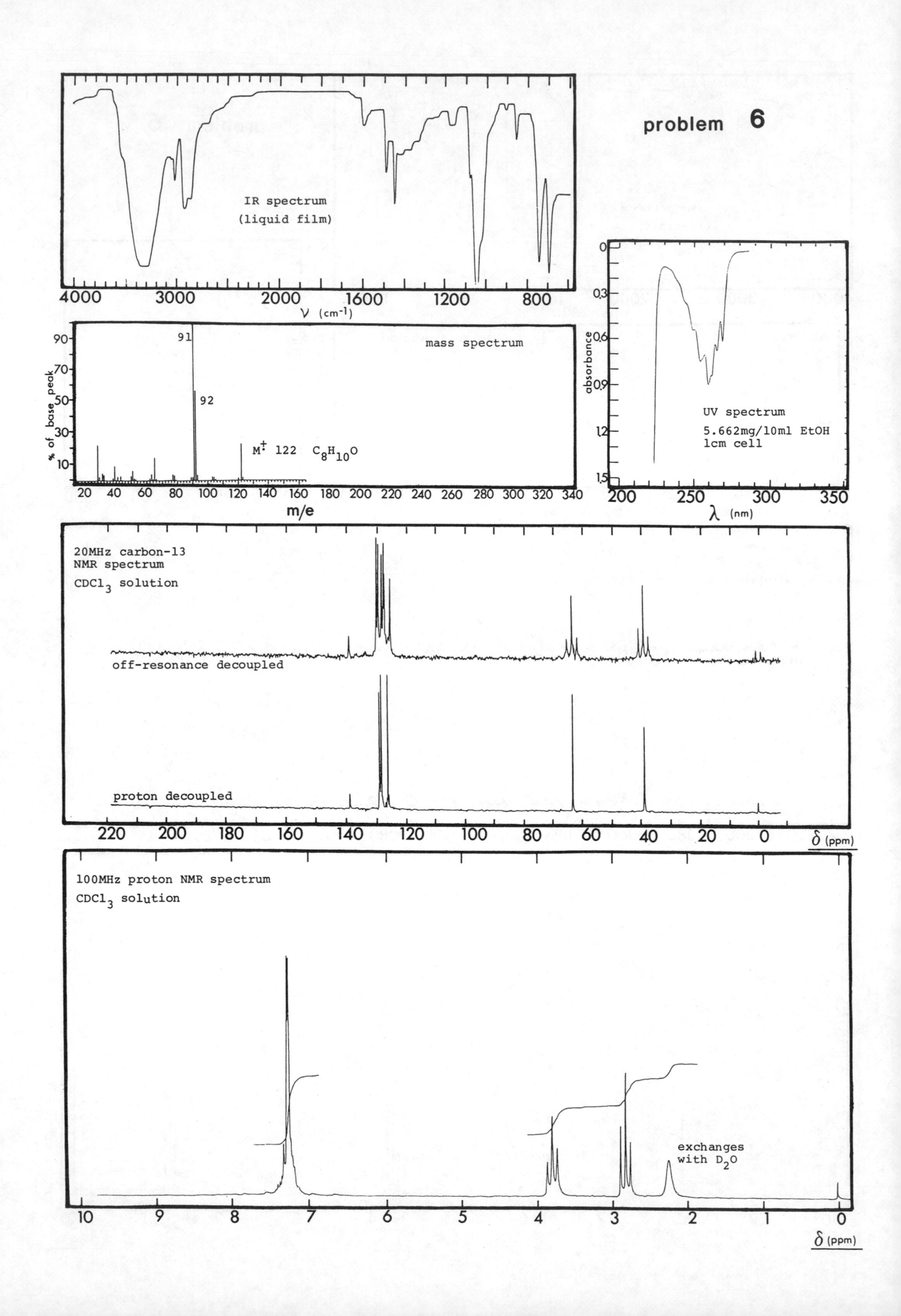

problem 6
IR spectrum
(liquid film)
4000
3000
2000
1600
1200
800
ν (cm-1)
mass spectrum
% of base peak
91
92
M+ 122 C8H10O
m/e
UV spectrum
5.662mg/10ml EtOH
1cm cell
absorbance
λ (nm)
20MHz carbon-13
NMR spectrum
CDCl3 solution
off-resonance decoupled
proton decoupled
δ (ppm)
100MHz proton NMR spectrum
CDCl3 solution
exchanges
with D2O
δ (ppm)

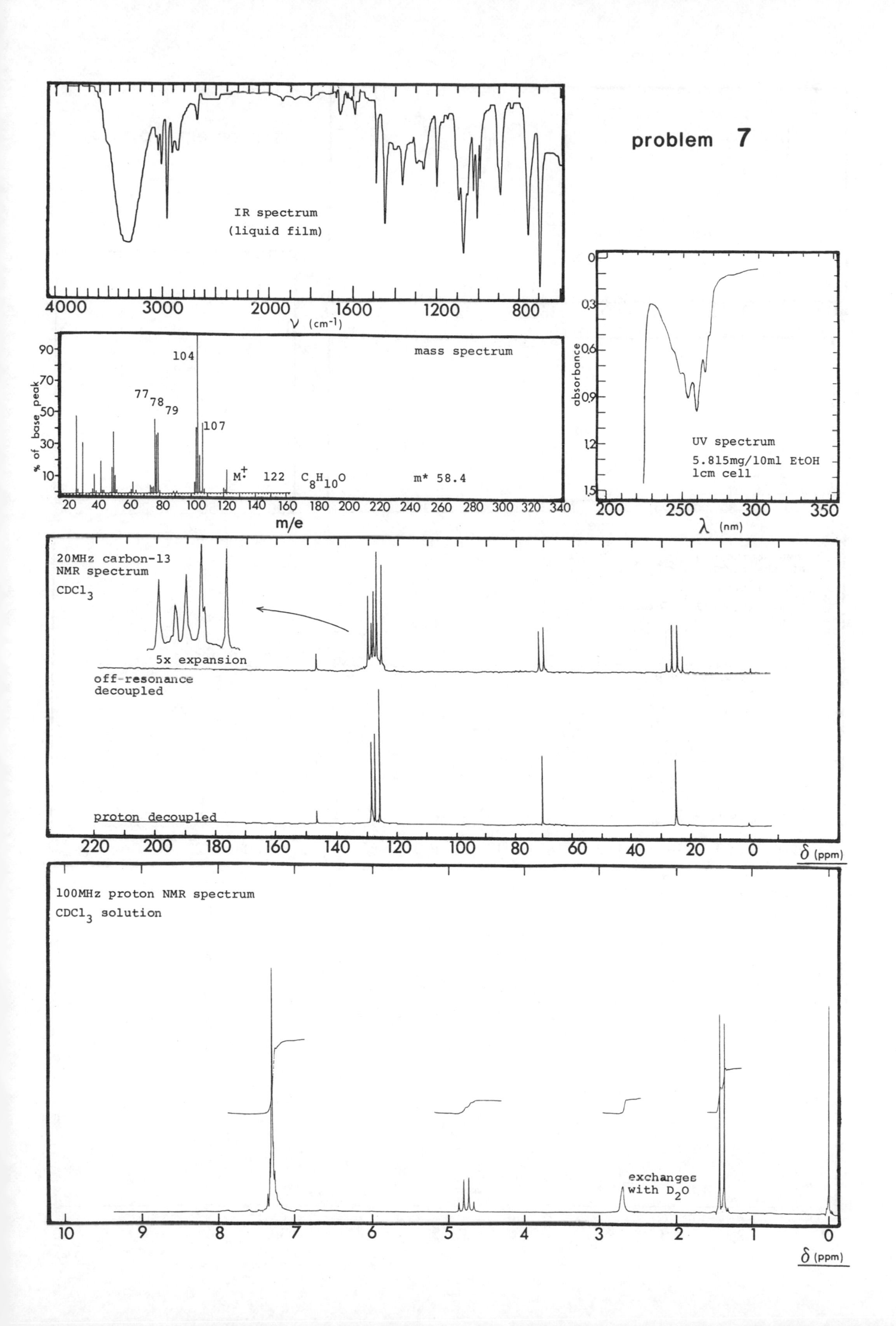
problem 7
IR spectrum
(liquid film)
4000
3000
2000
1600
1200
800
ν (cm-1)
mass spectrum
104
77 78 79
107
M+· 122 C8H10O
m* 58.4
% of base peak
m/e
UV spectrum
5.815mg/10ml EtOH
1cm cell
absorbance
λ (nm)
20MHz carbon-13
NMR spectrum
CDCl3
5x expansion
off-resonance
decoupled
proton decoupled
δ (ppm)
100MHz proton NMR spectrum
CDCl3 solution
exchanges
with D2O
δ (ppm)

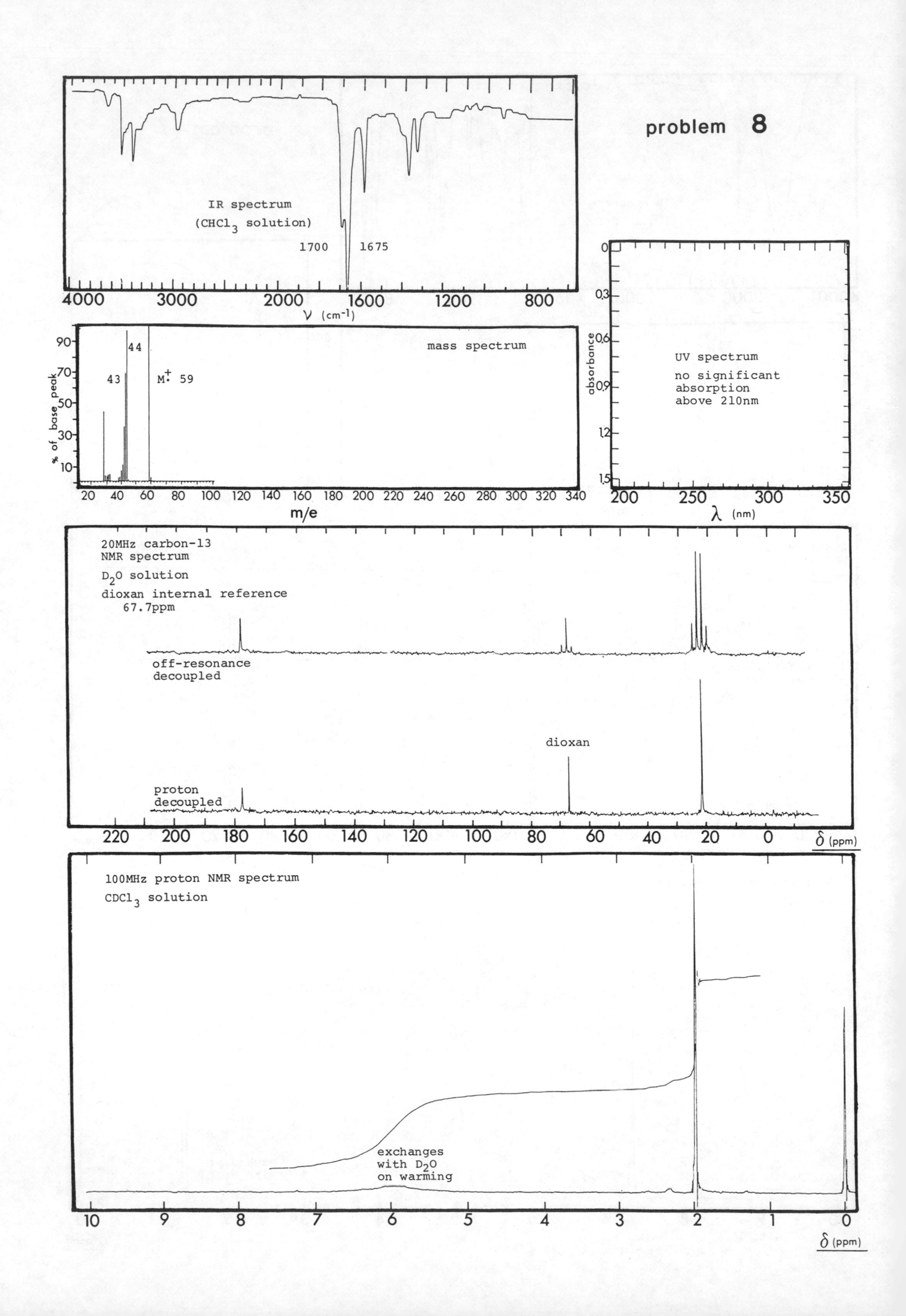

problem 8
IR spectrum
(CHCl3 solution)
1700
1675
4000
3000
2000
1600
1200
800
ν (cm-1)
mass spectrum
44
43
M+· 59
% of base peak
m/e
UV spectrum
no significant absorption above 210nm
absorbance
λ (nm)
20MHz carbon-13 NMR spectrum
D2O solution
dioxan internal reference 67.7ppm
off-resonance decoupled
dioxan
proton decoupled
δ (ppm)
100MHz proton NMR spectrum
CDCl3 solution
exchanges with D2O on warming
δ (ppm)

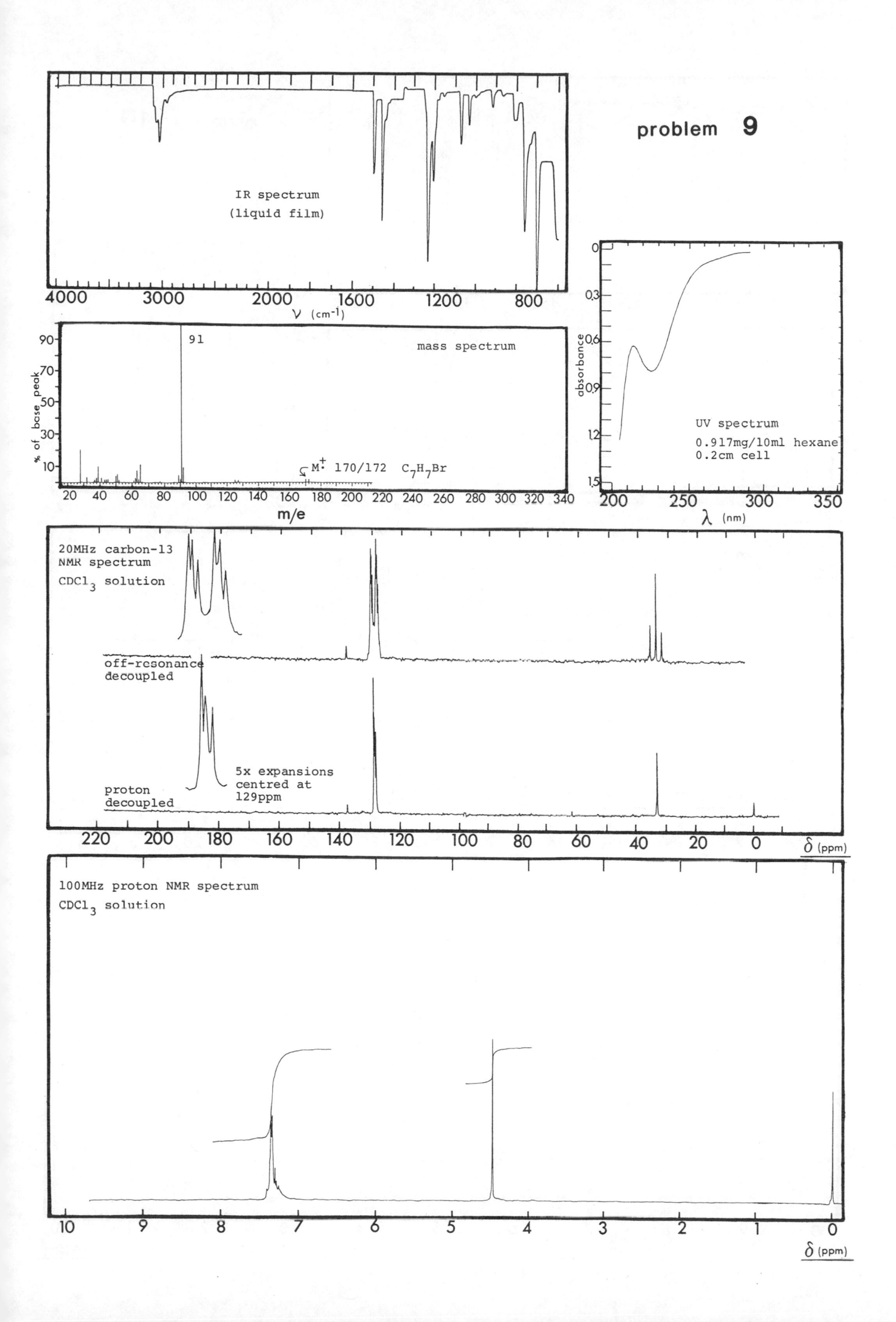

problem 9
IR spectrum
(liquid film)
4000
3000
2000
1600
1200
800
ν (cm-1)
mass spectrum
91
M+ 170/172 C7H7Br
% of base peak
m/e
UV spectrum
0.917mg/10ml hexane
0.2cm cell
absorbance
λ (nm)
20MHz carbon-13
NMR spectrum
CDCl3 solution
off-resonance
decoupled
proton
decoupled
5x expansions
centred at
129ppm
δ (ppm)
100MHz proton NMR spectrum
CDCl3 solution
δ (ppm)

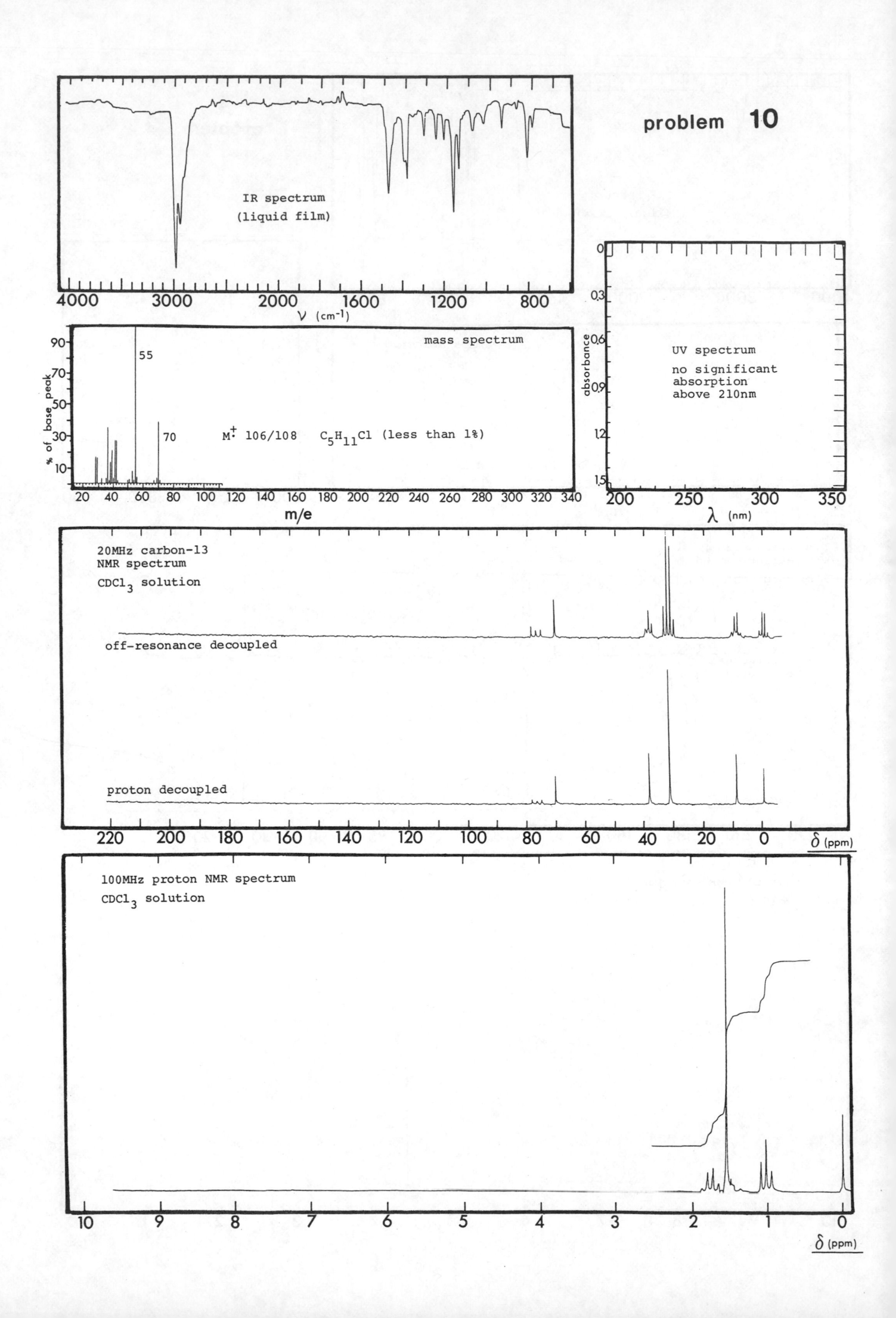
problem 10
IR spectrum
(liquid film)
4000
3000
2000
1600
1200
800
ν (cm⁻¹)
mass spectrum
55
70
M⁺· 106/108 C5H11Cl (less than 1%)
% of base peak
m/e
UV spectrum
no significant absorption above 210nm
absorbance
λ (nm)
20MHz carbon-13 NMR spectrum
CDCl3 solution
off-resonance decoupled
proton decoupled
δ (ppm)
100MHz proton NMR spectrum
CDCl3 solution
δ (ppm)

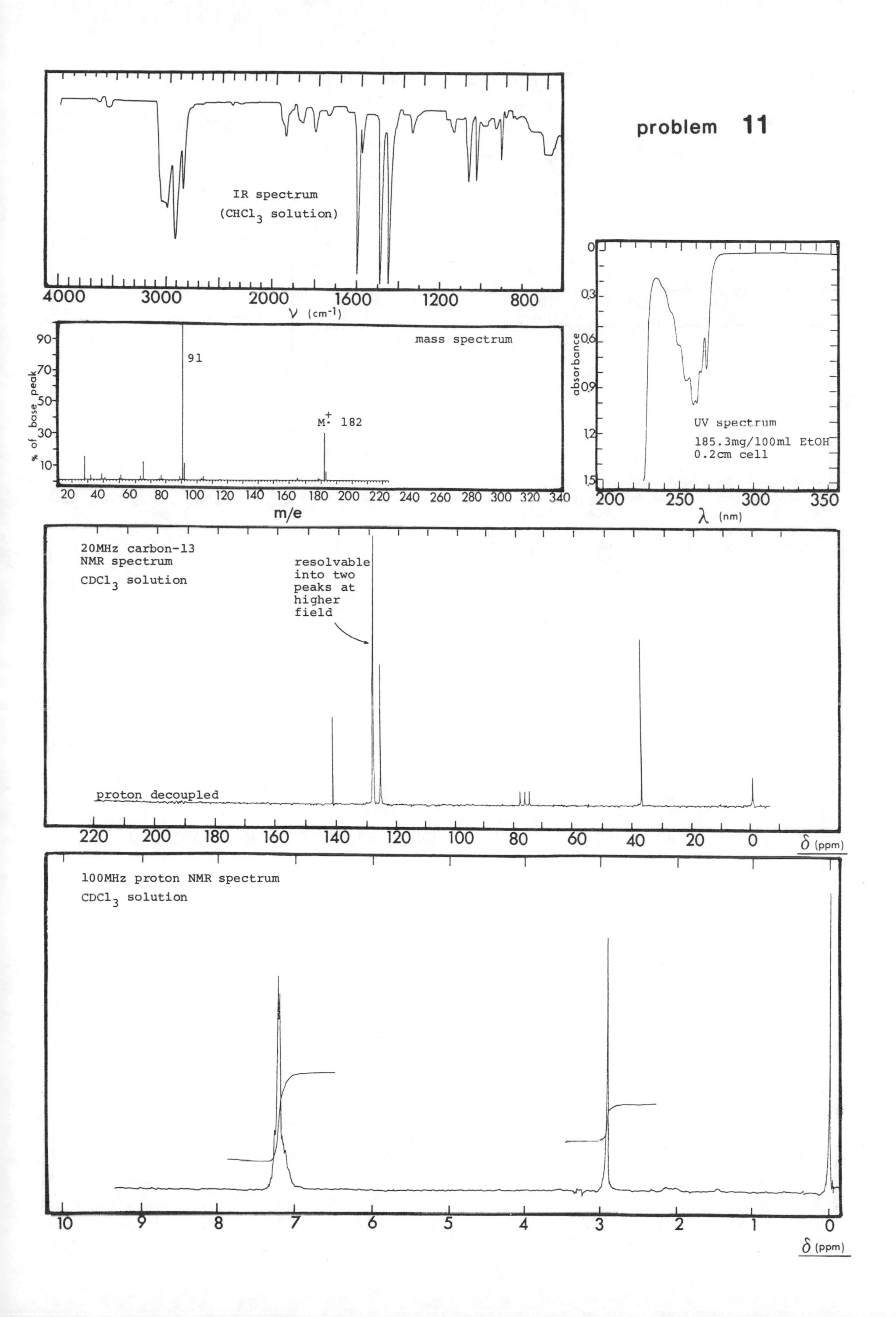

problem 11
IR spectrum
(CHCl3 solution)
4000
3000
2000
1600
1200
800
ν (cm-1)
mass spectrum
% of base peak
91
M+· 182
m/e
UV spectrum
185.3mg/100ml EtOH
0.2cm cell
absorbance
λ (nm)
20MHz carbon-13
NMR spectrum
CDCl3 solution
resolvable
into two
peaks at
higher
field
proton decoupled
δ (ppm)
100MHz proton NMR spectrum
CDCl3 solution
δ (ppm)

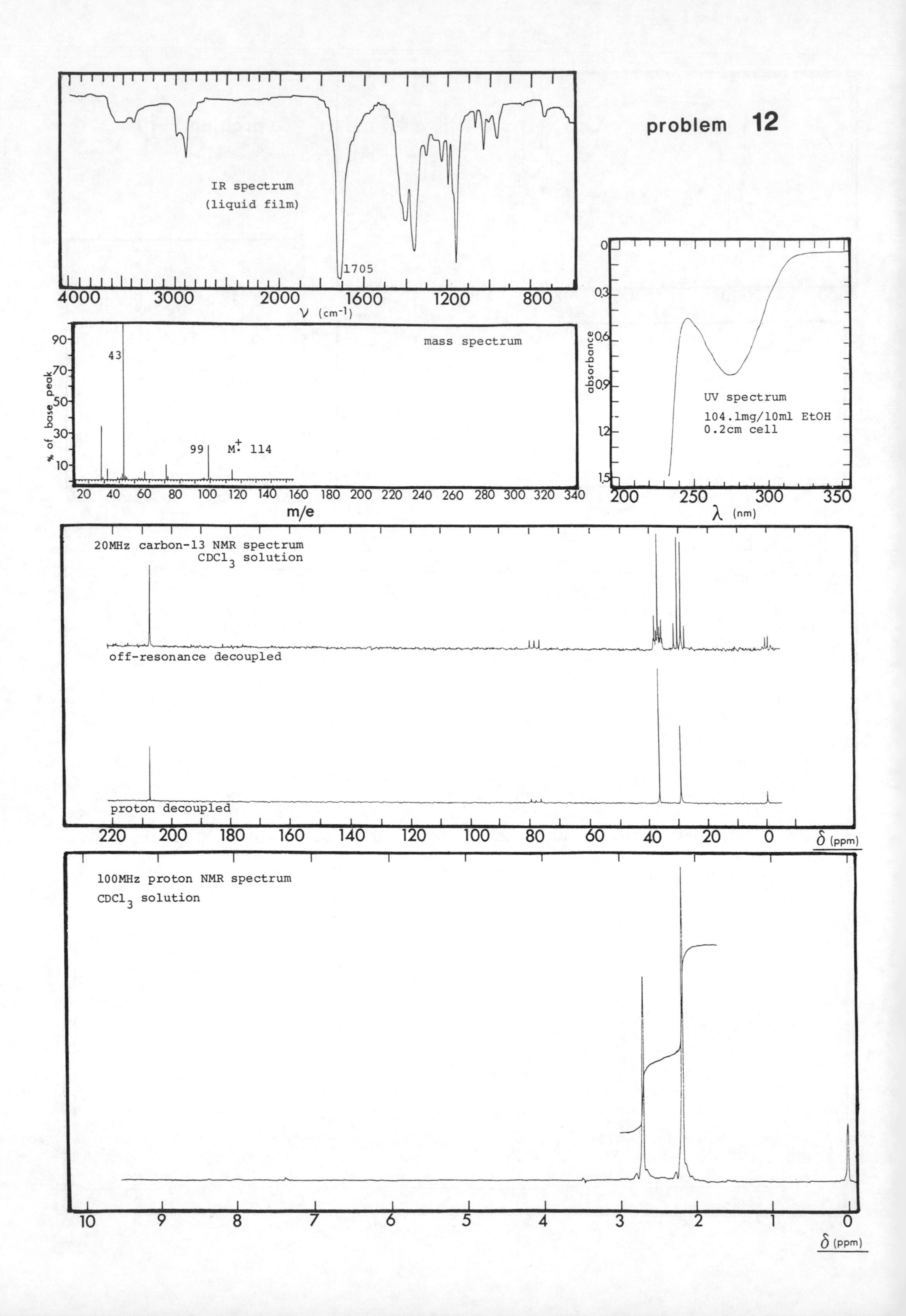
problem 12
IR spectrum
(liquid film)
1705
4000
3000
2000
1600
1200
800
ν (cm⁻¹)
mass spectrum
43
99
M⁺· 114
% of base peak
m/e
UV spectrum
104.1mg/10ml EtOH
0.2cm cell
absorbance
λ (nm)
20MHz carbon-13 NMR spectrum
CDCl₃ solution
off-resonance decoupled
proton decoupled
δ (ppm)
100MHz proton NMR spectrum
CDCl₃ solution
δ (ppm)

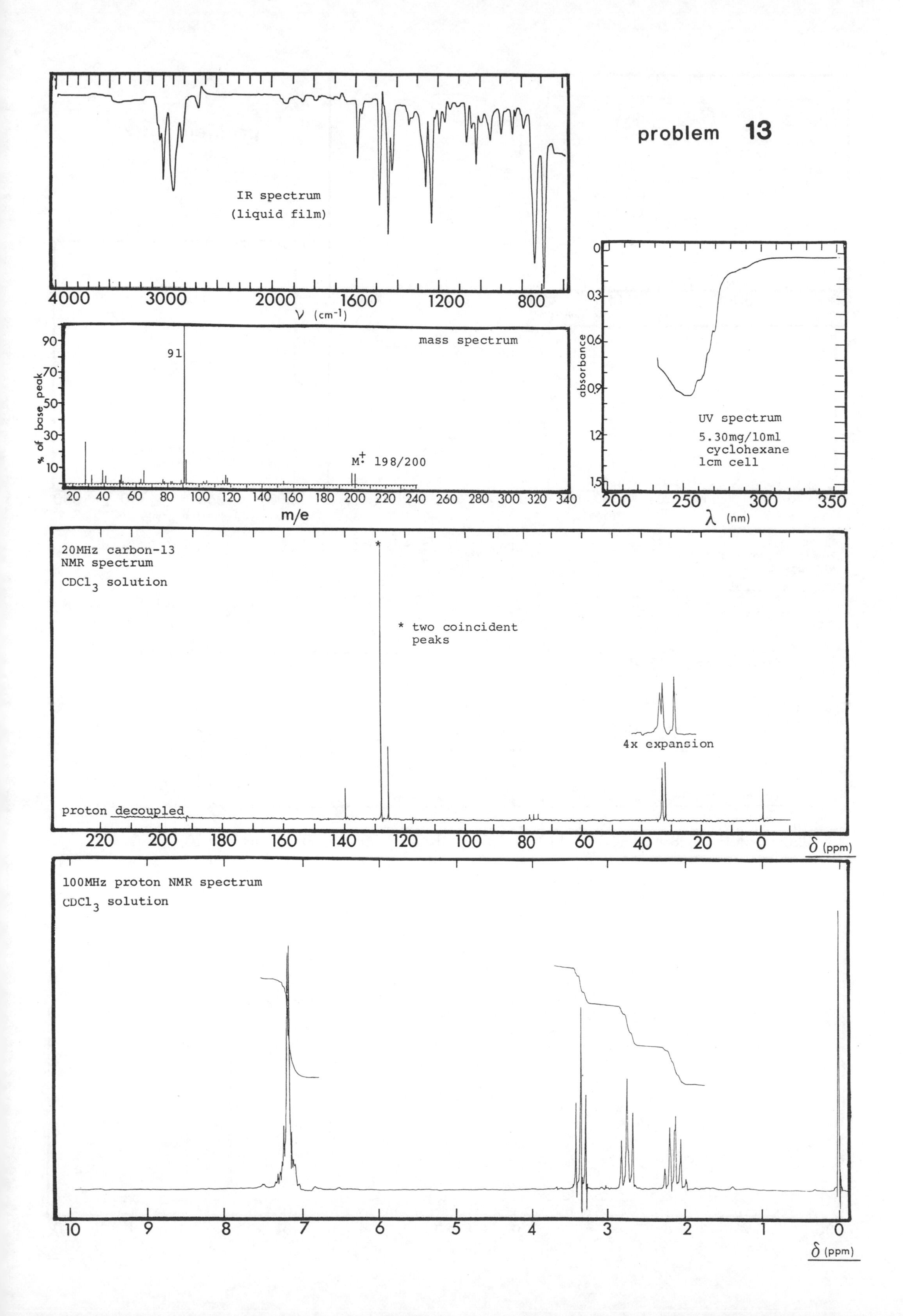

problem 13
IR spectrum
(liquid film)
4000
3000
2000
1600
1200
800
ν (cm⁻¹)
mass spectrum
% of base peak
91
M⁺· 198/200
m/e
absorbance
UV spectrum
5.30mg/10ml
cyclohexane
1cm cell
λ (nm)
20MHz carbon-13
NMR spectrum
CDCl3 solution
* two coincident
peaks
4x expansion
proton decoupled
δ (ppm)
100MHz proton NMR spectrum
CDCl3 solution
δ (ppm)

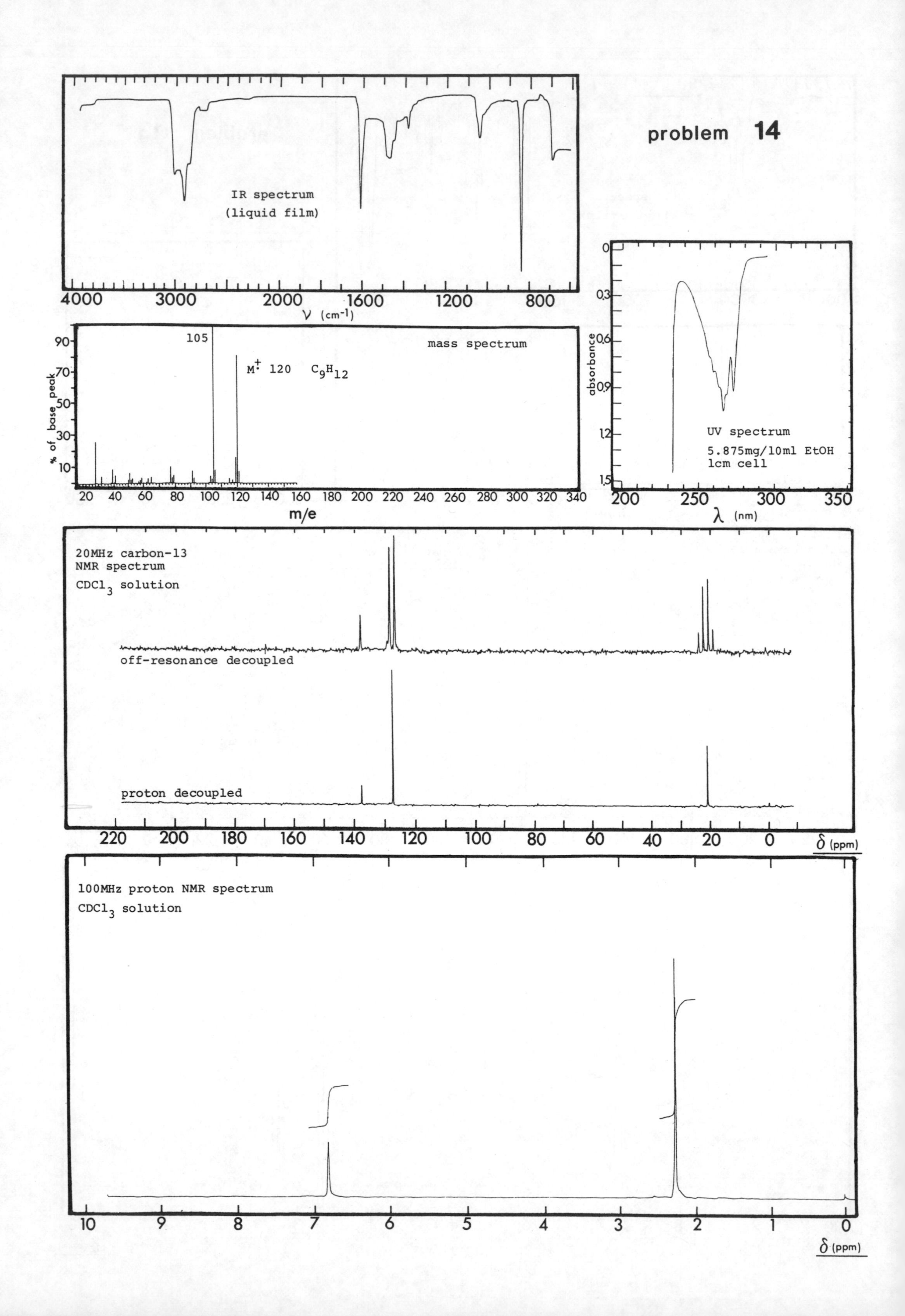
problem 14
IR spectrum
(liquid film)
4000
3000
2000
1600
1200
800
ν (cm-1)
mass spectrum
105
M+ 120 C9H12
% of base peak
m/e
UV spectrum
5.875mg/10ml EtOH
1cm cell
absorbance
λ (nm)
20MHz carbon-13
NMR spectrum
CDCl3 solution
off-resonance decoupled
proton decoupled
δ (ppm)
100MHz proton NMR spectrum
CDCl3 solution
δ (ppm)

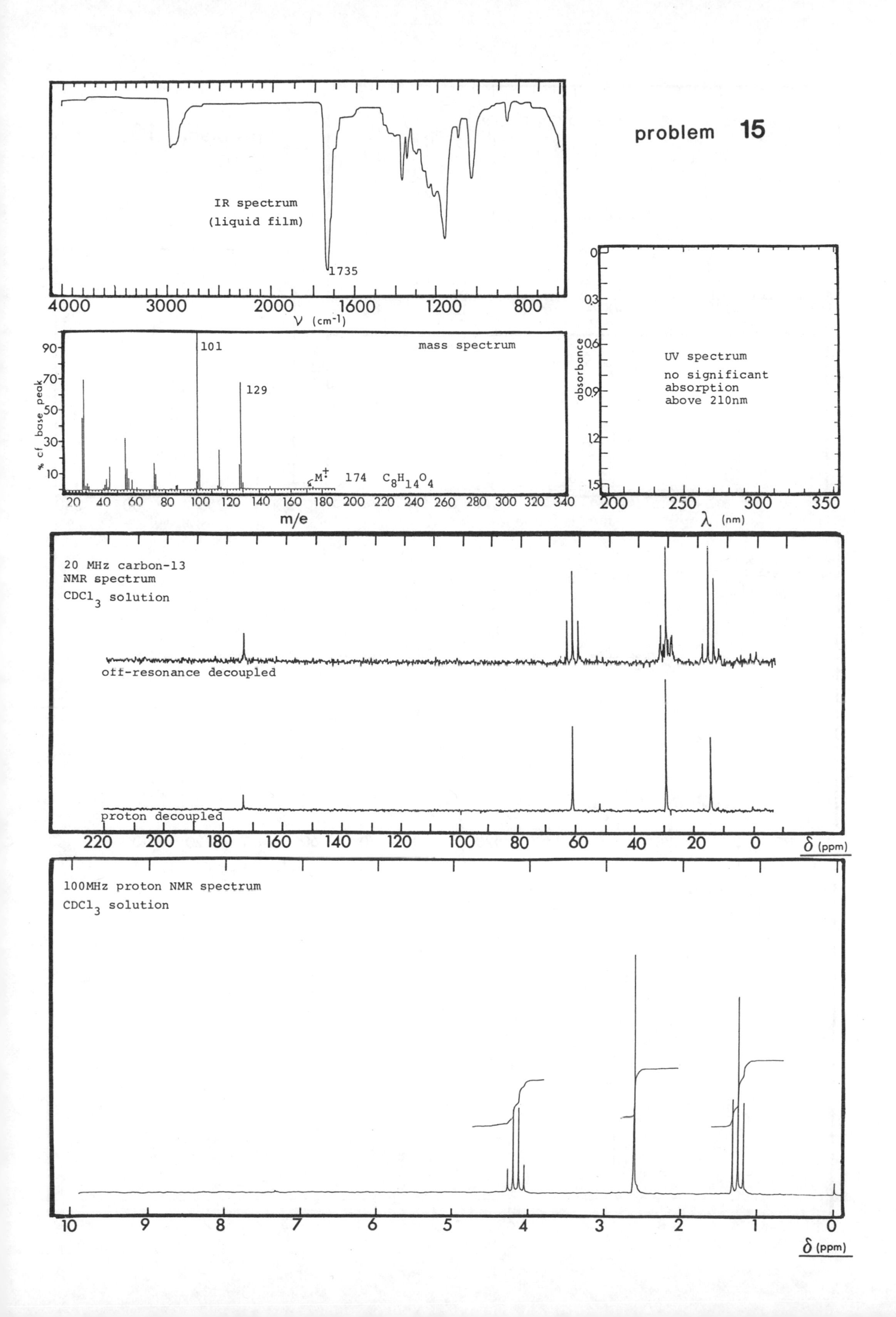

problem 15
IR spectrum
(liquid film)
1735
4000
3000
2000
1600
1200
800
ν (cm-1)
mass spectrum
101
129
M+ 174 C8H14O4
% of base peak
m/e
UV spectrum
no significant absorption above 210nm
absorbance
λ (nm)
20 MHz carbon-13 NMR spectrum
CDCl3 solution
off-resonance decoupled
proton decoupled
δ (ppm)
100MHz proton NMR spectrum
CDCl3 solution
δ (ppm)

problem **16**

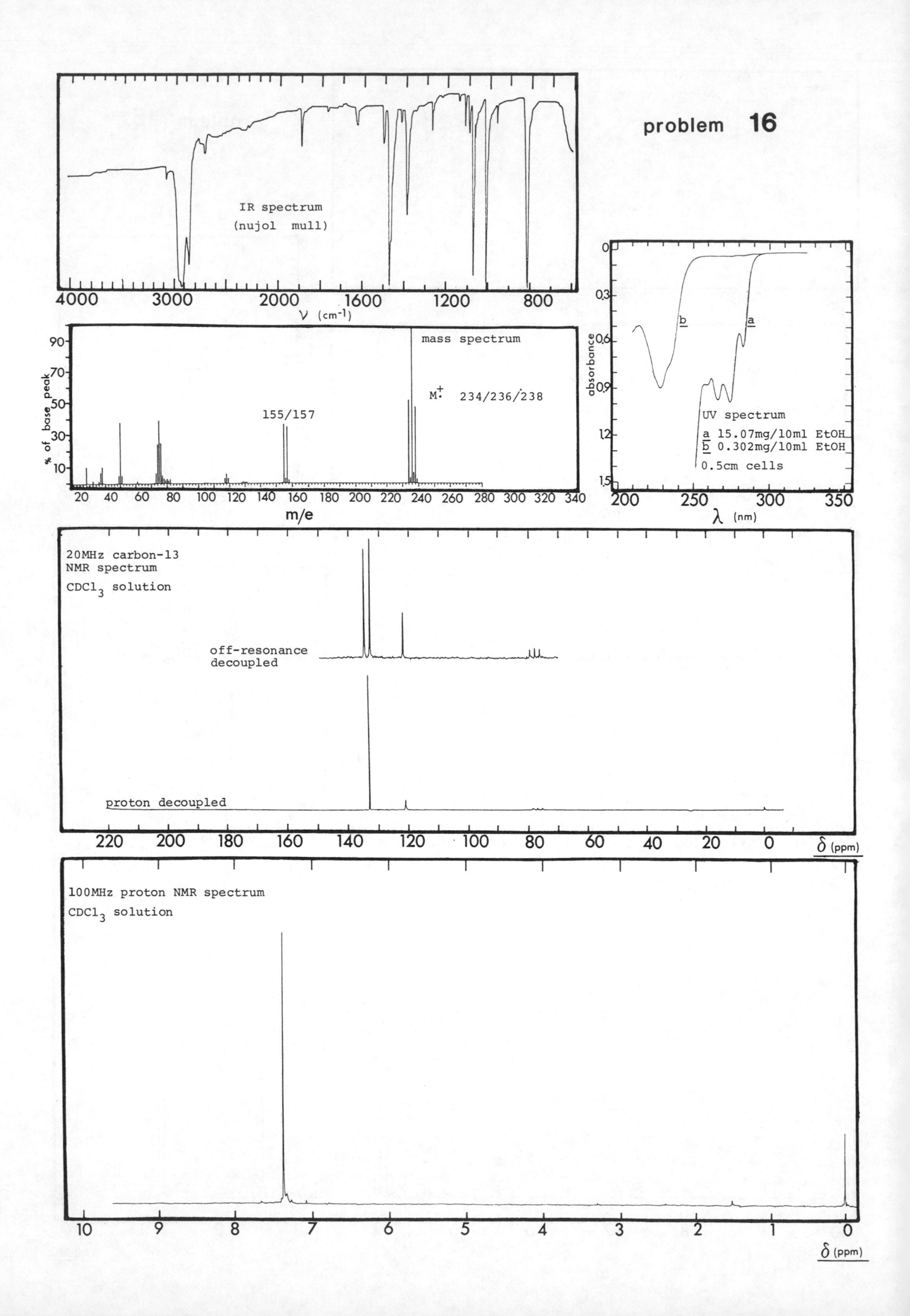

IR spectrum
(nujol mull)
4000
3000
2000
1600
1200
800
ν (cm⁻¹)
mass spectrum
M+· 234/236/238
155/157
% of base peak
m/e
UV spectrum
a 15.07mg/10ml EtOH
b 0.302mg/10ml EtOH
0.5cm cells
absorbance
λ (nm)
20MHz carbon-13
NMR spectrum
CDCl3 solution
off-resonance
decoupled
proton decoupled
δ (ppm)
100MHz proton NMR spectrum
CDCl3 solution
δ (ppm)

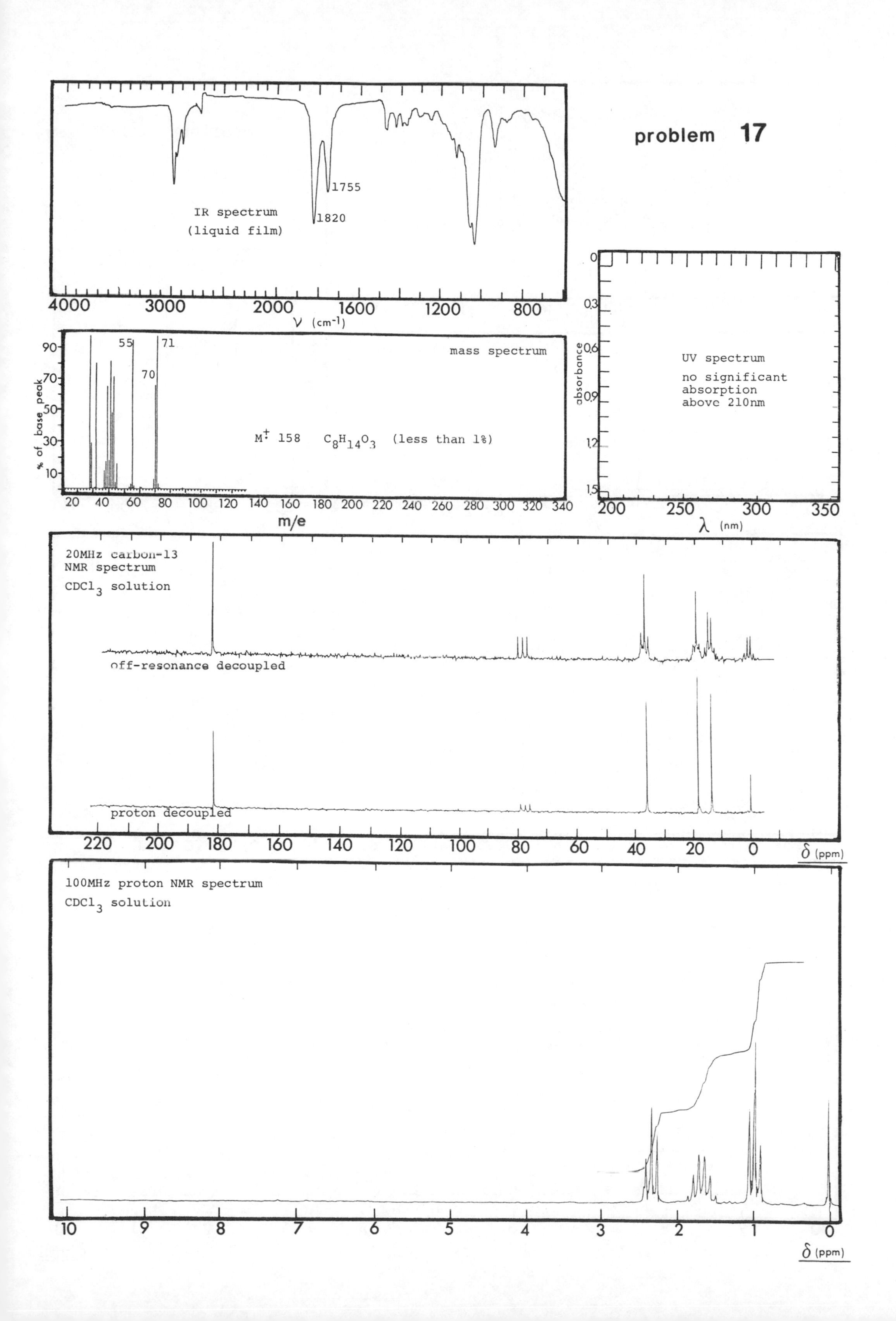

problem 17
IR spectrum
(liquid film)
1755
1820
4000
3000
2000
1600
1200
800
ν (cm-1)
mass spectrum
55
71
70
M+ 158 C8H14O3 (less than 1%)
% of base peak
m/e
UV spectrum
no significant
absorption
above 210nm
absorbance
λ (nm)
20MHz carbon-13
NMR spectrum
CDCl3 solution
off-resonance decoupled
proton decoupled
δ (ppm)
100MHz proton NMR spectrum
CDCl3 solution
δ (ppm)

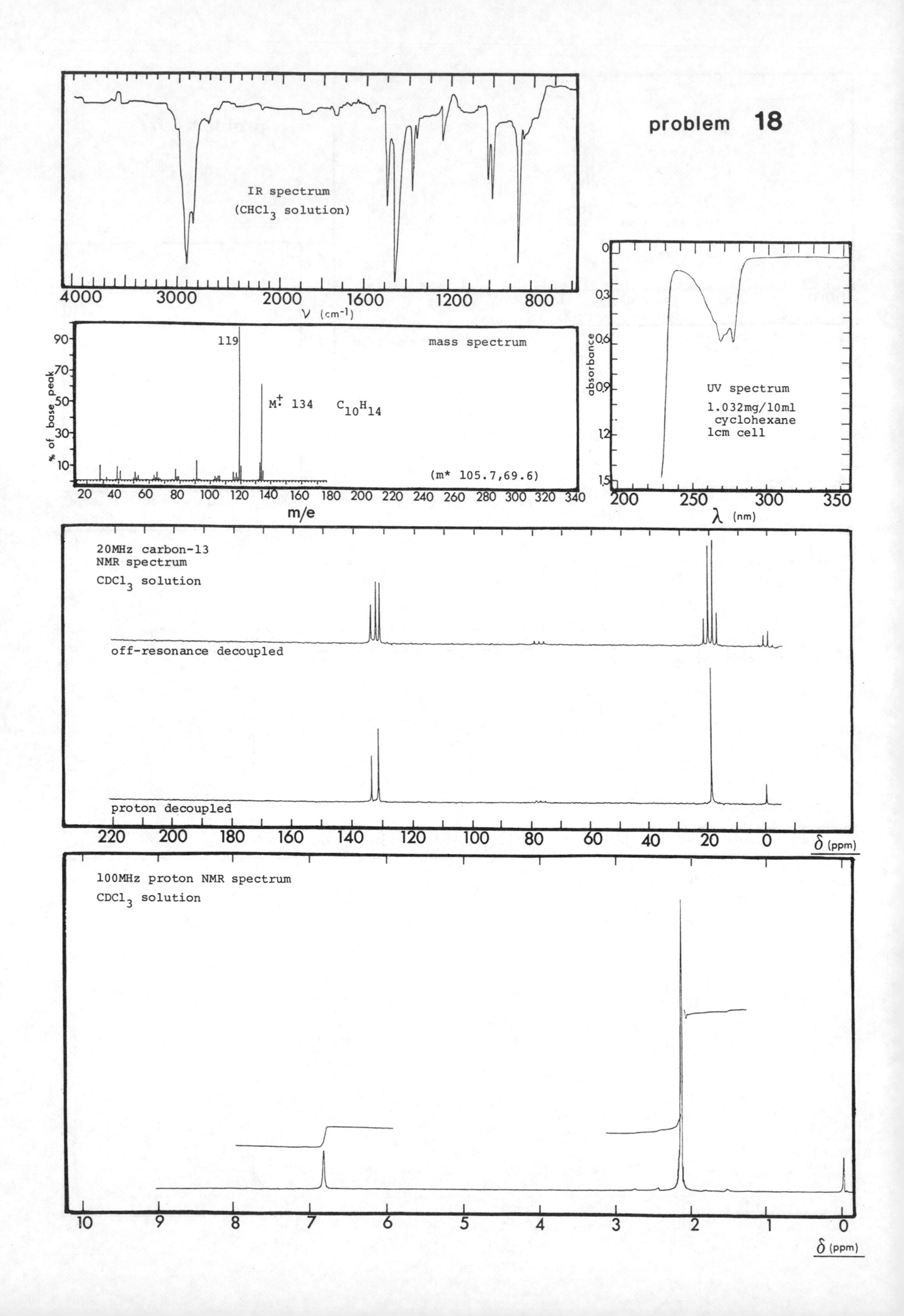
problem 18
IR spectrum
(CHCl3 solution)
4000
3000
2000
1600
1200
800
ν (cm-1)
mass spectrum
119
M+· 134
C10H14
(m* 105.7,69.6)
% of base peak
m/e
UV spectrum
1.032mg/10ml
cyclohexane
1cm cell
absorbance
λ (nm)
20MHz carbon-13
NMR spectrum
CDCl3 solution
off-resonance decoupled
proton decoupled
δ (ppm)
100MHz proton NMR spectrum
CDCl3 solution
δ (ppm)

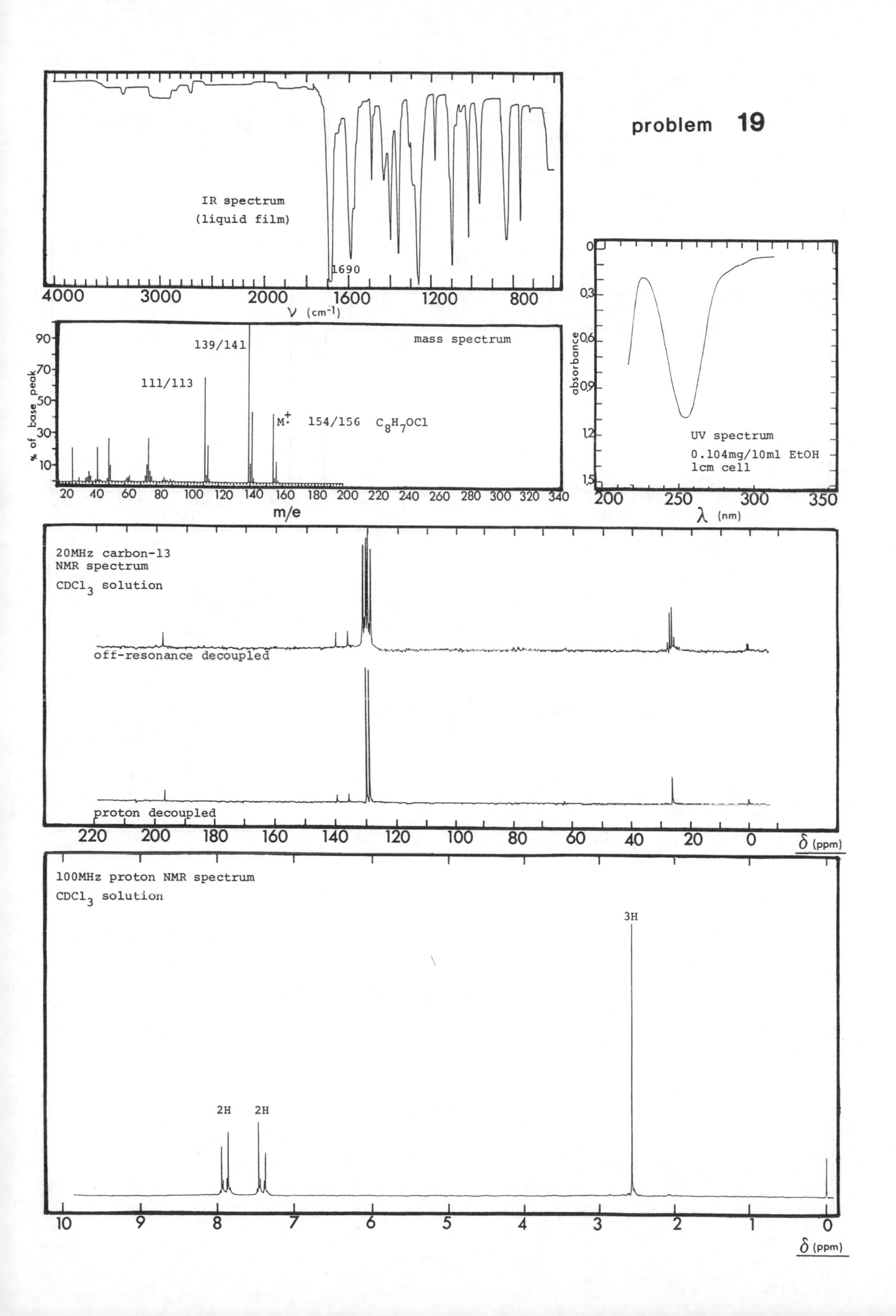
problem 19
IR spectrum
(liquid film)
1690
4000
3000
2000
1600
1200
800
ν (cm⁻¹)
mass spectrum
139/141
111/113
M+· 154/156 C_8H_7OCl
% of base peak
m/e
UV spectrum
0.104mg/10ml EtOH
1cm cell
absorbance
λ (nm)
20MHz carbon-13
NMR spectrum
$CDCl_3$ solution
off-resonance decoupled
proton decoupled
δ (ppm)
100MHz proton NMR spectrum
$CDCl_3$ solution
3H
2H
2H
δ (ppm)

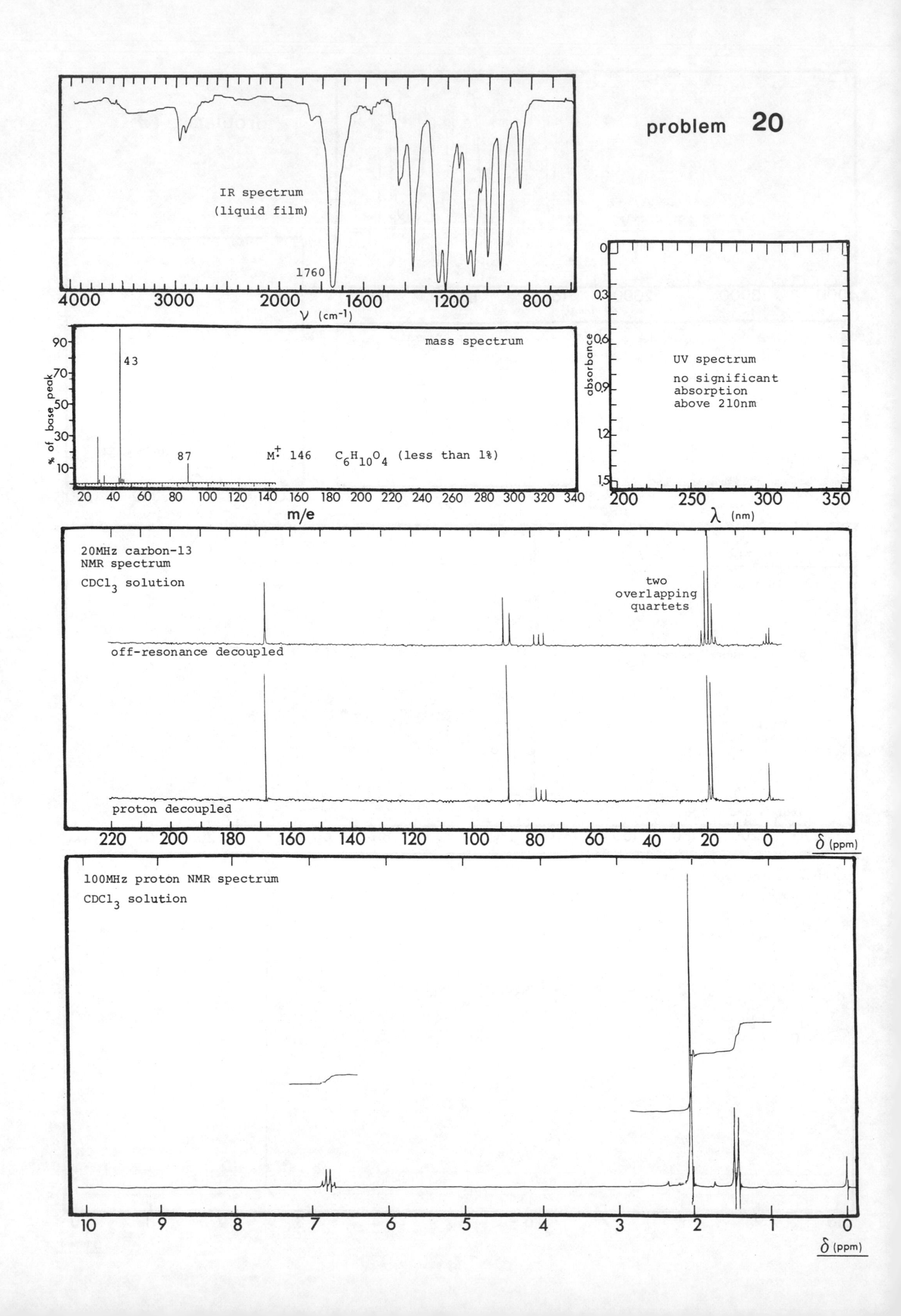

problem 20
IR spectrum
(liquid film)
1760
4000
3000
2000
1600
1200
800
ν (cm-1)
mass spectrum
43
87
M+ 146 C6H10O4 (less than 1%)
% of base peak
90
70
50
30
10
20 40 60 80 100 120 140 160 180 200 220 240 260 280 300 320 340
m/e
UV spectrum
no significant
absorption
above 210nm
absorbance
0
0.3
0.6
0.9
1.2
1.5
200 250 300 350
λ (nm)
20MHz carbon-13
NMR spectrum
CDCl3 solution
two
overlapping
quartets
off-resonance decoupled
proton decoupled
220 200 180 160 140 120 100 80 60 40 20 0
δ (ppm)
100MHz proton NMR spectrum
CDCl3 solution
10 9 8 7 6 5 4 3 2 1 0
δ (ppm)

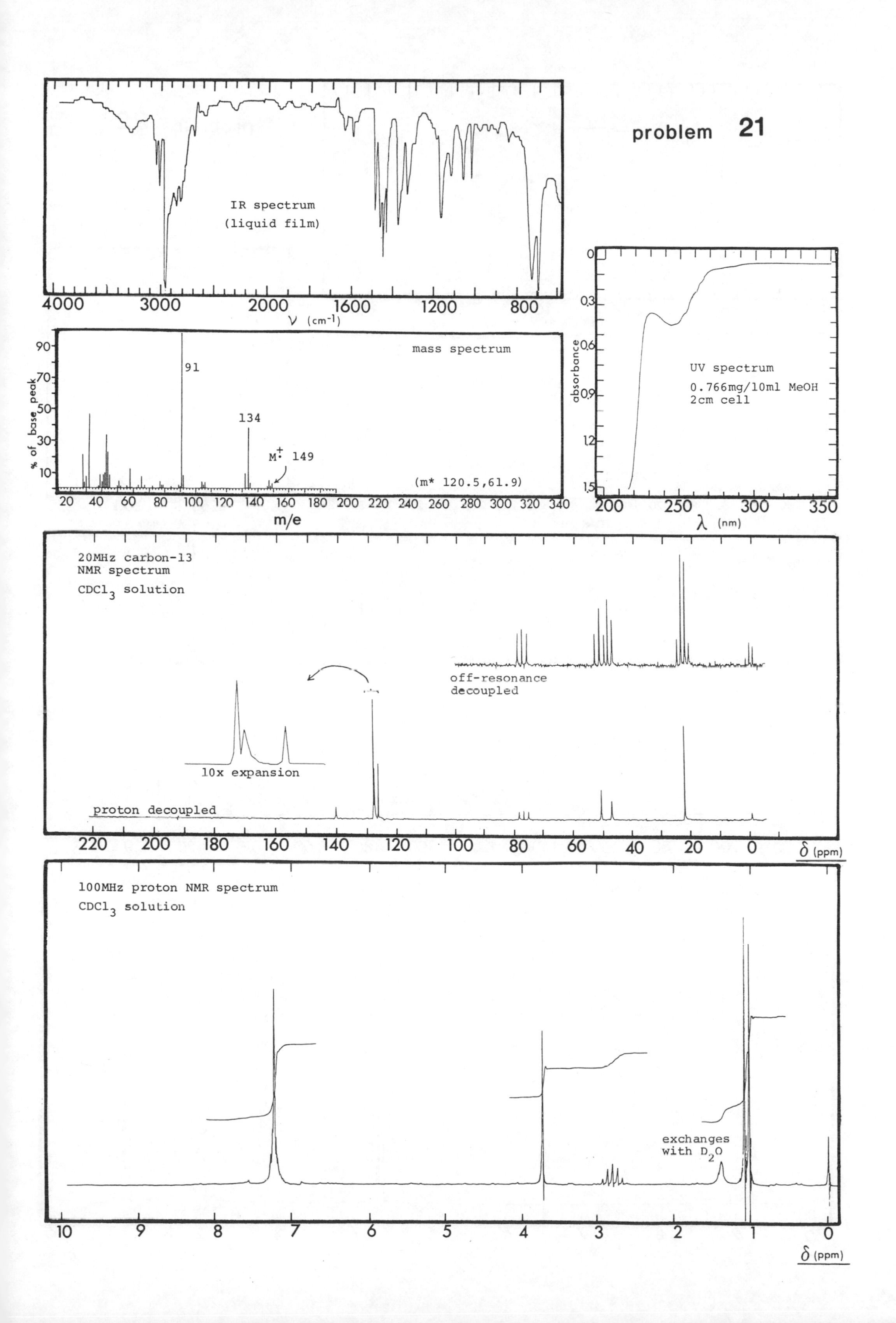
problem 21
IR spectrum
(liquid film)
4000
3000
2000
1600
1200
800
ν (cm-1)
mass spectrum
91
134
M+ 149
(m* 120.5,61.9)
% of base peak
m/e
UV spectrum
0.766mg/10ml MeOH
2cm cell
absorbance
λ (nm)
20MHz carbon-13
NMR spectrum
CDCl3 solution
off-resonance
decoupled
10x expansion
proton decoupled
δ (ppm)
100MHz proton NMR spectrum
CDCl3 solution
exchanges
with D2O
δ (ppm)

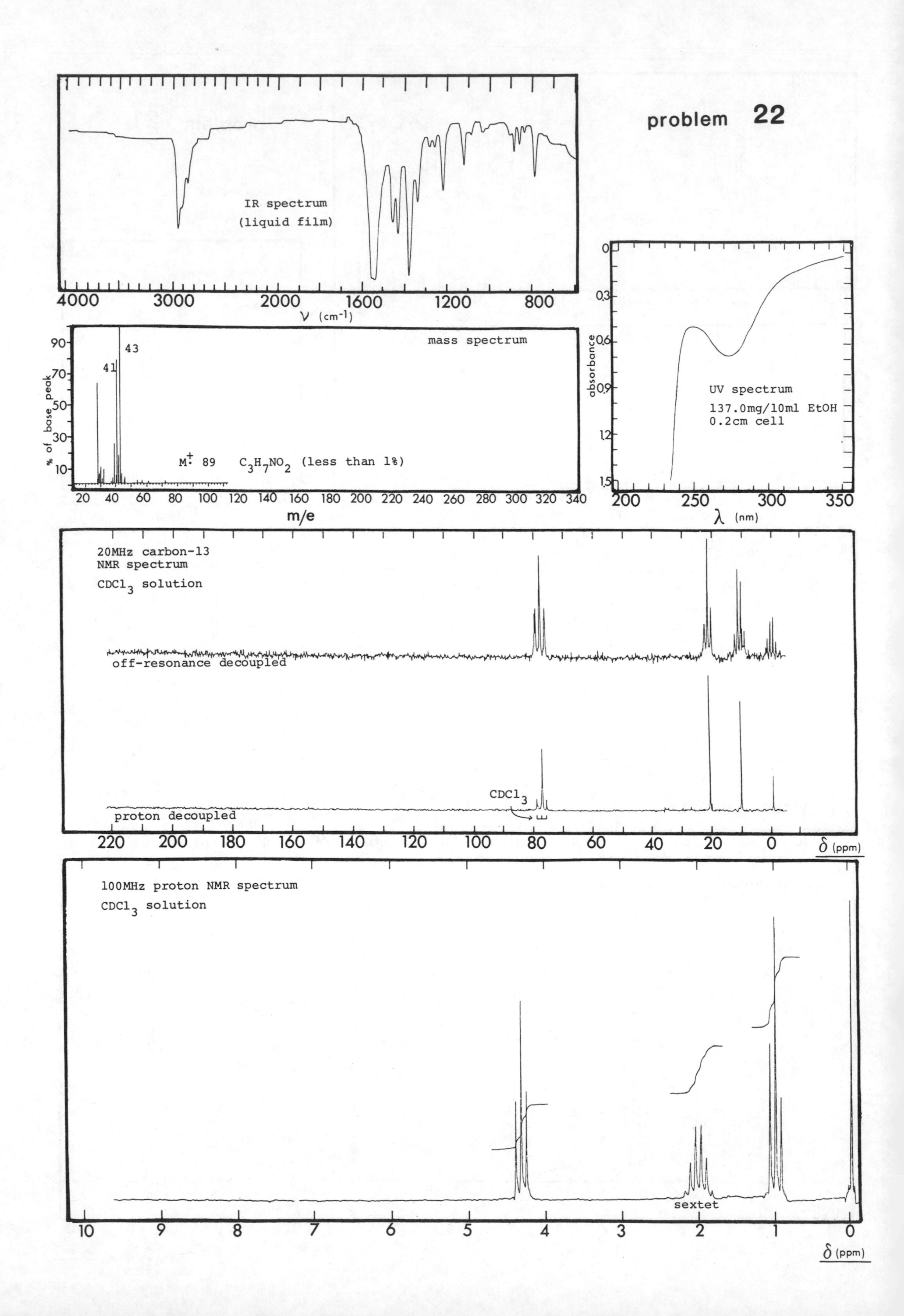
problem 22
IR spectrum
(liquid film)
4000
3000
2000
1600
1200
800
ν (cm⁻¹)
mass spectrum
% of base peak
43
41
M⁺· 89 C3H7NO2 (less than 1%)
m/e
UV spectrum
137.0mg/10ml EtOH
0.2cm cell
absorbance
λ (nm)
20MHz carbon-13
NMR spectrum
CDCl3 solution
off-resonance decoupled
proton decoupled
CDCl3
δ (ppm)
100MHz proton NMR spectrum
CDCl3 solution
sextet
δ (ppm)

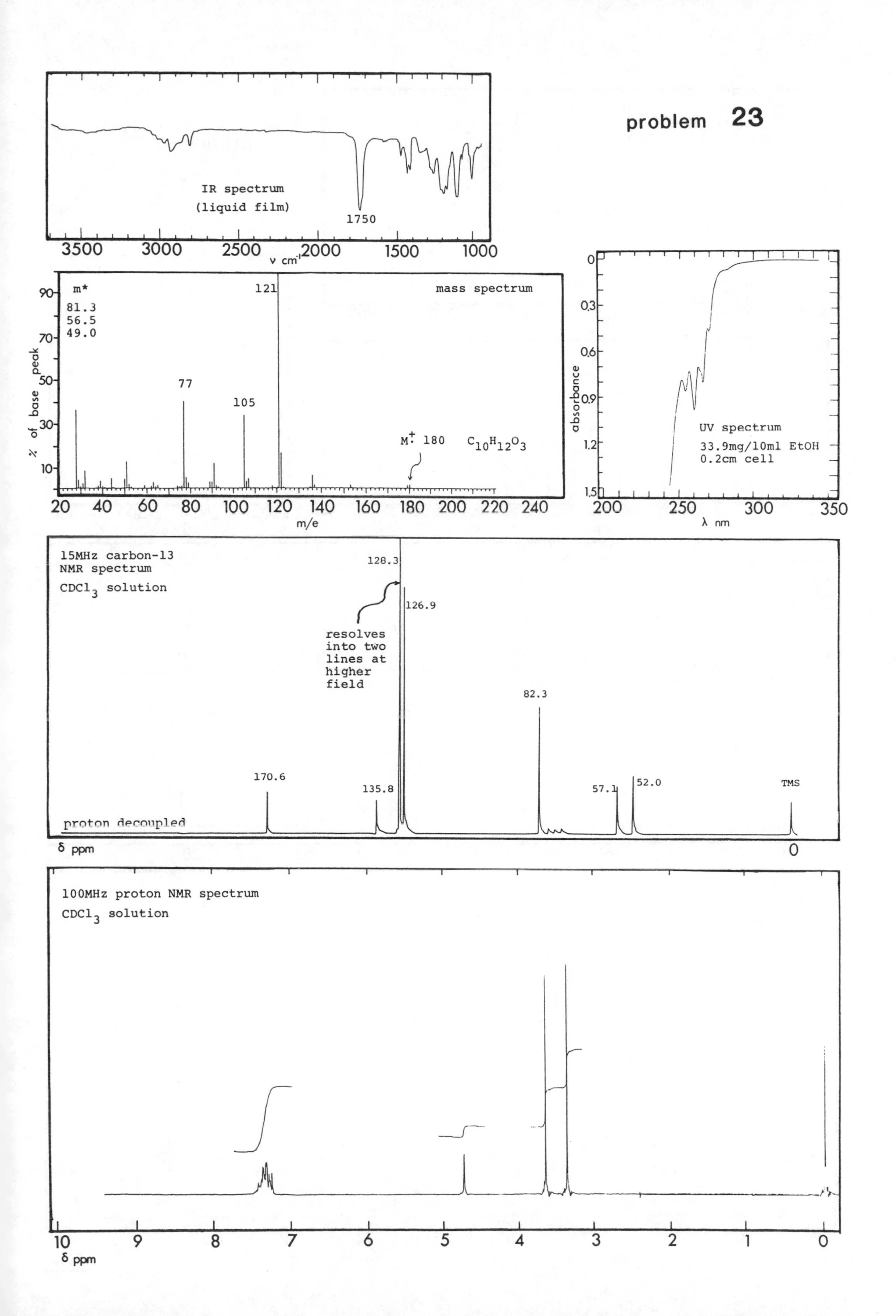
problem 23
IR spectrum
(liquid film)
1750
3500
3000
2500
2000
1500
1000
ν cm⁻¹
mass spectrum
m*
81.3
56.5
49.0
121
77
105
M⁺· 180
C10H12O3
% of base peak
m/e
UV spectrum
33.9mg/10ml EtOH
0.2cm cell
absorbance
λ nm
15MHz carbon-13
NMR spectrum
CDCl3 solution
128.3
126.9
resolves
into two
lines at
higher
field
82.3
170.6
135.8
57.1
52.0
TMS
proton decoupled
δ ppm
100MHz proton NMR spectrum
CDCl3 solution
δ ppm

problem 24

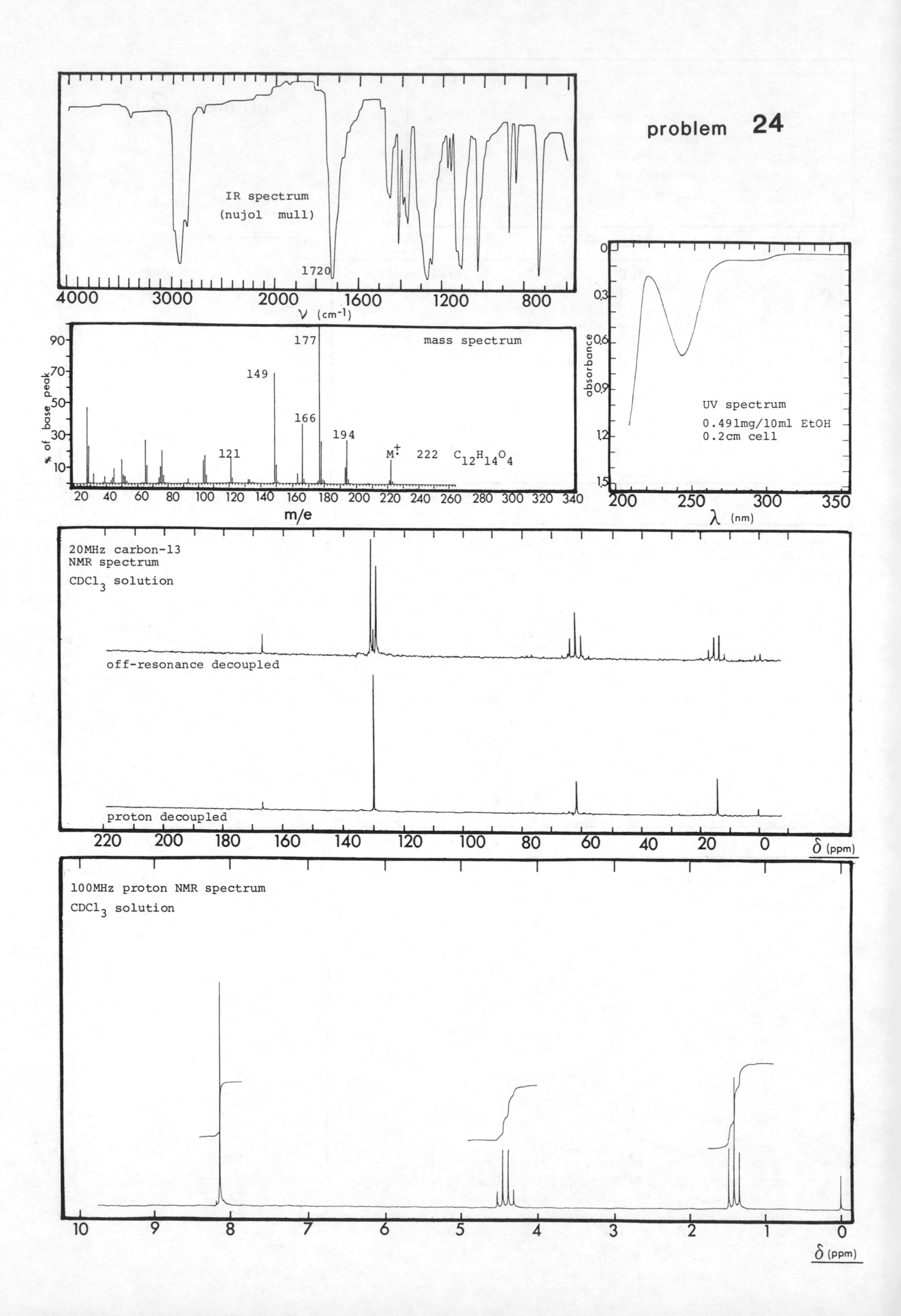

IR spectrum
(nujol mull)
1720
4000
3000
2000
1600
1200
800
ν (cm⁻¹)
mass spectrum
177
149
166
194
121
M⁺ 222 C12H14O4
% of base peak
m/e
UV spectrum
0.491mg/10ml EtOH
0.2cm cell
absorbance
λ (nm)
20MHz carbon-13
NMR spectrum
CDCl3 solution
off-resonance decoupled
proton decoupled
δ (ppm)
100MHz proton NMR spectrum
CDCl3 solution
δ (ppm)

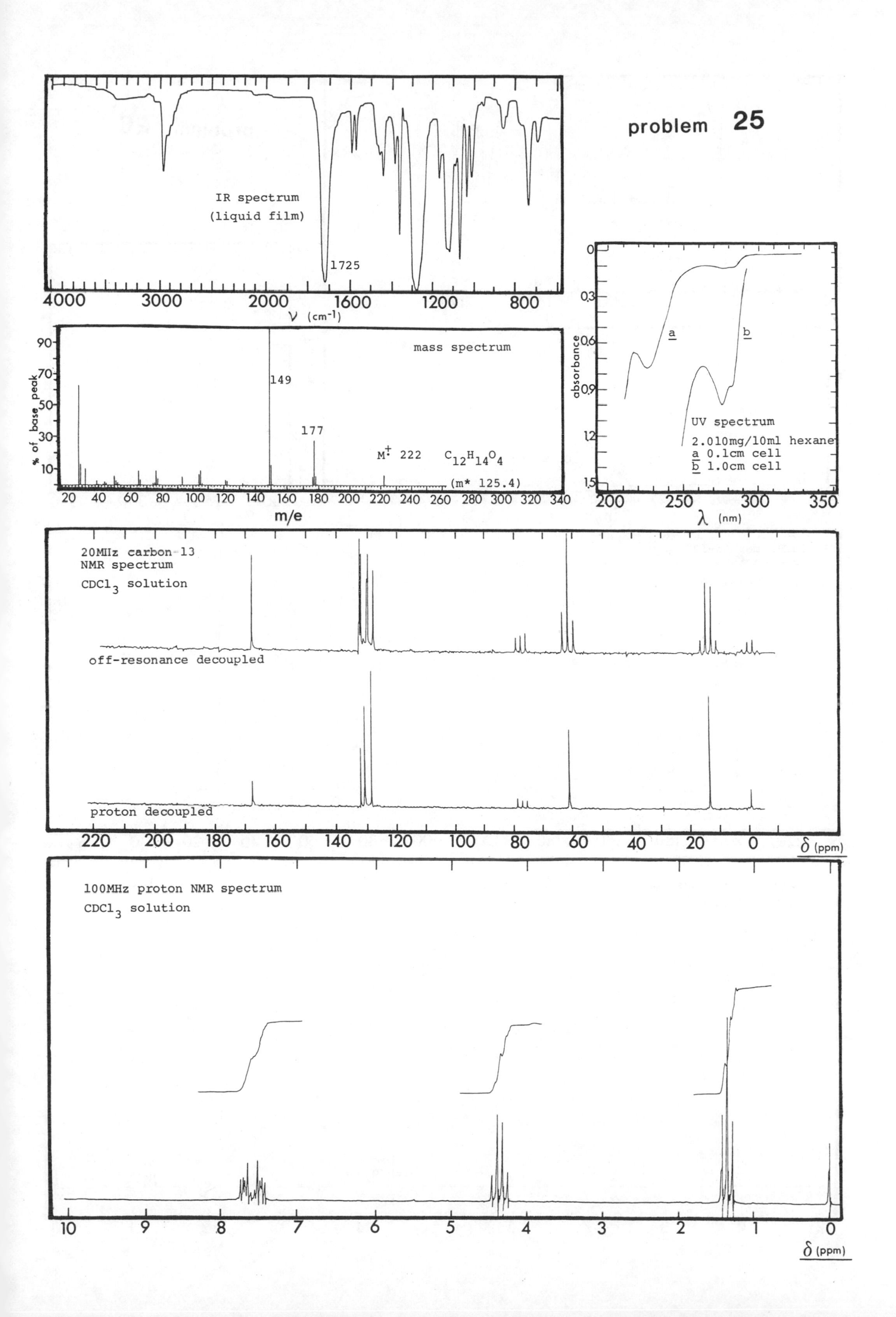
problem 25
IR spectrum
(liquid film)
1725
4000
3000
2000
1600
1200
800
ν (cm-1)
mass spectrum
149
177
M+ 222
C12H14O4
(m* 125.4)
% of base peak
m/e
UV spectrum
2.010mg/10ml hexane
a 0.1cm cell
b 1.0cm cell
absorbance
λ (nm)
20MHz carbon-13
NMR spectrum
CDCl3 solution
off-resonance decoupled
proton decoupled
δ (ppm)
100MHz proton NMR spectrum
CDCl3 solution
δ (ppm)

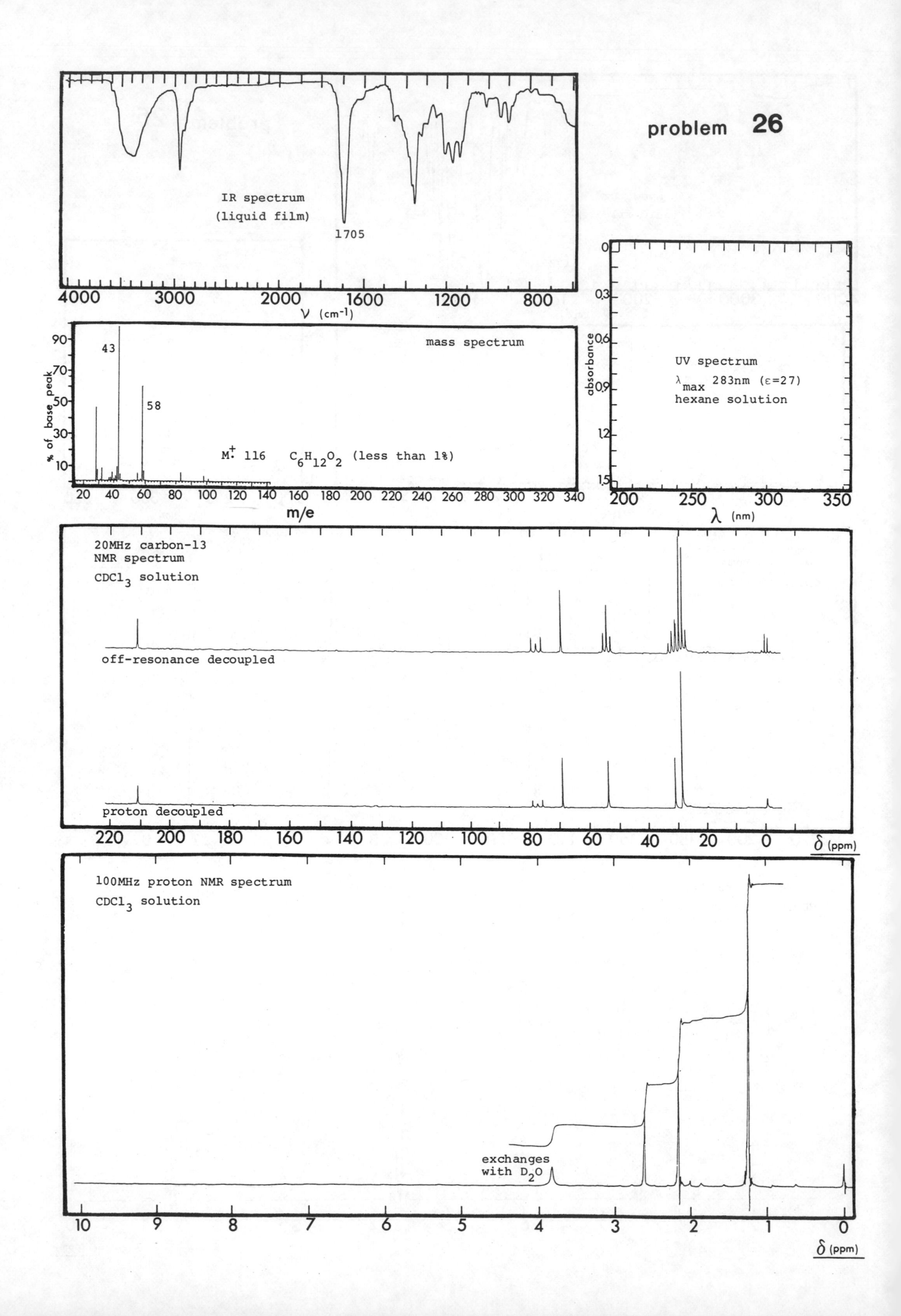

problem 26
IR spectrum
(liquid film)
1705
4000
3000
2000
1600
1200
800
ν (cm⁻¹)
mass spectrum
43
58
M⁺ 116 C6H12O2 (less than 1%)
% of base peak
m/e
UV spectrum
λmax 283nm (ε=27)
hexane solution
absorbance
λ (nm)
20MHz carbon-13
NMR spectrum
CDCl3 solution
off-resonance decoupled
proton decoupled
δ (ppm)
100MHz proton NMR spectrum
CDCl3 solution
exchanges
with D2O
δ (ppm)

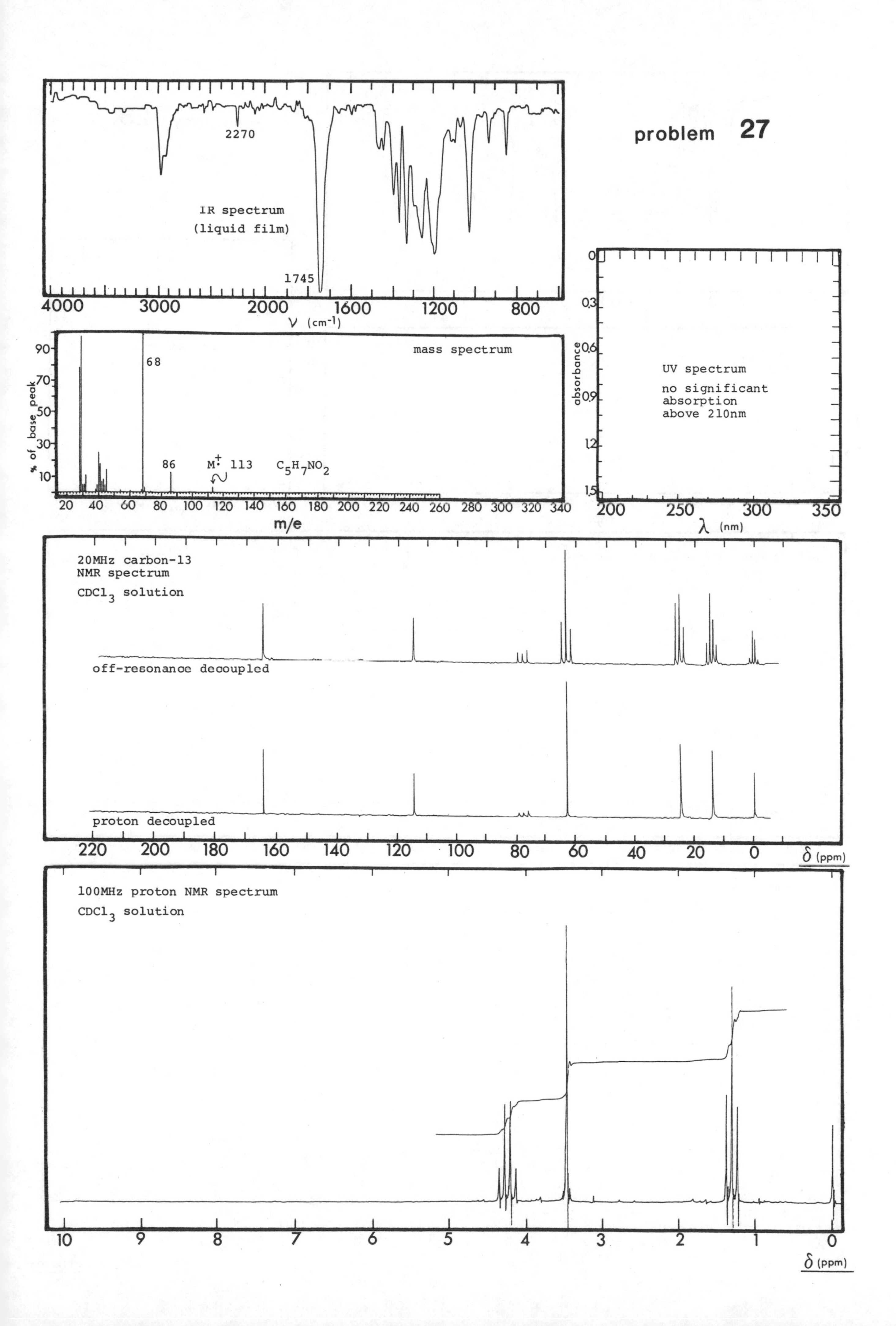
problem 27
IR spectrum
(liquid film)
2270
1745
4000
3000
2000
1600
1200
800
ν (cm⁻¹)
mass spectrum
68
86
M⁺ 113
$C_5H_7NO_2$
% of base peak
m/e
UV spectrum
no significant absorption above 210nm
absorbance
λ (nm)
20MHz carbon-13 NMR spectrum
$CDCl_3$ solution
off-resonance decoupled
proton decoupled
δ (ppm)
100MHz proton NMR spectrum
$CDCl_3$ solution
δ (ppm)

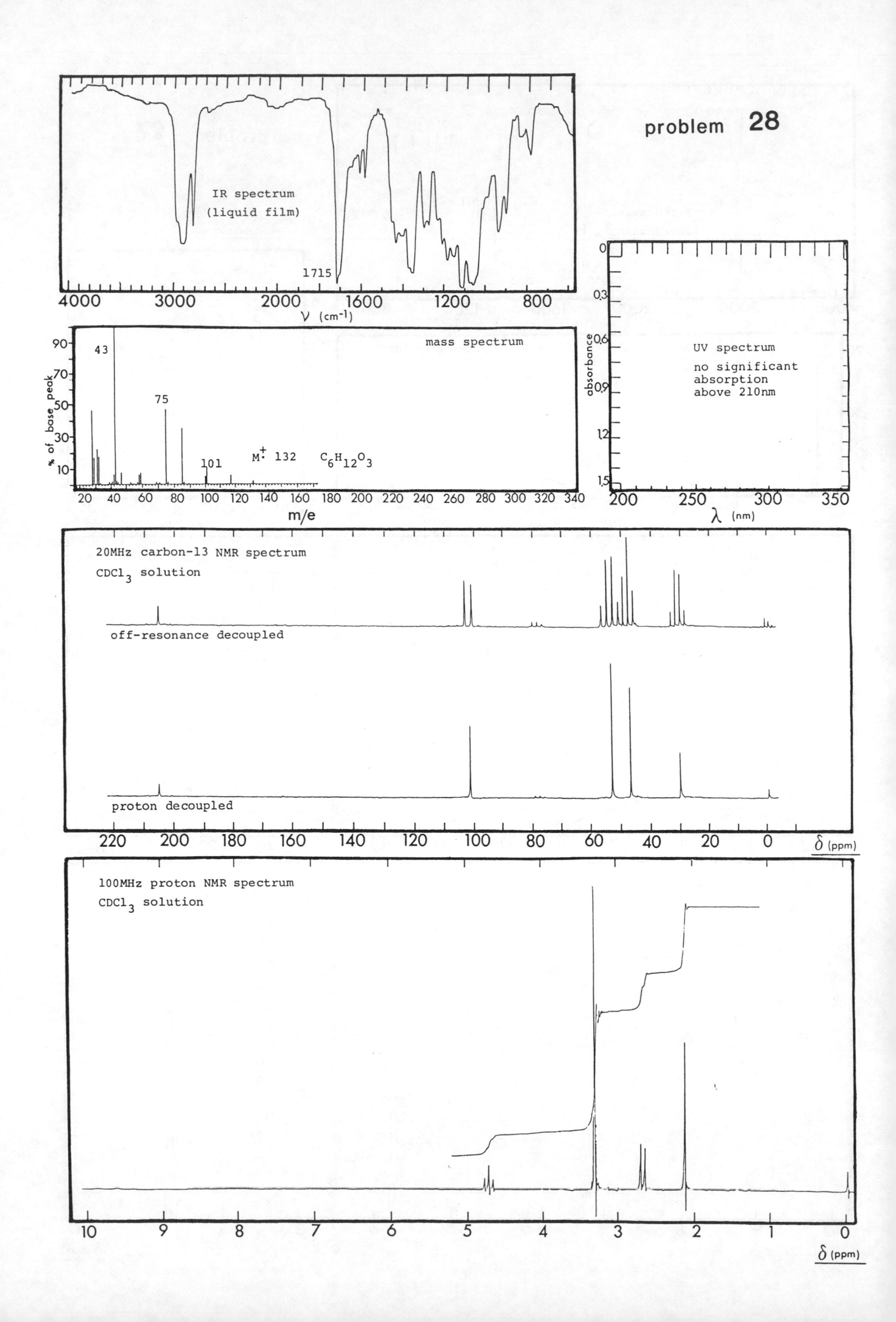
problem 28
IR spectrum
(liquid film)
1715
4000
3000
2000
1600
1200
800
ν (cm-1)
mass spectrum
43
75
101
M+ 132
C6H12O3
% of base peak
m/e
UV spectrum
no significant
absorption
above 210nm
absorbance
λ (nm)
20MHz carbon-13 NMR spectrum
CDCl3 solution
off-resonance decoupled
proton decoupled
δ (ppm)
100MHz proton NMR spectrum
CDCl3 solution
δ (ppm)

problem **29**

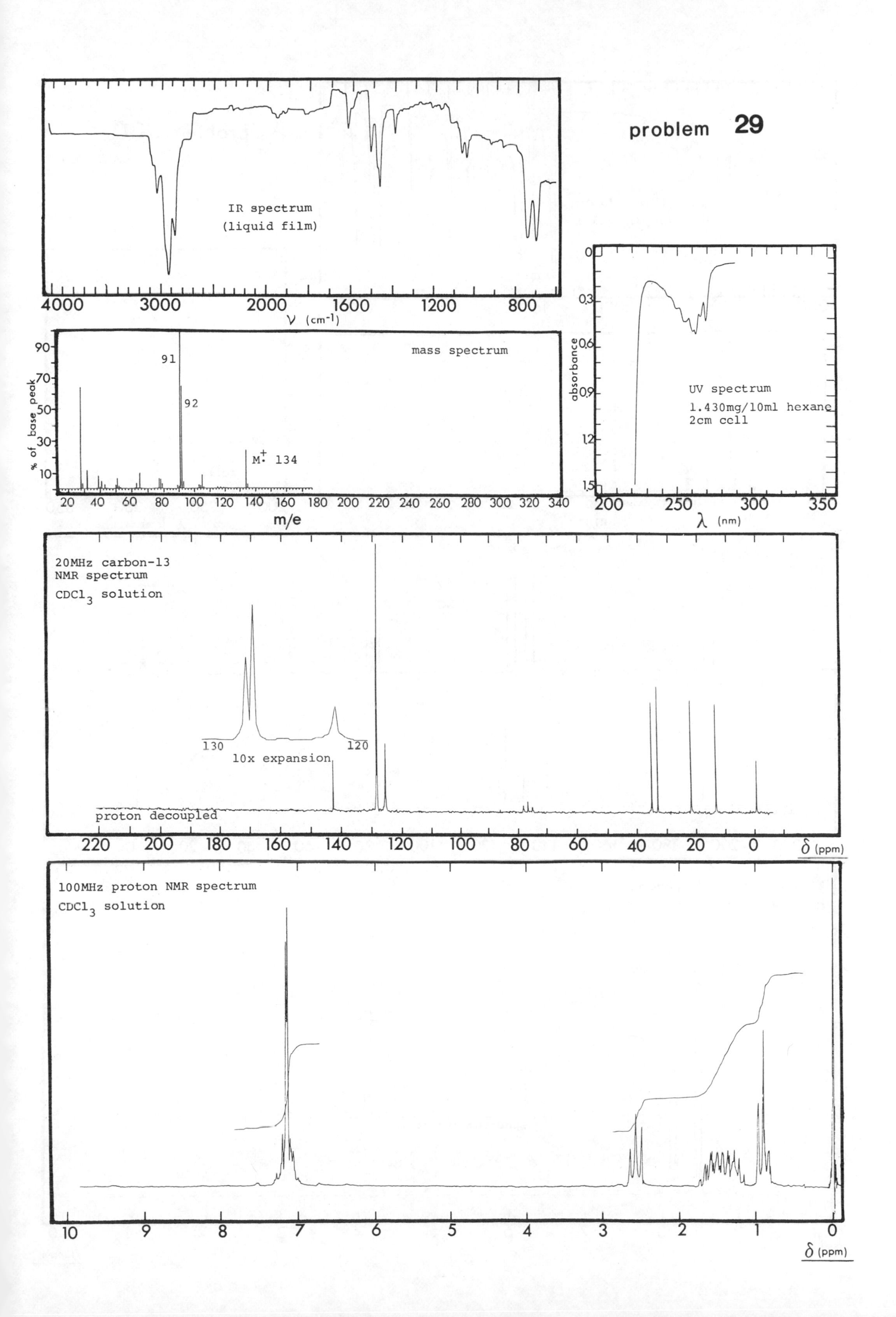

IR spectrum
(liquid film)
4000
3000
2000
1600
1200
800
ν (cm-1)
mass spectrum
91
92
M+. 134
% of base peak
m/e
UV spectrum
1.430mg/10ml hexane
2cm cell
absorbance
λ (nm)
20MHz carbon-13
NMR spectrum
CDCl3 solution
10x expansion
proton decoupled
δ (ppm)
100MHz proton NMR spectrum
CDCl3 solution
δ (ppm)

problem 30

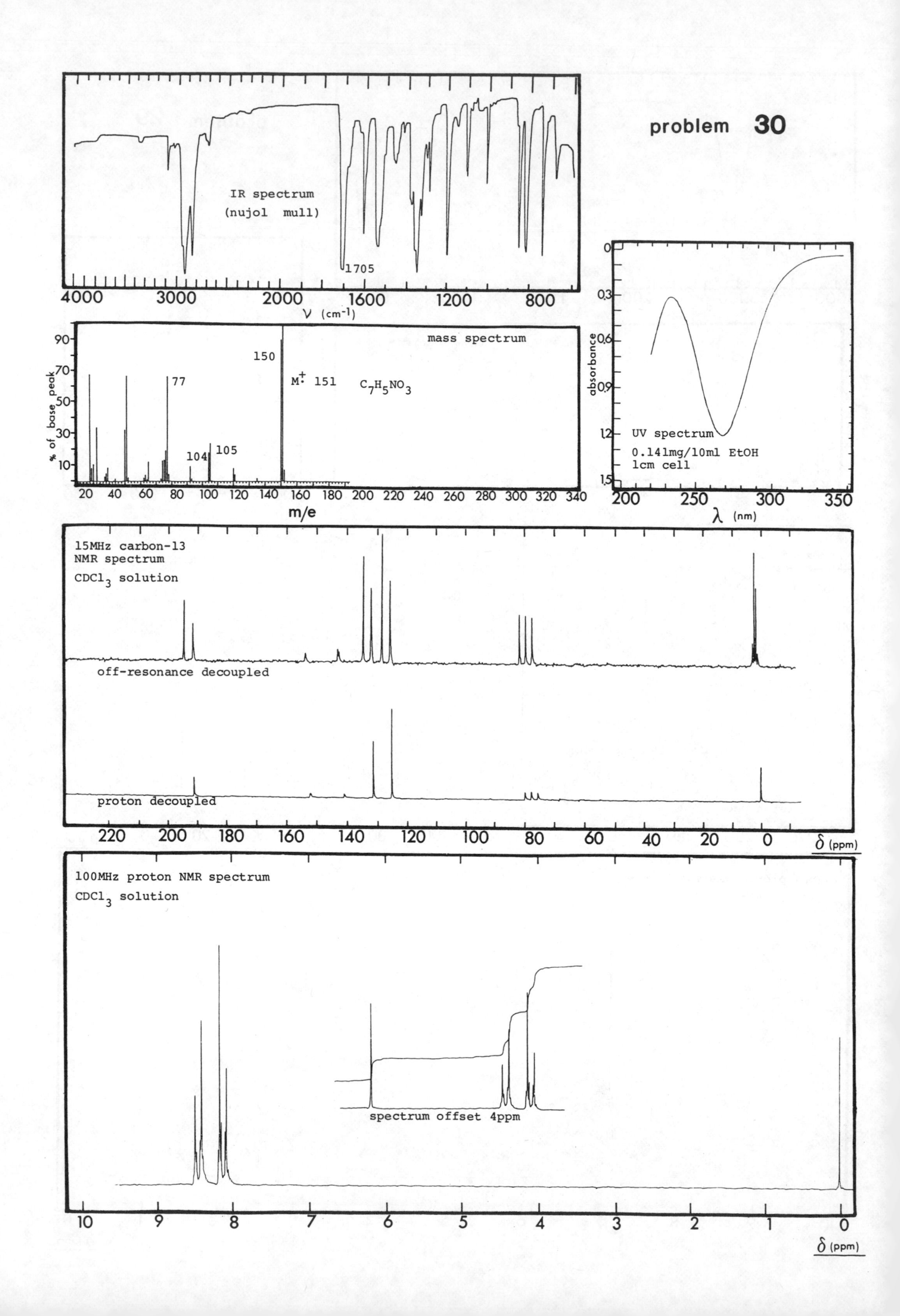

IR spectrum
(nujol mull)
1705
4000
3000
2000
1600
1200
800
ν (cm-1)
mass spectrum
% of base peak
150
77
M+· 151
C7H5NO3
104
105
m/e
UV spectrum
0.141mg/10ml EtOH
1cm cell
absorbance
λ (nm)
15MHz carbon-13
NMR spectrum
CDCl3 solution
off-resonance decoupled
proton decoupled
δ (ppm)
100MHz proton NMR spectrum
CDCl3 solution
spectrum offset 4ppm
δ (ppm)

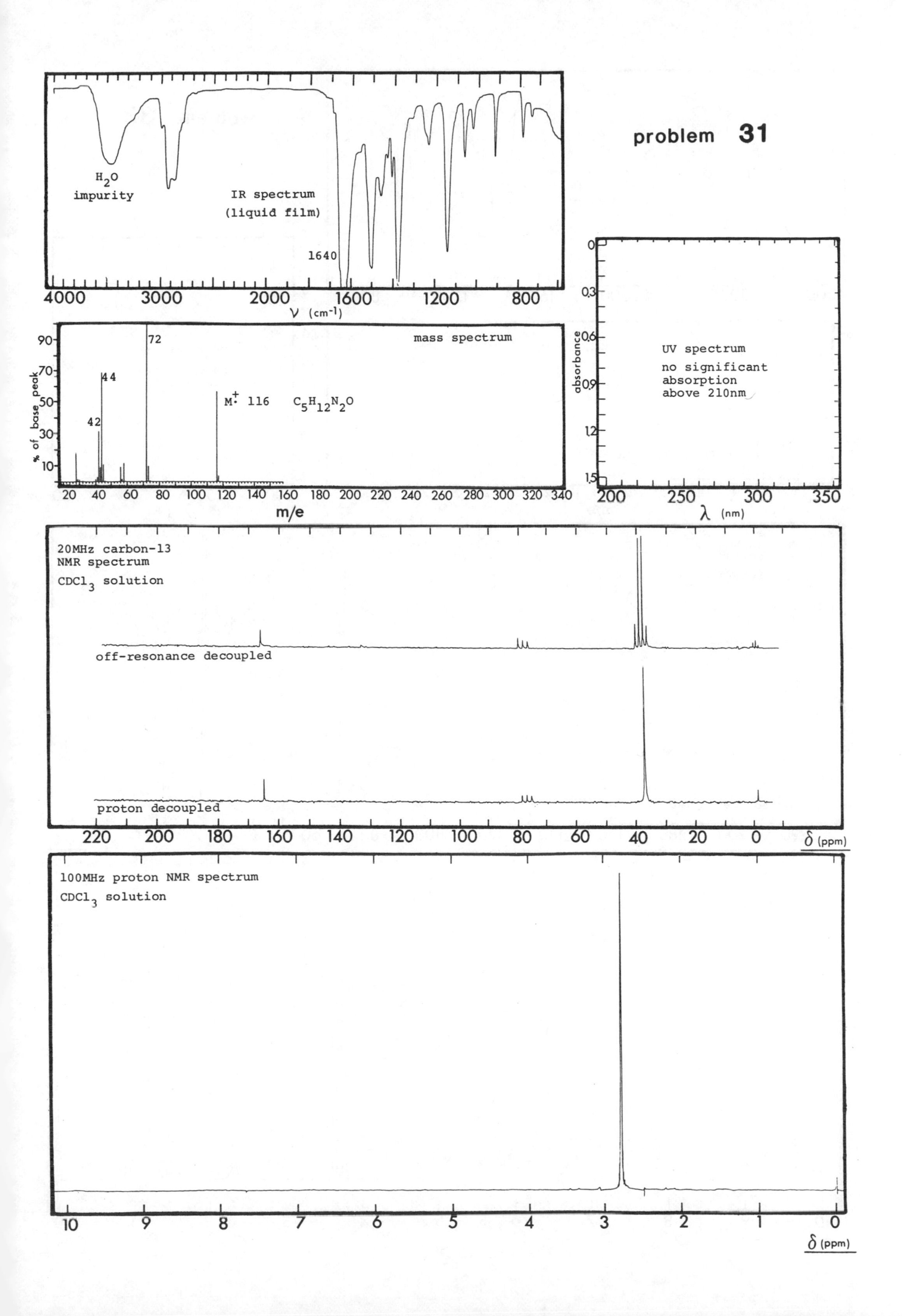

problem 31
IR spectrum
(liquid film)
H_2O
impurity
1640
4000
3000
2000
1600
1200
800
ν (cm-1)
mass spectrum
72
44
42
M⁺ 116
$C_5H_{12}N_2O$
% of base peak
m/e
UV spectrum
no significant
absorption
above 210nm
absorbance
λ (nm)
20MHz carbon-13
NMR spectrum
$CDCl_3$ solution
off-resonance decoupled
proton decoupled
δ (ppm)
100MHz proton NMR spectrum
$CDCl_3$ solution
δ (ppm)

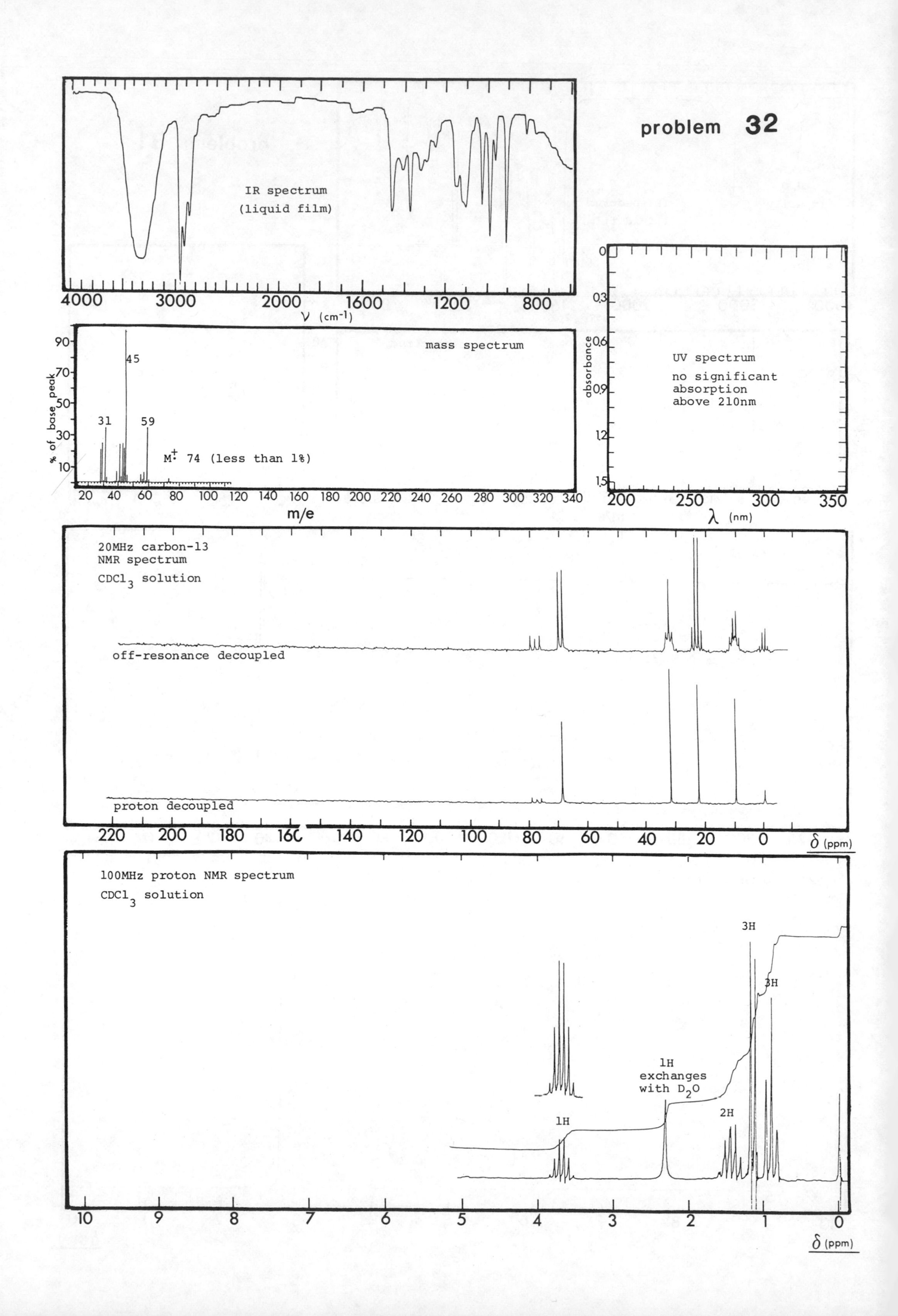

problem 32
IR spectrum
(liquid film)
4000
3000
2000
1600
1200
800
ν (cm-1)
mass spectrum
% of base peak
45
31
59
M+ 74 (less than 1%)
m/e
UV spectrum
no significant
absorption
above 210nm
absorbance
λ (nm)
20MHz carbon-13
NMR spectrum
CDCl3 solution
off-resonance decoupled
proton decoupled
δ (ppm)
100MHz proton NMR spectrum
CDCl3 solution
1H
1H
exchanges
with D2O
2H
3H
3H
δ (ppm)

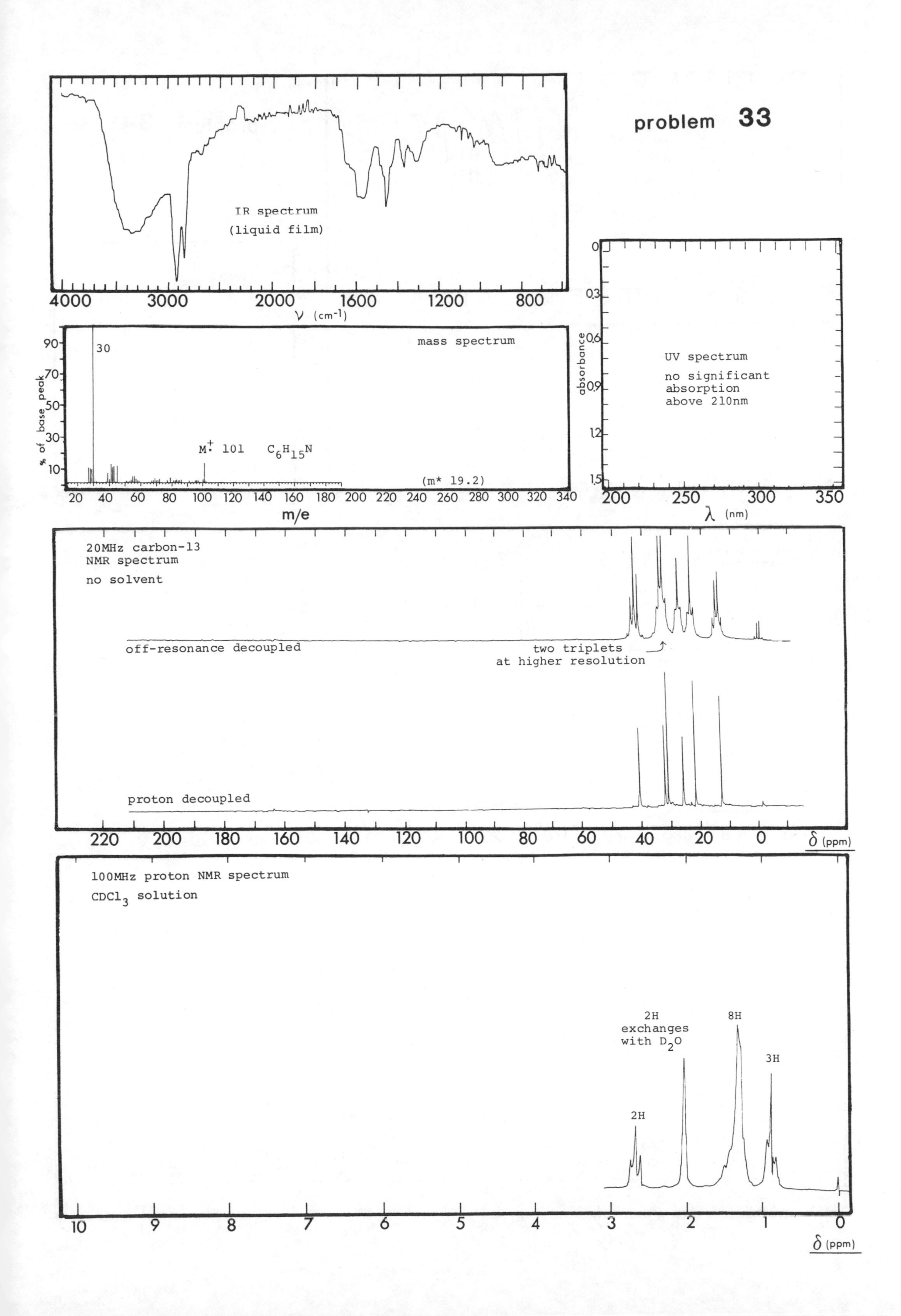

problem 33
IR spectrum
(liquid film)
4000
3000
2000
1600
1200
800
ν (cm-1)
mass spectrum
30
M⁺· 101 C6H15N
(m* 19.2)
% of base peak
m/e
UV spectrum
no significant
absorption
above 210nm
absorbance
λ (nm)
20MHz carbon-13
NMR spectrum
no solvent
off-resonance decoupled
two triplets
at higher resolution
proton decoupled
δ (ppm)
100MHz proton NMR spectrum
CDCl3 solution
2H
exchanges
with D2O
8H
3H
2H
δ (ppm)

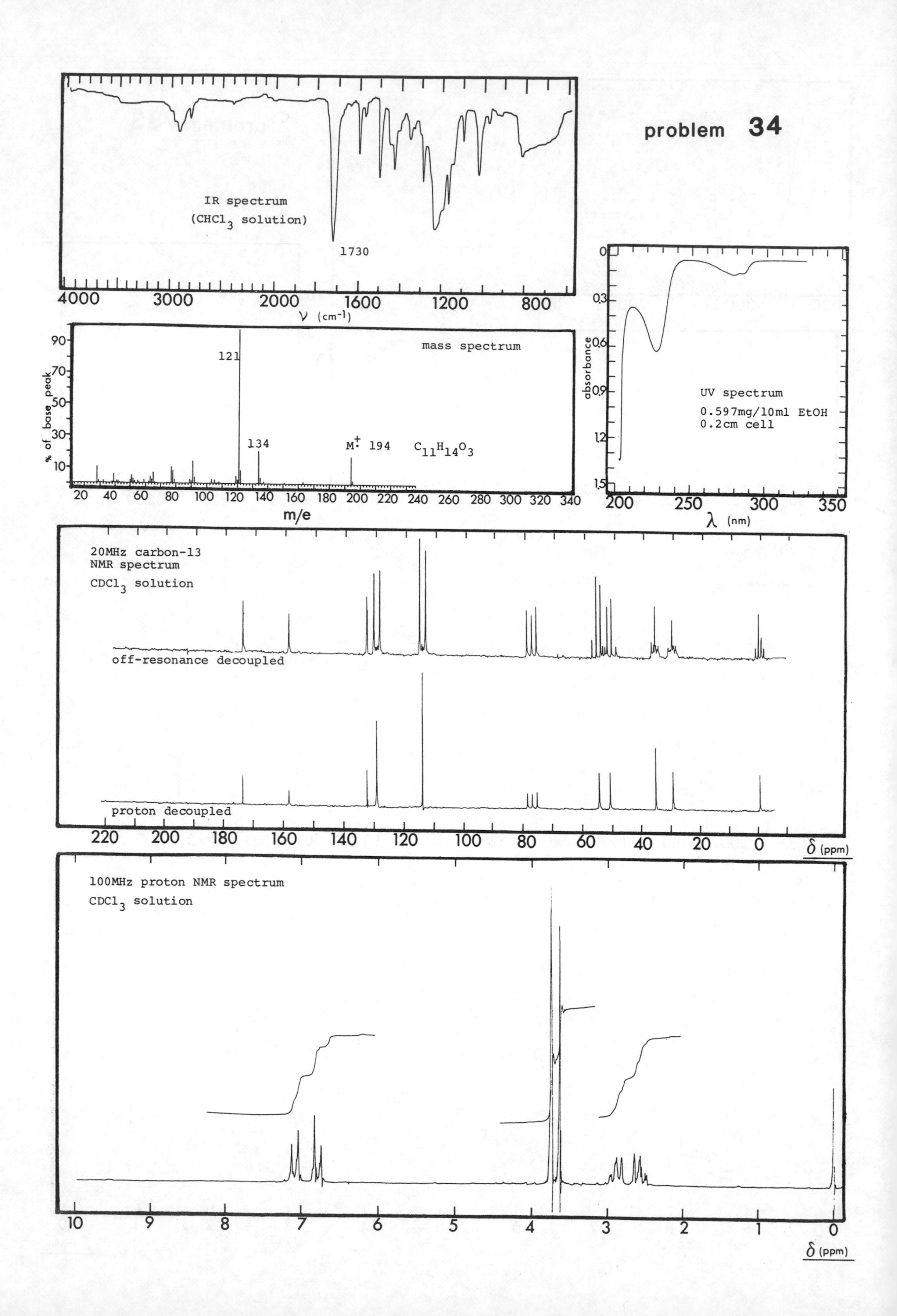

problem 34
IR spectrum
(CHCl3 solution)
1730
4000
3000
2000
1600
1200
800
ν (cm-1)
mass spectrum
121
134
M+· 194
C11H14O3
% of base peak
m/e
UV spectrum
0.597mg/10ml EtOH
0.2cm cell
absorbance
λ (nm)
20MHz carbon-13
NMR spectrum
CDCl3 solution
off-resonance decoupled
proton decoupled
δ (ppm)
100MHz proton NMR spectrum
CDCl3 solution
δ (ppm)

problem **35**

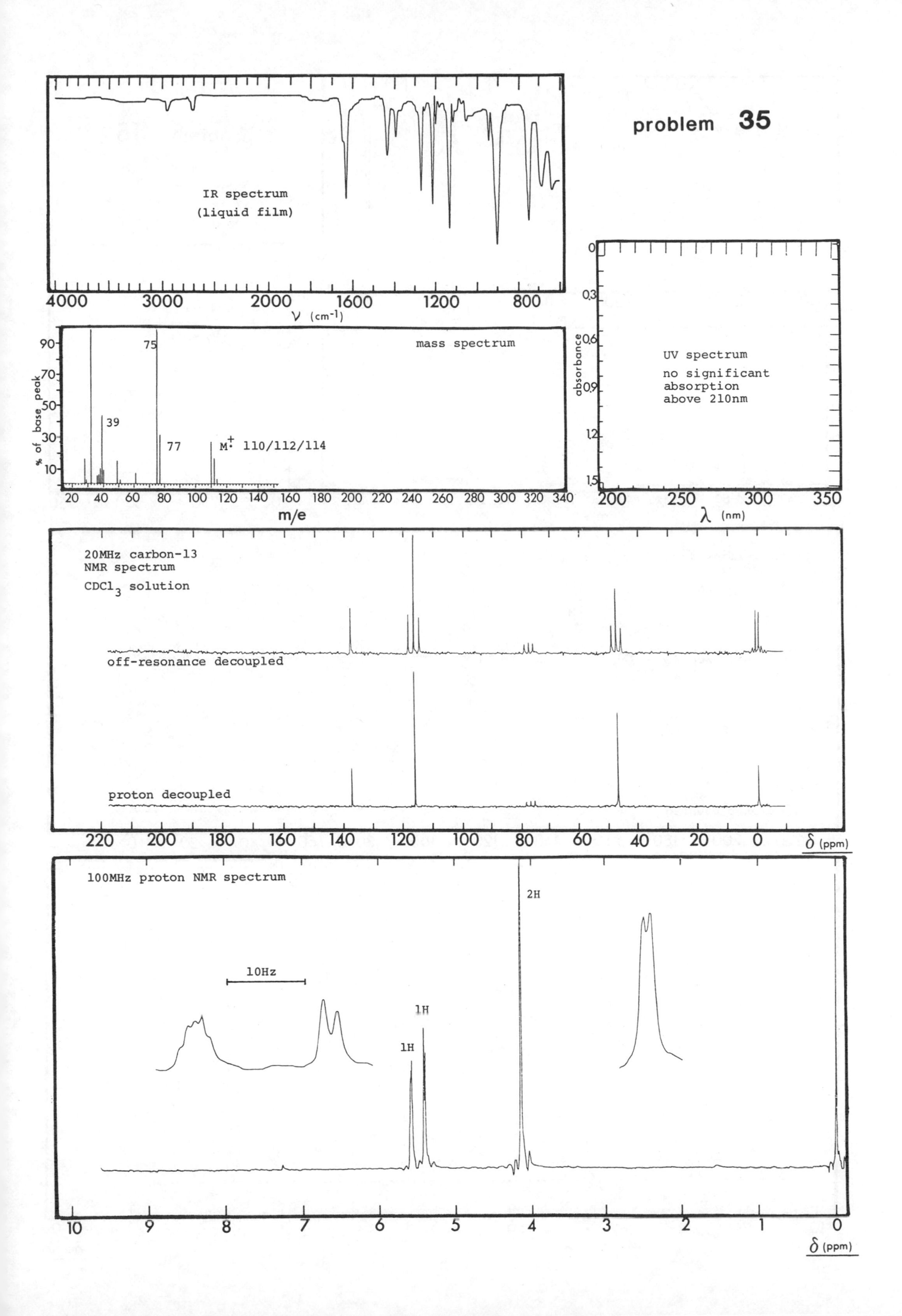

IR spectrum
(liquid film)
4000
3000
2000
1600
1200
800
ν (cm-1)
mass spectrum
% of base peak
90
70
50
30
10
75
39
77
M⁺· 110/112/114
20 40 60 80 100 120 140 160 180 200 220 240 260 280 300 320 340
m/e
UV spectrum
no significant
absorption
above 210nm
absorbance
0
0.3
0.6
0.9
1.2
1.5
200 250 300 350
λ (nm)
20MHz carbon-13
NMR spectrum
CDCl3 solution
off-resonance decoupled
proton decoupled
220 200 180 160 140 120 100 80 60 40 20 0
δ (ppm)
100MHz proton NMR spectrum
10Hz
2H
1H
1H
10 9 8 7 6 5 4 3 2 1 0
δ (ppm)

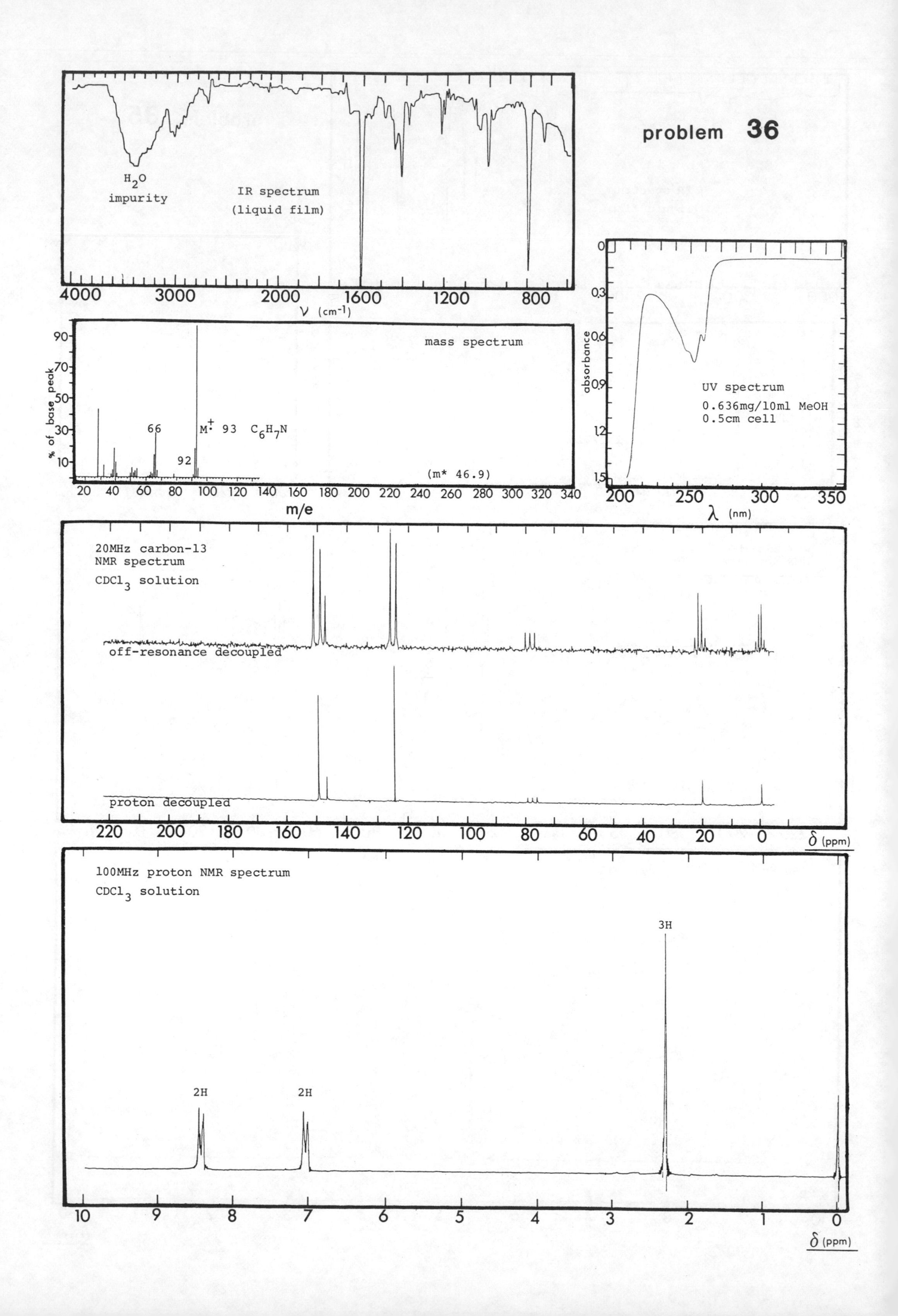
problem 36
H2O impurity
IR spectrum (liquid film)
4000 3000 2000 1600 1200 800
ν (cm-1)
mass spectrum
% of base peak
90 70 50 30 10
66
M+· 93 C6H7N
92
(m* 46.9)
20 40 60 80 100 120 140 160 180 200 220 240 260 280 300 320 340
m/e
UV spectrum
0.636mg/10ml MeOH
0.5cm cell
absorbance
0 0.3 0.6 0.9 1.2 1.5
200 250 300 350
λ (nm)
20MHz carbon-13
NMR spectrum
CDCl3 solution
off-resonance decoupled
proton decoupled
220 200 180 160 140 120 100 80 60 40 20 0
δ (ppm)
100MHz proton NMR spectrum
CDCl3 solution
3H
2H
2H
10 9 8 7 6 5 4 3 2 1 0
δ (ppm)

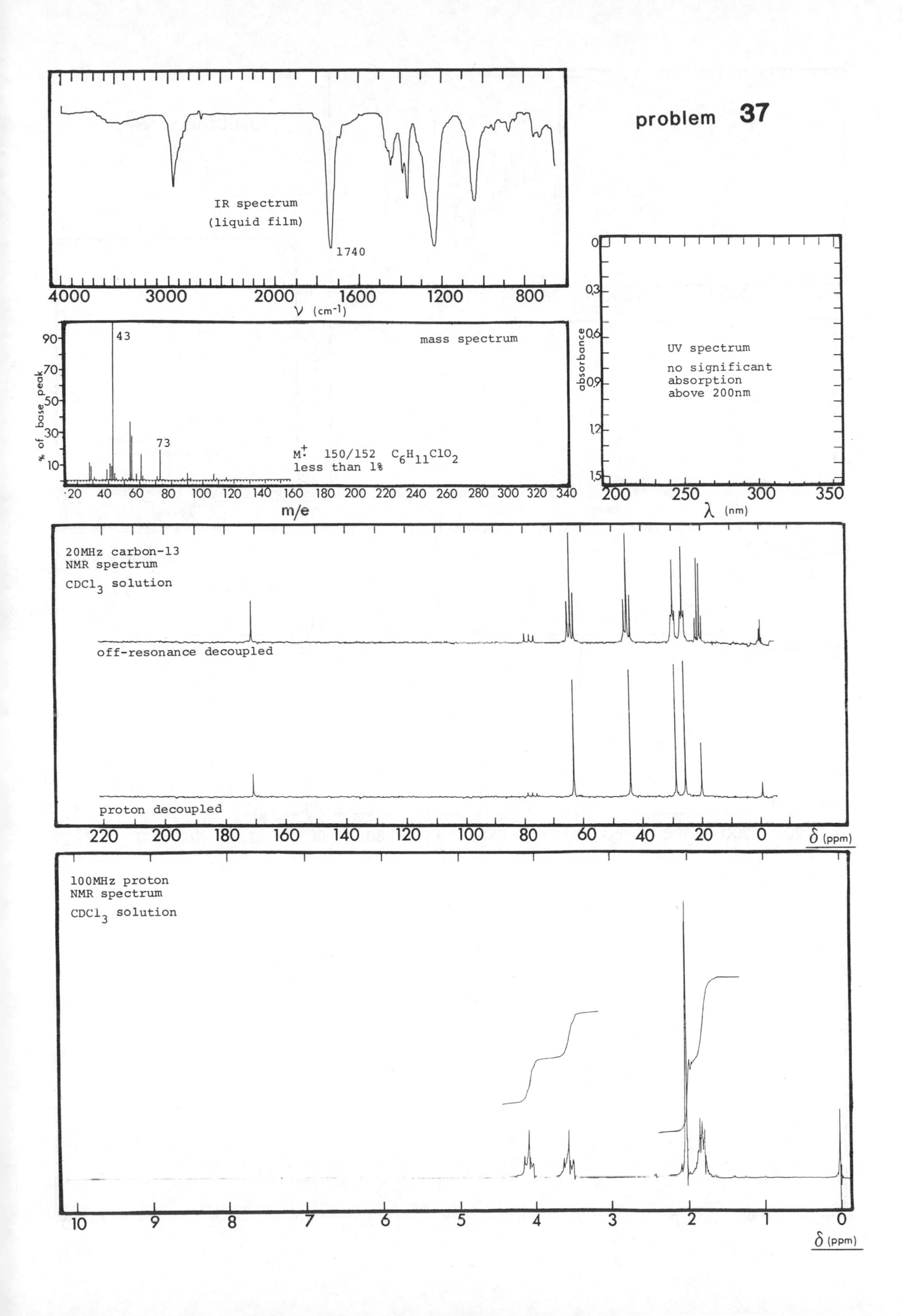

problem 37
IR spectrum
(liquid film)
1740
4000
3000
2000
1600
1200
800
ν (cm⁻¹)
mass spectrum
43
73
M⁺· 150/152 C6H11ClO2
less than 1%
% of base peak
m/e
UV spectrum
no significant
absorption
above 200nm
absorbance
λ (nm)
20MHz carbon-13
NMR spectrum
CDCl3 solution
off-resonance decoupled
proton decoupled
δ (ppm)
100MHz proton
NMR spectrum
CDCl3 solution
δ (ppm)

problem **38**

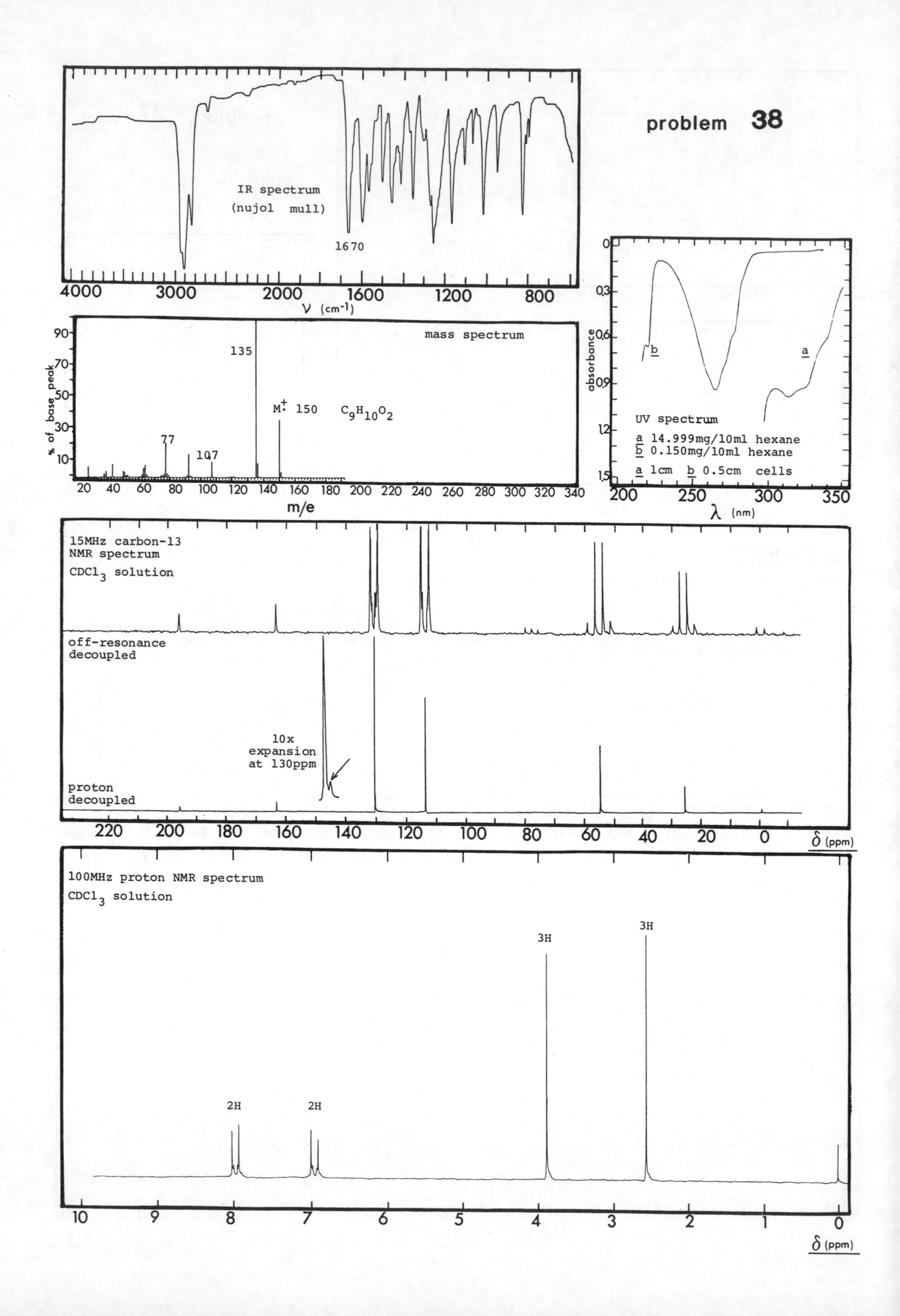

IR spectrum
(nujol mull)
1670
4000
3000
2000
1600
1200
800
ν (cm⁻¹)
mass spectrum
135
M⁺· 150
C9H10O2
77
107
% of base peak
m/e
UV spectrum
a 14.999mg/10ml hexane
b 0.150mg/10ml hexane
a 1cm b 0.5cm cells
absorbance
λ (nm)
15MHz carbon-13
NMR spectrum
CDCl3 solution
off-resonance
decoupled
10x
expansion
at 130ppm
proton
decoupled
δ (ppm)
100MHz proton NMR spectrum
CDCl3 solution
3H
3H
2H
2H
δ (ppm)

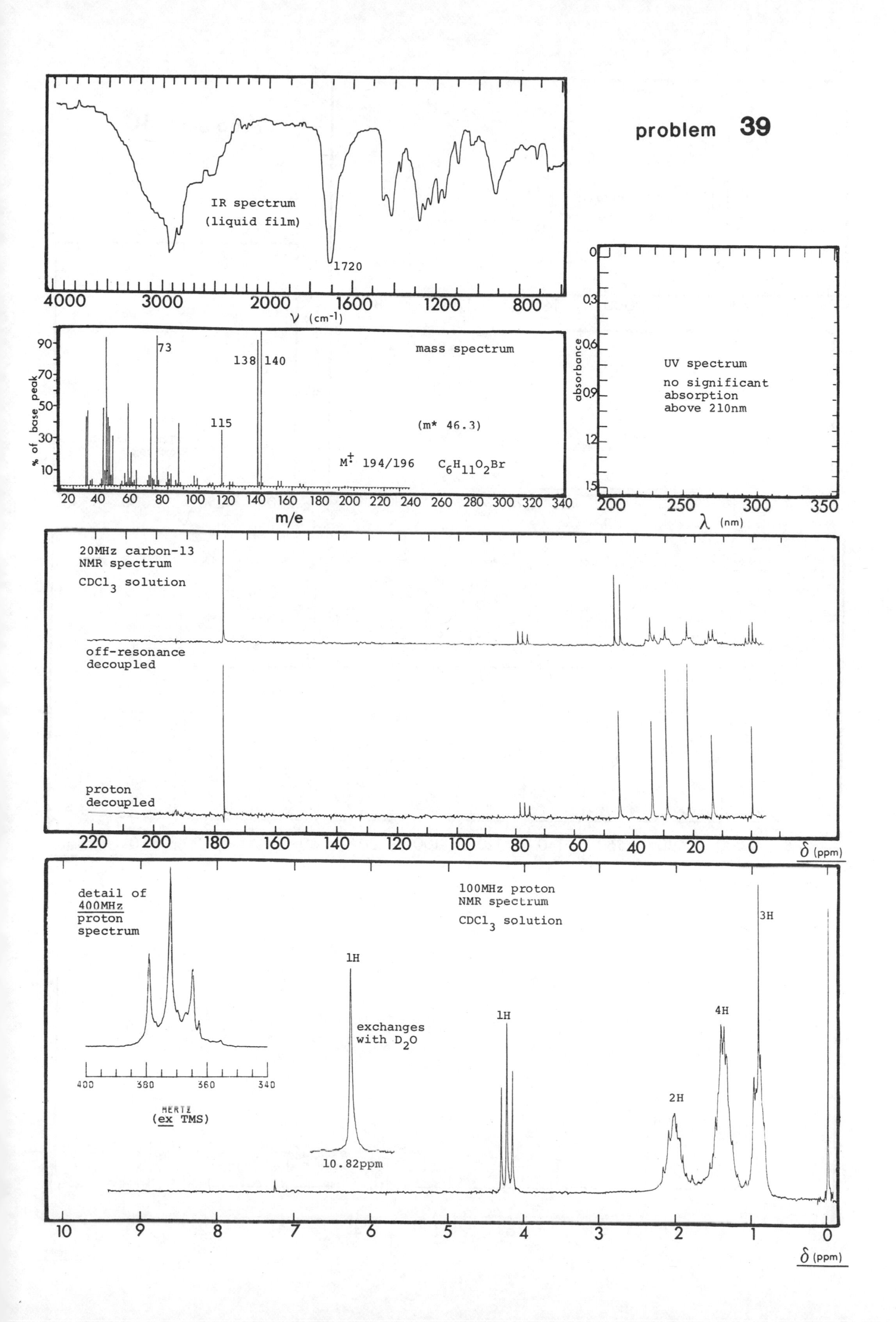

problem 39
IR spectrum
(liquid film)
1720
4000
3000
2000
1600
1200
800
ν (cm⁻¹)
mass spectrum
73
138
140
115
(m* 46.3)
M+· 194/196
C6H11O2Br
% of base peak
m/e
UV spectrum
no significant absorption above 210nm
absorbance
λ (nm)
20MHz carbon-13 NMR spectrum
CDCl3 solution
off-resonance decoupled
proton decoupled
δ (ppm)
detail of 400MHz proton spectrum
HERTZ (ex TMS)
100MHz proton NMR spectrum
CDCl3 solution
3H
1H
exchanges with D2O
10.82ppm
1H
4H
2H
δ (ppm)

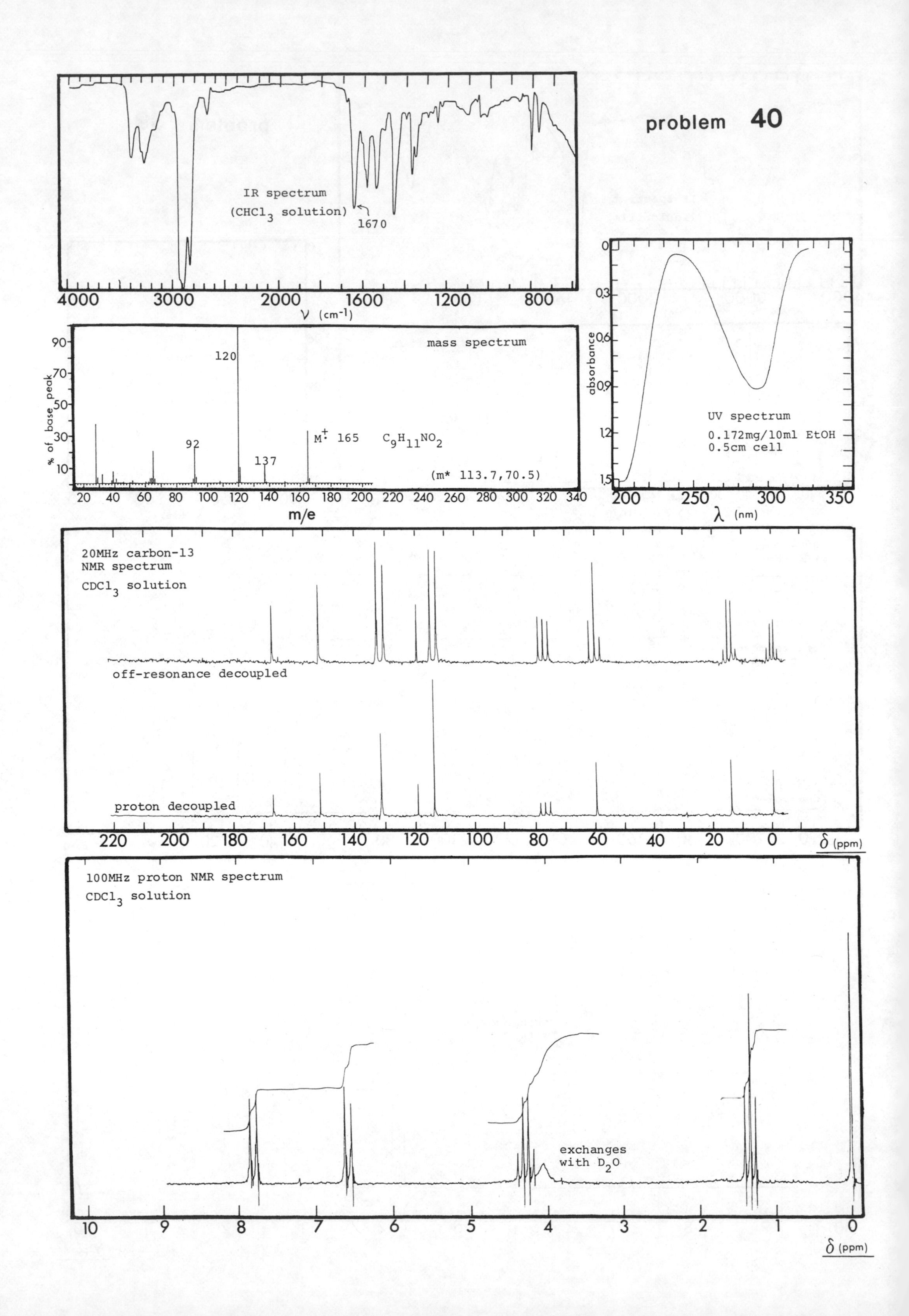
problem 40
IR spectrum
(CHCl3 solution)
1670
4000
3000
2000
1600
1200
800
ν (cm-1)
mass spectrum
120
92
137
M+ 165
C9H11NO2
(m* 113.7,70.5)
% of base peak
m/e
UV spectrum
0.172mg/10ml EtOH
0.5cm cell
absorbance
λ (nm)
20MHz carbon-13
NMR spectrum
CDCl3 solution
off-resonance decoupled
proton decoupled
δ (ppm)
100MHz proton NMR spectrum
CDCl3 solution
exchanges
with D2O
δ (ppm)

problem 41

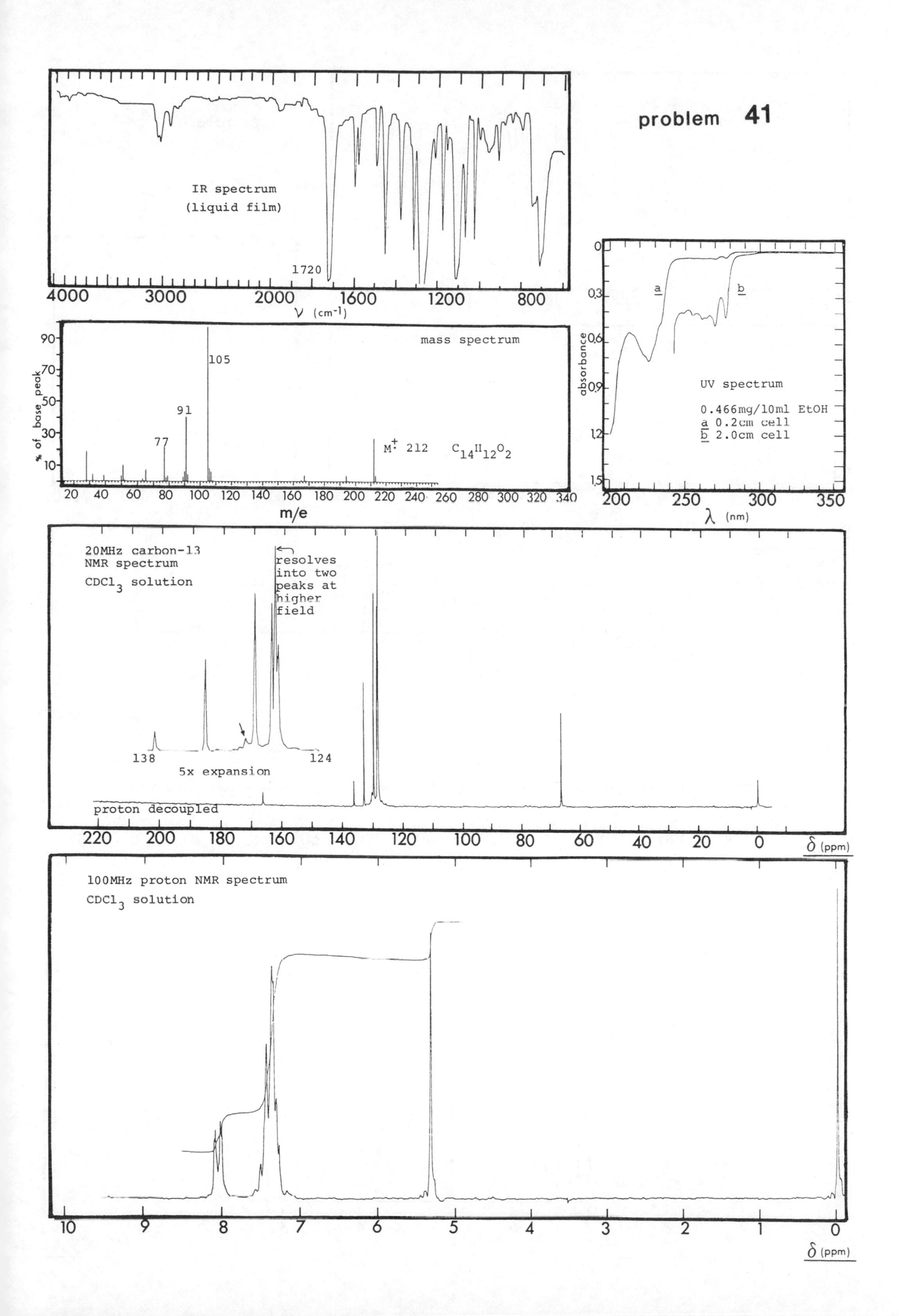

IR spectrum
(liquid film)
1720
4000
3000
2000
1600
1200
800
ν (cm-1)
mass spectrum
105
91
77
M+ 212
C14H12O2
% of base peak
m/e
UV spectrum
0.466mg/10ml EtOH
a 0.2cm cell
b 2.0cm cell
absorbance
λ (nm)
20MHz carbon-13
NMR spectrum
CDCl3 solution
resolves
into two
peaks at
higher
field
138
124
5x expansion
proton decoupled
δ (ppm)
100MHz proton NMR spectrum
CDCl3 solution
δ (ppm)

problem **42**

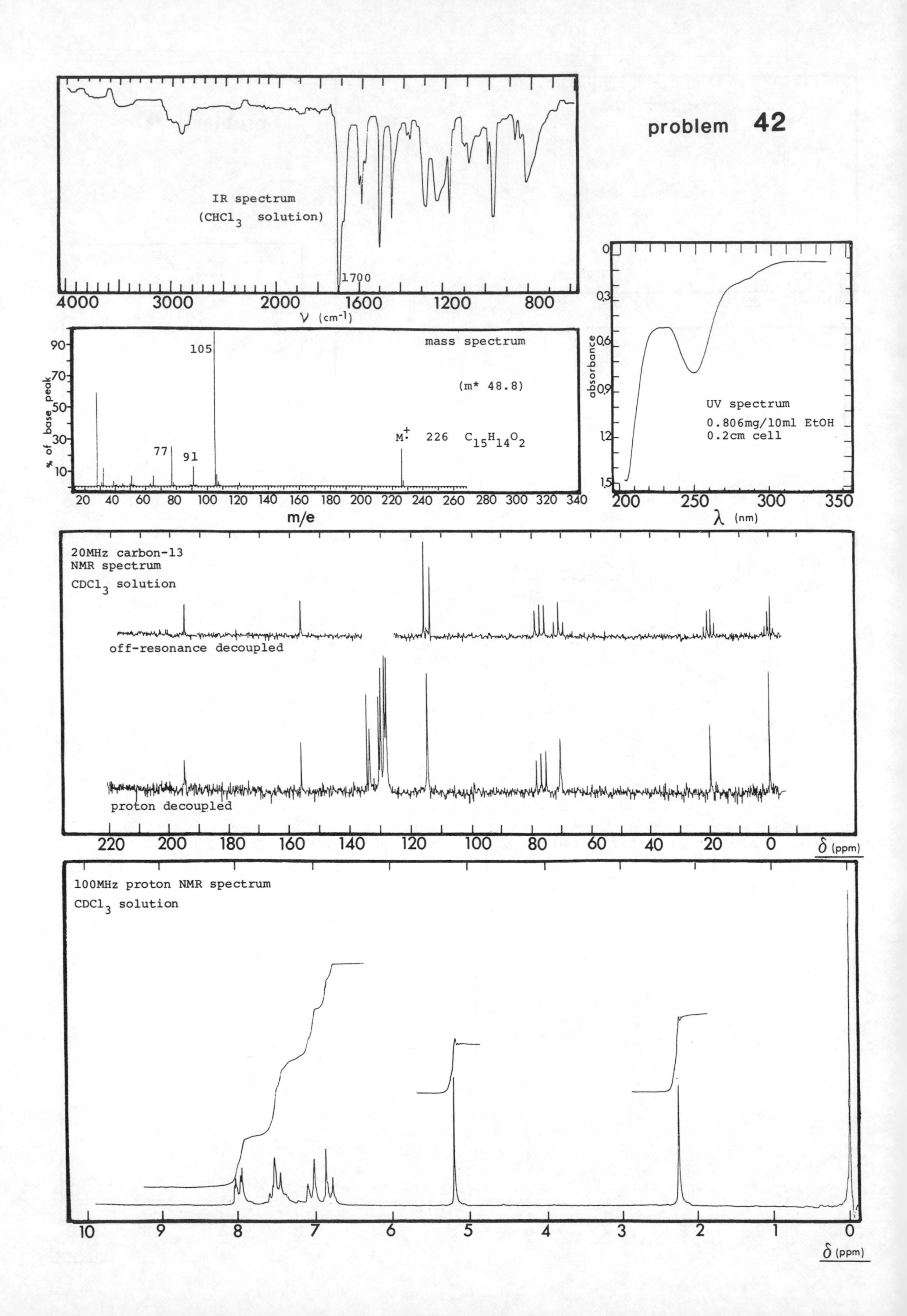

IR spectrum
(CHCl3 solution)
1700
4000
3000
2000
1600
1200
800
ν (cm-1)
mass spectrum
105
(m* 48.8)
77
91
M+ 226 C15H14O2
% of base peak
m/e
UV spectrum
0.806mg/10ml EtOH
0.2cm cell
absorbance
λ (nm)
20MHz carbon-13
NMR spectrum
CDCl3 solution
off-resonance decoupled
proton decoupled
δ (ppm)
100MHz proton NMR spectrum
CDCl3 solution
δ (ppm)

problem 43

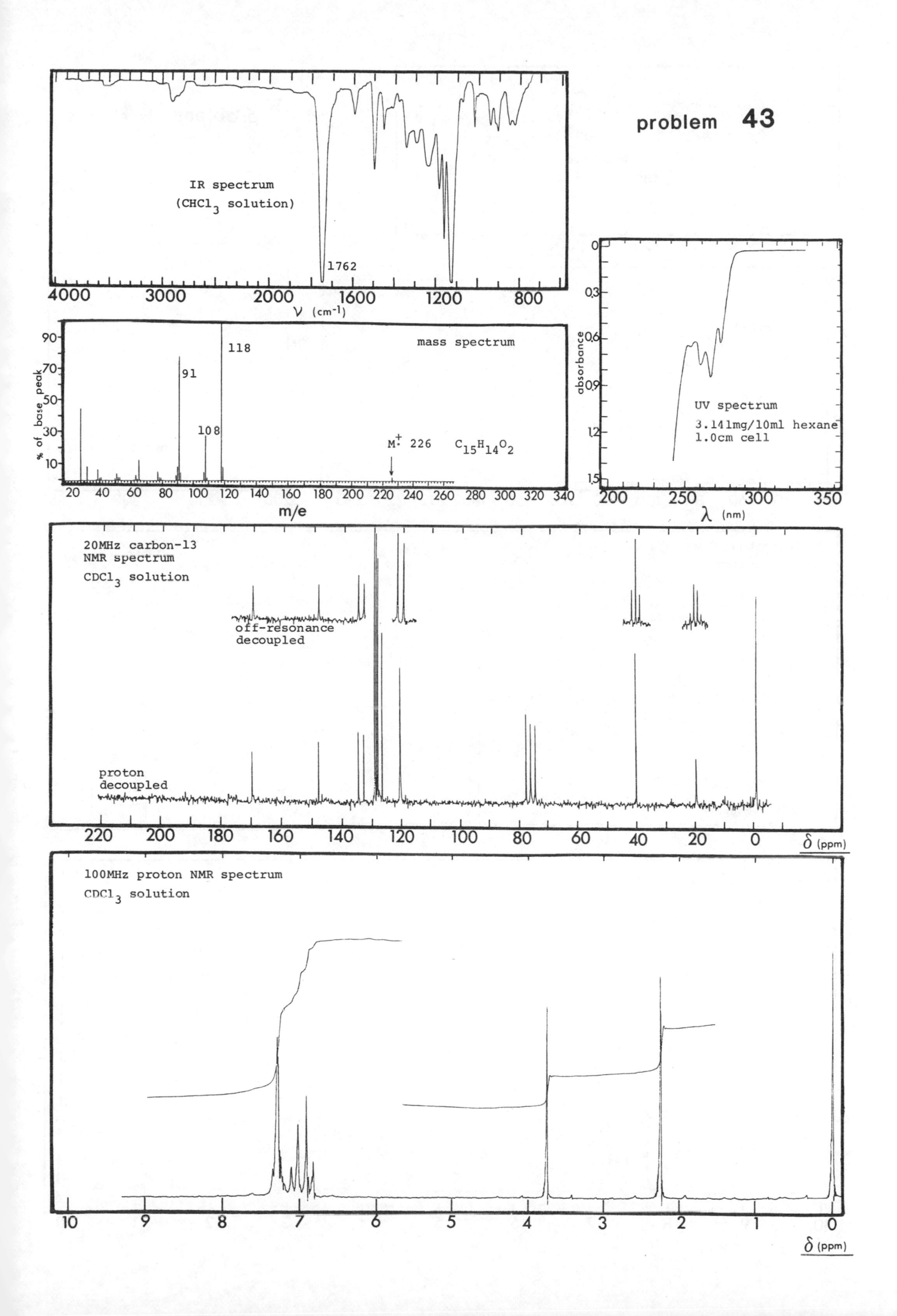

IR spectrum
(CHCl3 solution)
1762
4000
3000
2000
1600
1200
800
ν (cm-1)
mass spectrum
118
91
108
M+ 226
C15H14O2
% of base peak
m/e
UV spectrum
3.141mg/10ml hexane
1.0cm cell
absorbance
λ (nm)
20MHz carbon-13
NMR spectrum
CDCl3 solution
off-resonance
decoupled
proton
decoupled
δ (ppm)
100MHz proton NMR spectrum
CDCl3 solution
δ (ppm)

problem 44

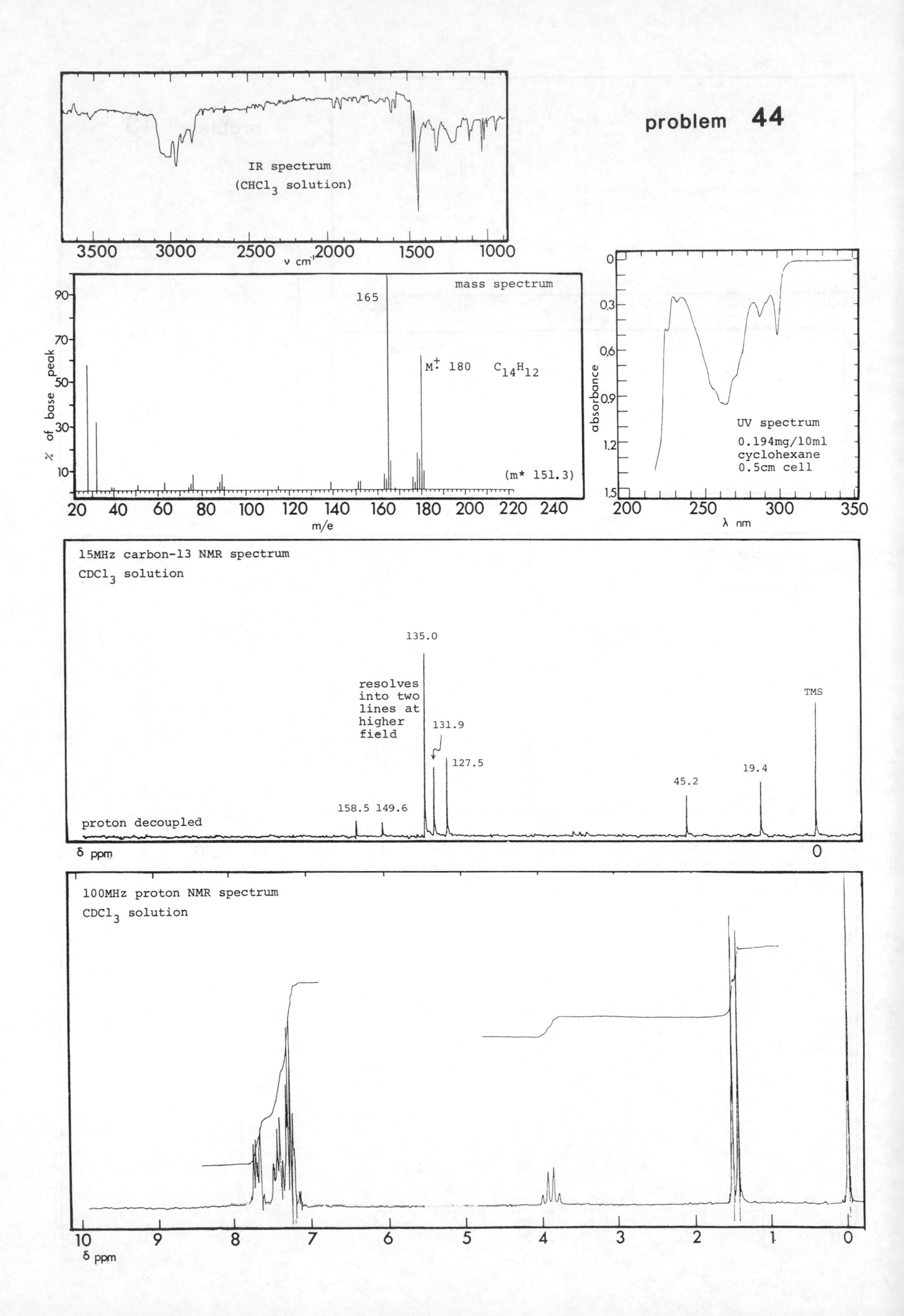

IR spectrum
(CHCl3 solution)
3500 3000 2500 2000 1500 1000
ν cm⁻¹
mass spectrum
165
M⁺ 180 C14H12
(m* 151.3)
% of base peak
90 70 50 30 10
20 40 60 80 100 120 140 160 180 200 220 240
m/e
absorbance
0 0.3 0.6 0.9 1.2 1.5
UV spectrum
0.194mg/10ml
cyclohexane
0.5cm cell
200 250 300 350
λ nm
15MHz carbon-13 NMR spectrum
CDCl3 solution
135.0
resolves into two lines at higher field
131.9
127.5
158.5 149.6
45.2
19.4
TMS
proton decoupled
δ ppm
0
100MHz proton NMR spectrum
CDCl3 solution
10 9 8 7 6 5 4 3 2 1 0
δ ppm

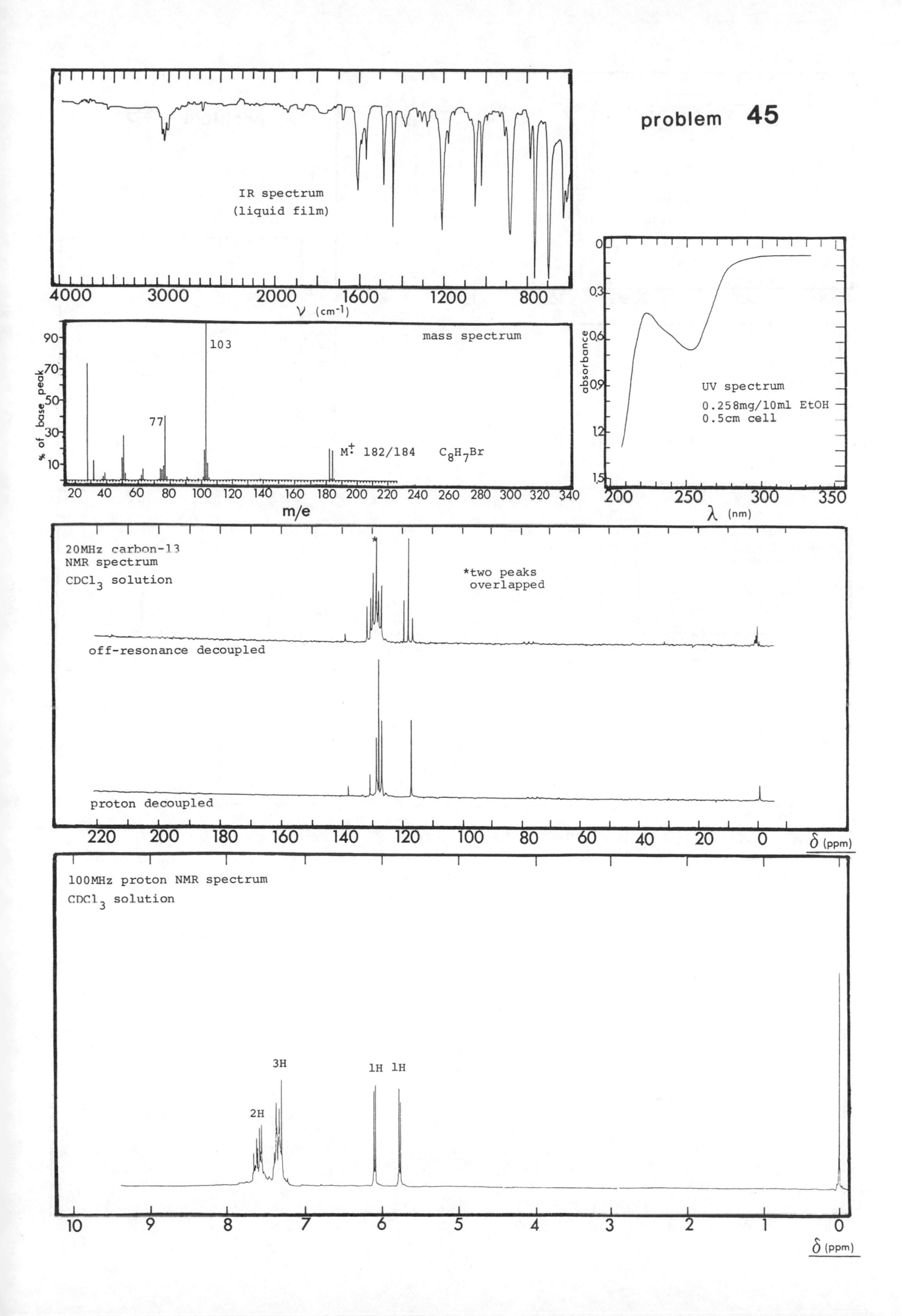
problem 45
IR spectrum
(liquid film)
4000
3000
2000
1600
1200
800
ν (cm-1)
mass spectrum
103
77
M+ 182/184 C8H7Br
% of base peak
m/e
UV spectrum
0.258mg/10ml EtOH
0.5cm cell
absorbance
λ (nm)
20MHz carbon-13
NMR spectrum
CDCl3 solution
*two peaks
overlapped
off-resonance decoupled
proton decoupled
δ (ppm)
100MHz proton NMR spectrum
CDCl3 solution
3H
2H
1H 1H
δ (ppm)

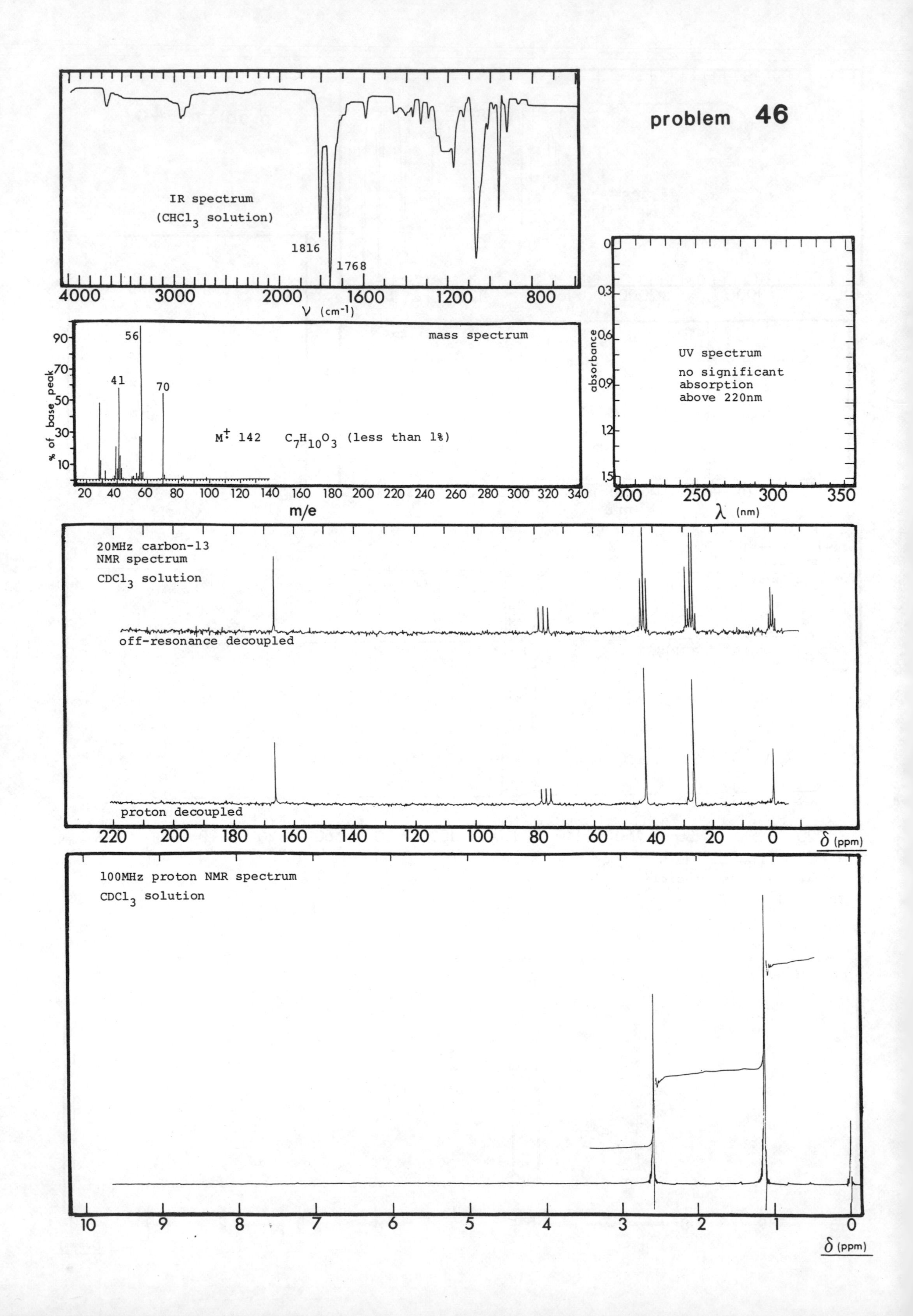
problem 46
IR spectrum
(CHCl3 solution)
1816
1768
4000
3000
2000
1600
1200
800
ν (cm-1)
mass spectrum
56
41
70
M+ 142 C7H10O3 (less than 1%)
% of base peak
m/e
UV spectrum
no significant
absorption
above 220nm
absorbance
λ (nm)
20MHz carbon-13
NMR spectrum
CDCl3 solution
off-resonance decoupled
proton decoupled
δ (ppm)
100MHz proton NMR spectrum
CDCl3 solution
δ (ppm)

problem **47**

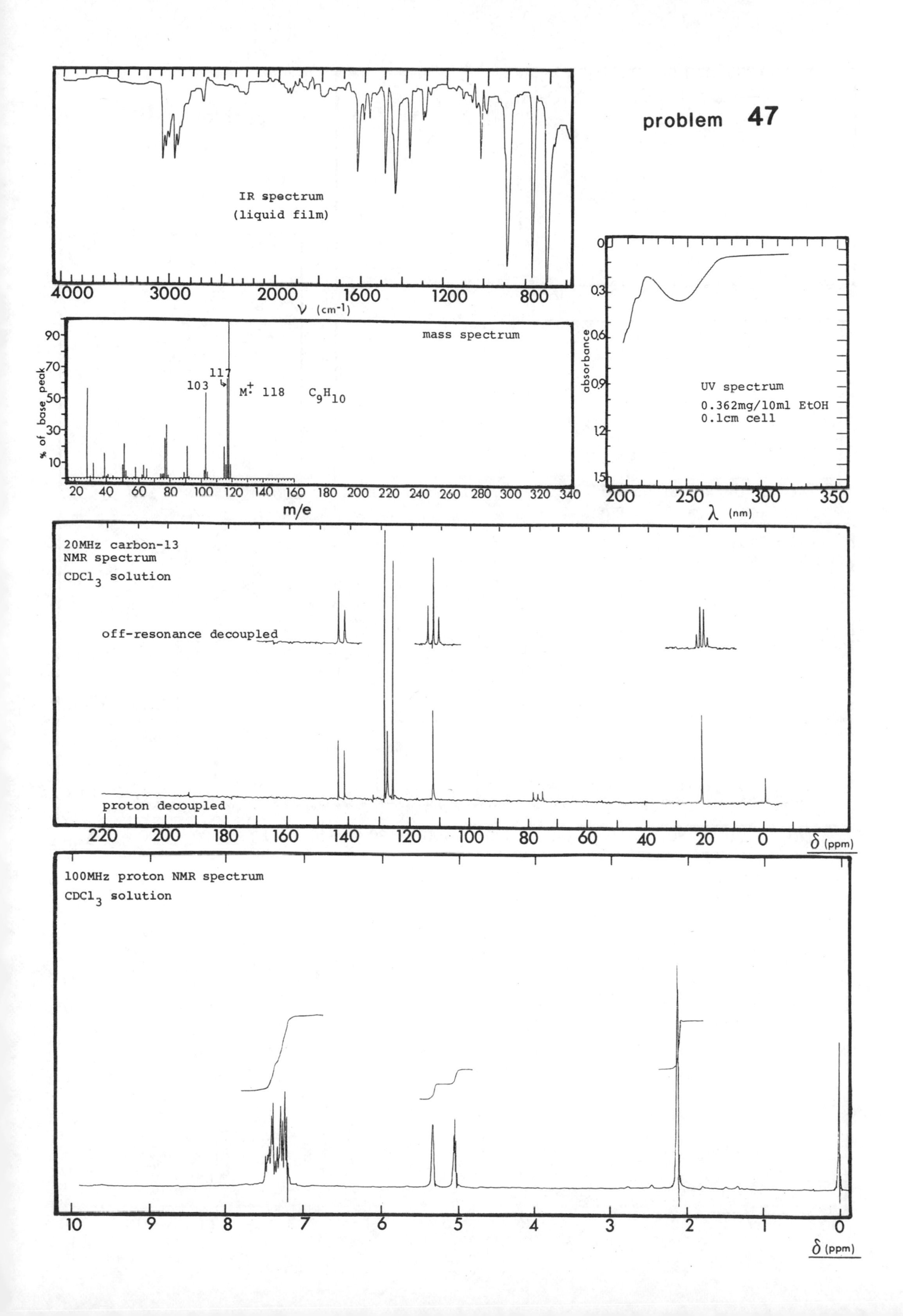

IR spectrum
(liquid film)
4000
3000
2000
1600
1200
800
ν (cm-1)
mass spectrum
% of base peak
117
103
M+ 118
C9H10
m/e
UV spectrum
0.362mg/10ml EtOH
0.1cm cell
absorbance
λ (nm)
20MHz carbon-13
NMR spectrum
CDCl3 solution
off-resonance decoupled
proton decoupled
δ (ppm)
100MHz proton NMR spectrum
CDCl3 solution
δ (ppm)

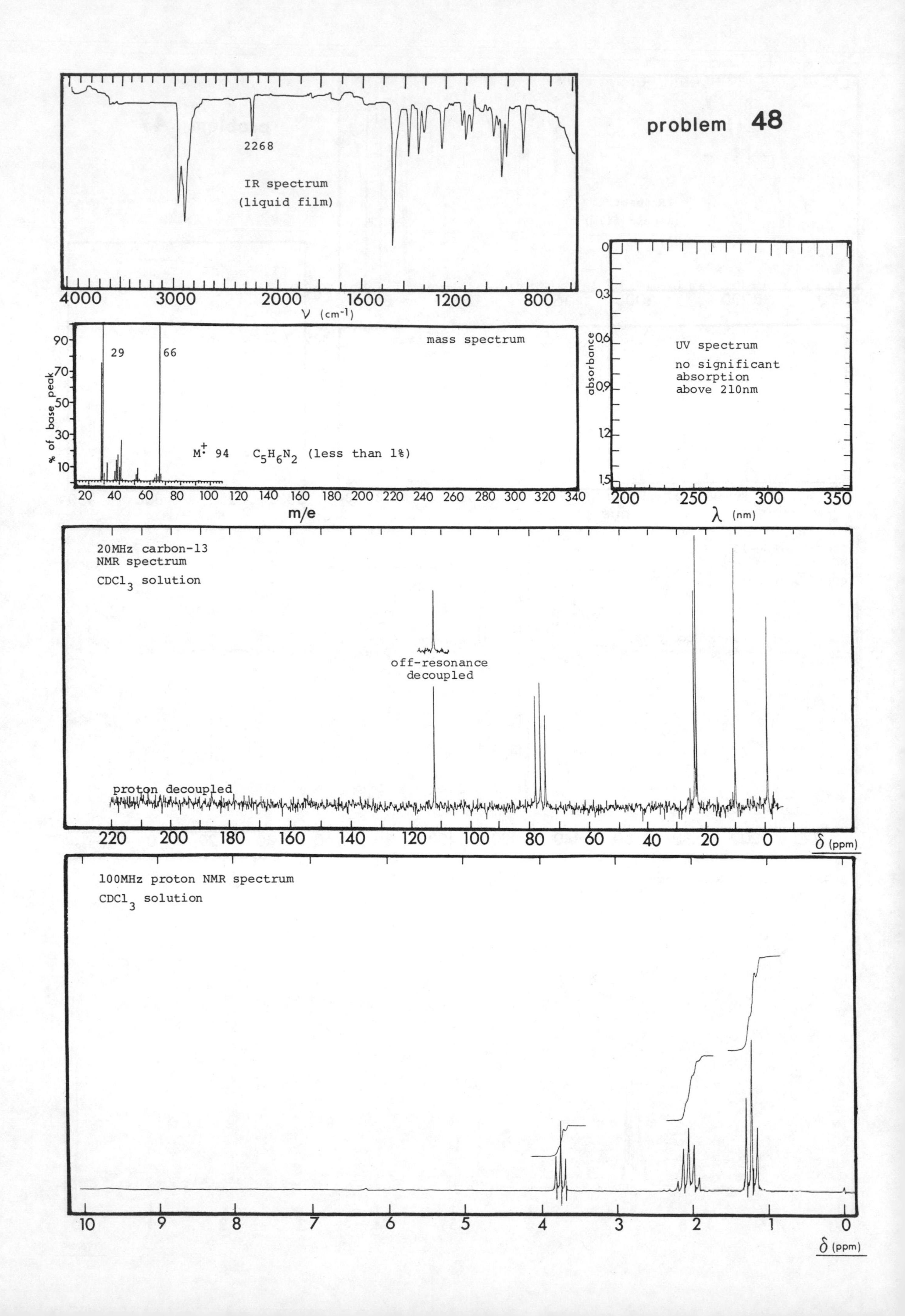

problem 48
2268
IR spectrum
(liquid film)
4000 3000 2000 1600 1200 800
ν (cm⁻¹)
mass spectrum
29
66
M⁺· 94 $C_5H_6N_2$ (less than 1%)
% of base peak
m/e
UV spectrum
no significant absorption above 210nm
absorbance
λ (nm)
20MHz carbon-13 NMR spectrum
$CDCl_3$ solution
off-resonance decoupled
proton decoupled
δ (ppm)
100MHz proton NMR spectrum
$CDCl_3$ solution
δ (ppm)

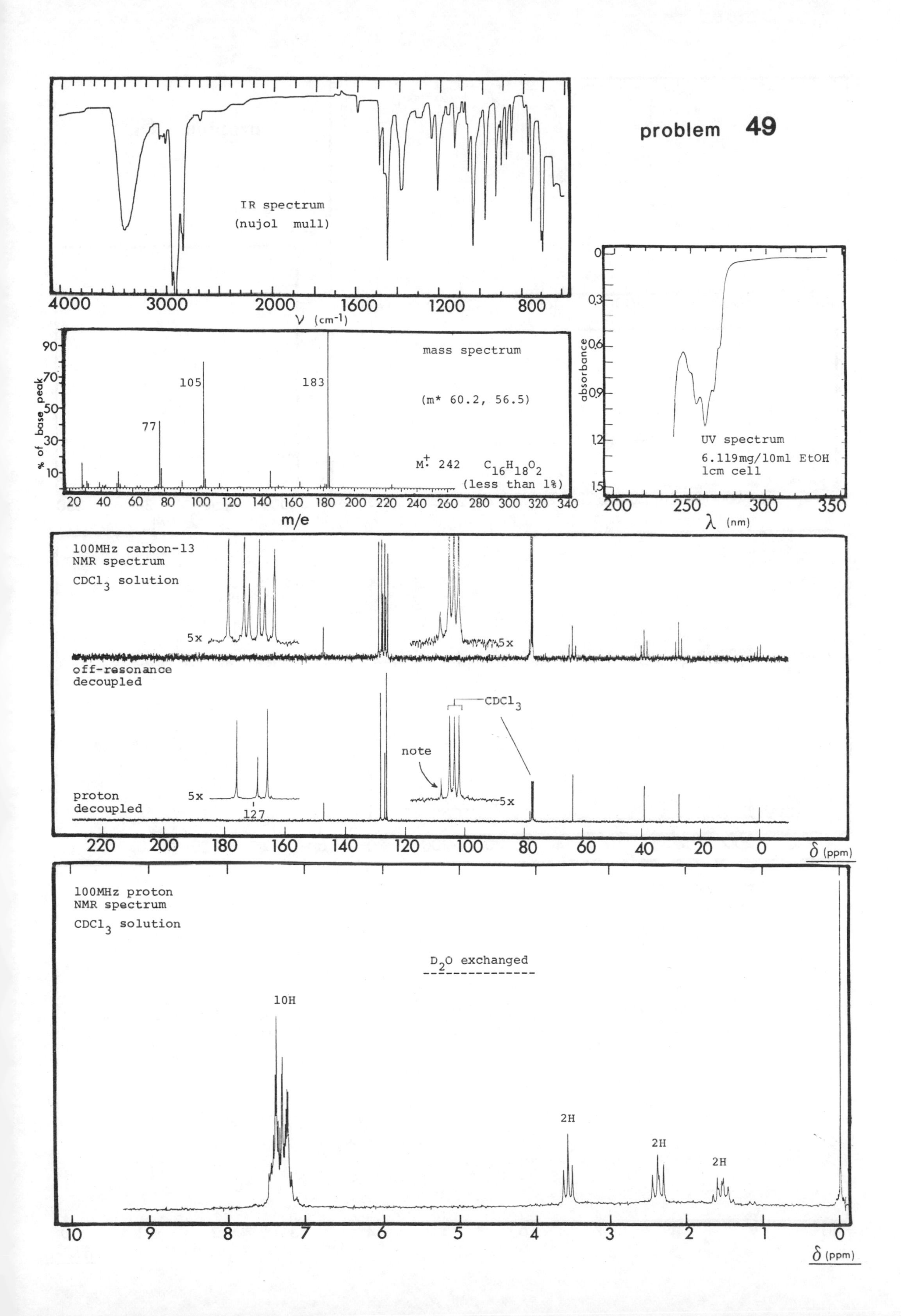
problem 49
IR spectrum
(nujol mull)
ν (cm⁻¹)
mass spectrum
(m* 60.2, 56.5)
77
105
183
M⁺ 242 C16H18O2
(less than 1%)
% of base peak
m/e
UV spectrum
6.119mg/10ml EtOH
1cm cell
absorbance
λ (nm)
100MHz carbon-13
NMR spectrum
CDCl3 solution
5x
off-resonance
decoupled
CDCl3
note
proton
decoupled
127
δ (ppm)
100MHz proton
NMR spectrum
CDCl3 solution
D2O exchanged
10H
2H
2H
2H
δ (ppm)

problem 50

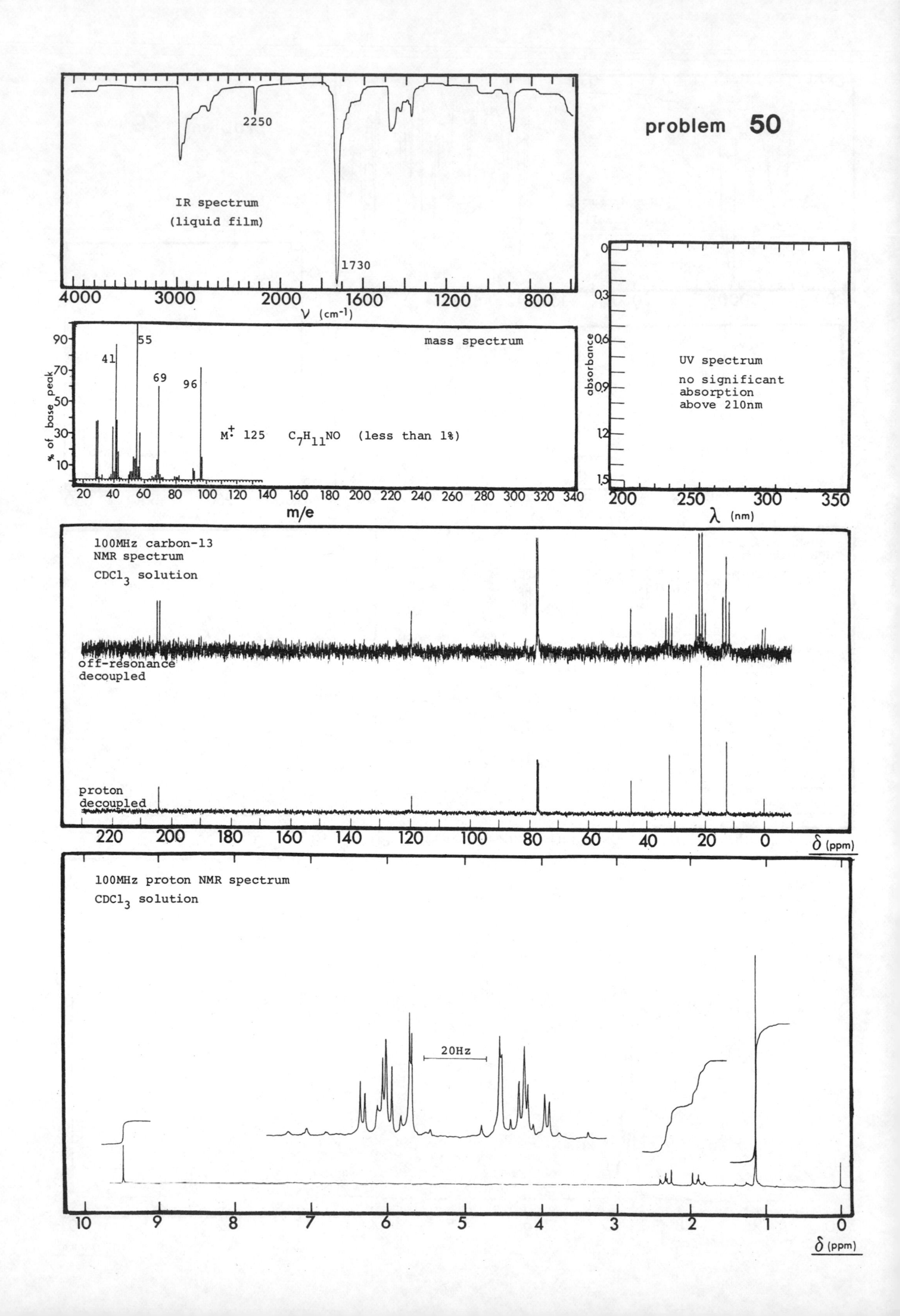

2250
IR spectrum
(liquid film)
1730
4000
3000
2000
1600
1200
800
ν (cm-1)
mass spectrum
90
70
50
30
10
% of base peak
41
55
69
96
M+ 125 C7H11NO (less than 1%)
20 40 60 80 100 120 140 160 180 200 220 240 260 280 300 320 340
m/e
0
0.3
0.6
0.9
1.2
1.5
absorbance
UV spectrum
no significant absorption above 210nm
200 250 300 350
λ (nm)
100MHz carbon-13 NMR spectrum
CDCl3 solution
off-resonance decoupled
proton decoupled
220 200 180 160 140 120 100 80 60 40 20 0
δ (ppm)
100MHz proton NMR spectrum
CDCl3 solution
20Hz
10 9 8 7 6 5 4 3 2 1 0
δ (ppm)

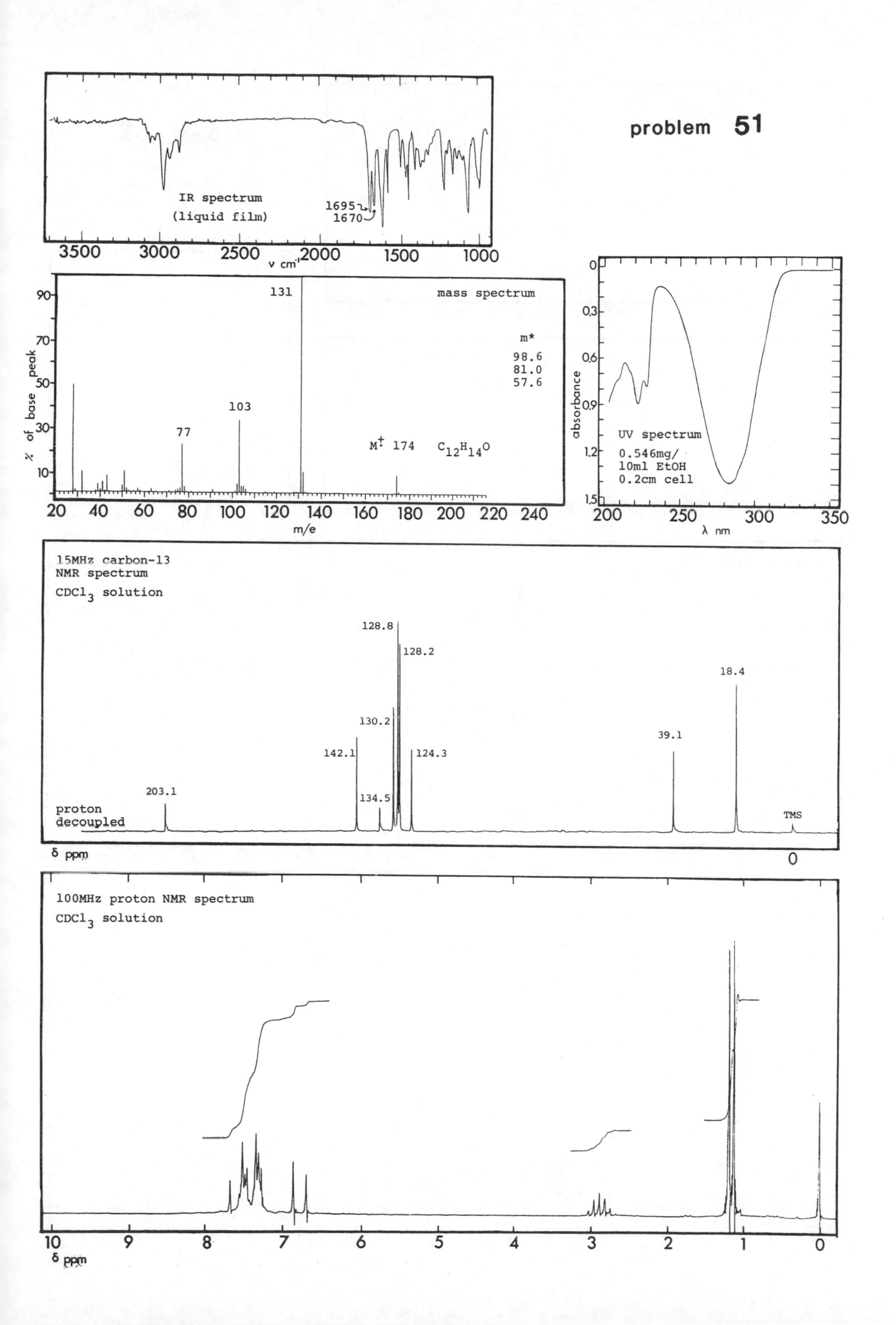

problem 51
IR spectrum
(liquid film)
1695
1670
3500
3000
2500
2000
1500
1000
ν cm⁻¹
mass spectrum
131
103
77
m*
98.6
81.0
57.6
M⁺· 174
C12H14O
% of base peak
m/e
UV spectrum
0.546mg/
10ml EtOH
0.2cm cell
absorbance
λ nm
15MHz carbon-13
NMR spectrum
CDCl3 solution
203.1
142.1
134.5
130.2
128.8
128.2
124.3
39.1
18.4
TMS
proton
decoupled
δ ppm
100MHz proton NMR spectrum
CDCl3 solution
δ ppm

problem 52

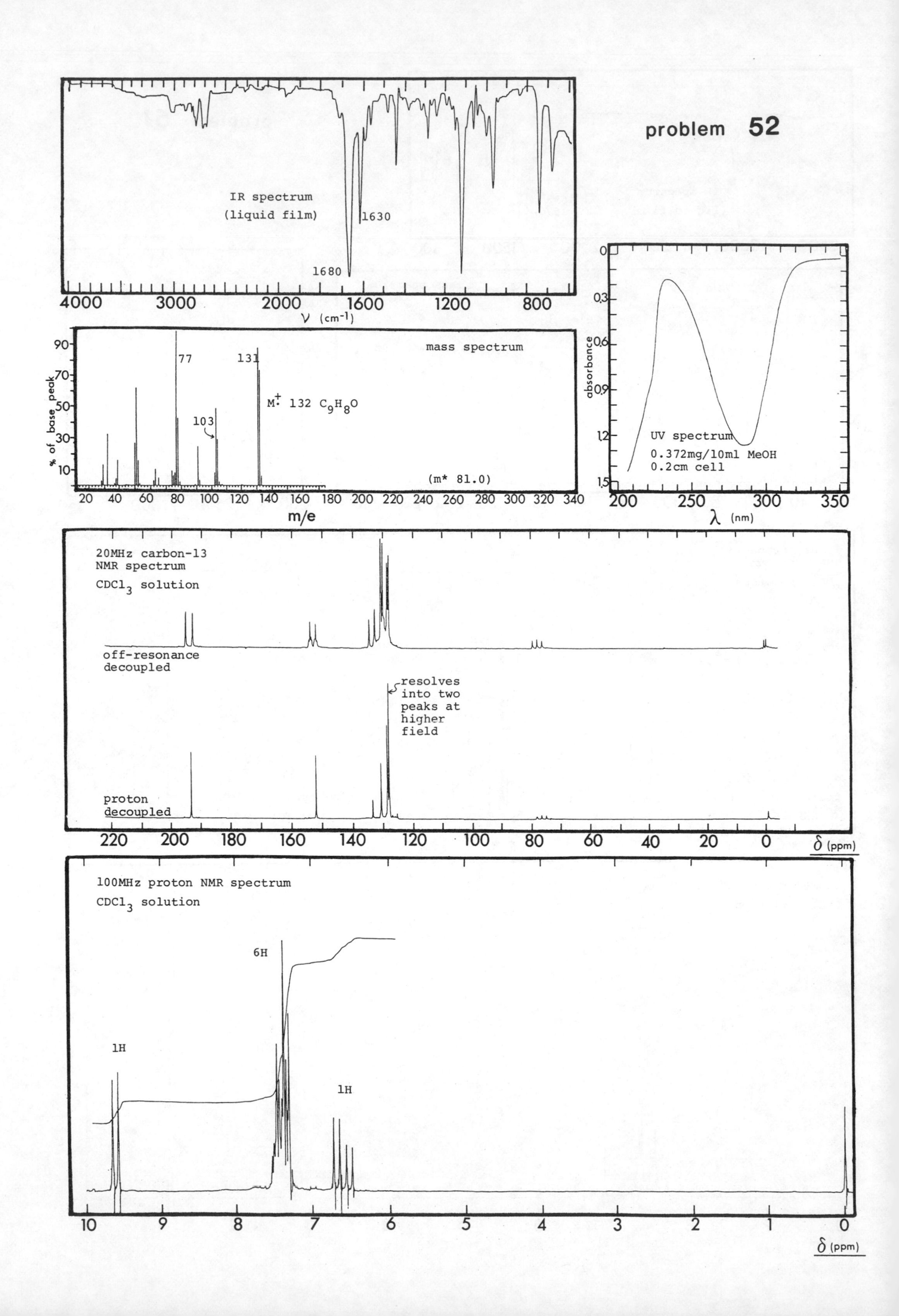

IR spectrum
(liquid film)
1630
1680
4000
3000
2000
1600
1200
800
ν (cm-1)
mass spectrum
77
131
103
M+· 132 C9H8O
(m* 81.0)
% of base peak
m/e
absorbance
UV spectrum
0.372mg/10ml MeOH
0.2cm cell
λ (nm)
20MHz carbon-13
NMR spectrum
CDCl3 solution
off-resonance
decoupled
resolves
into two
peaks at
higher
field
proton
decoupled
δ (ppm)
100MHz proton NMR spectrum
CDCl3 solution
6H
1H
1H
δ (ppm)

problem 53

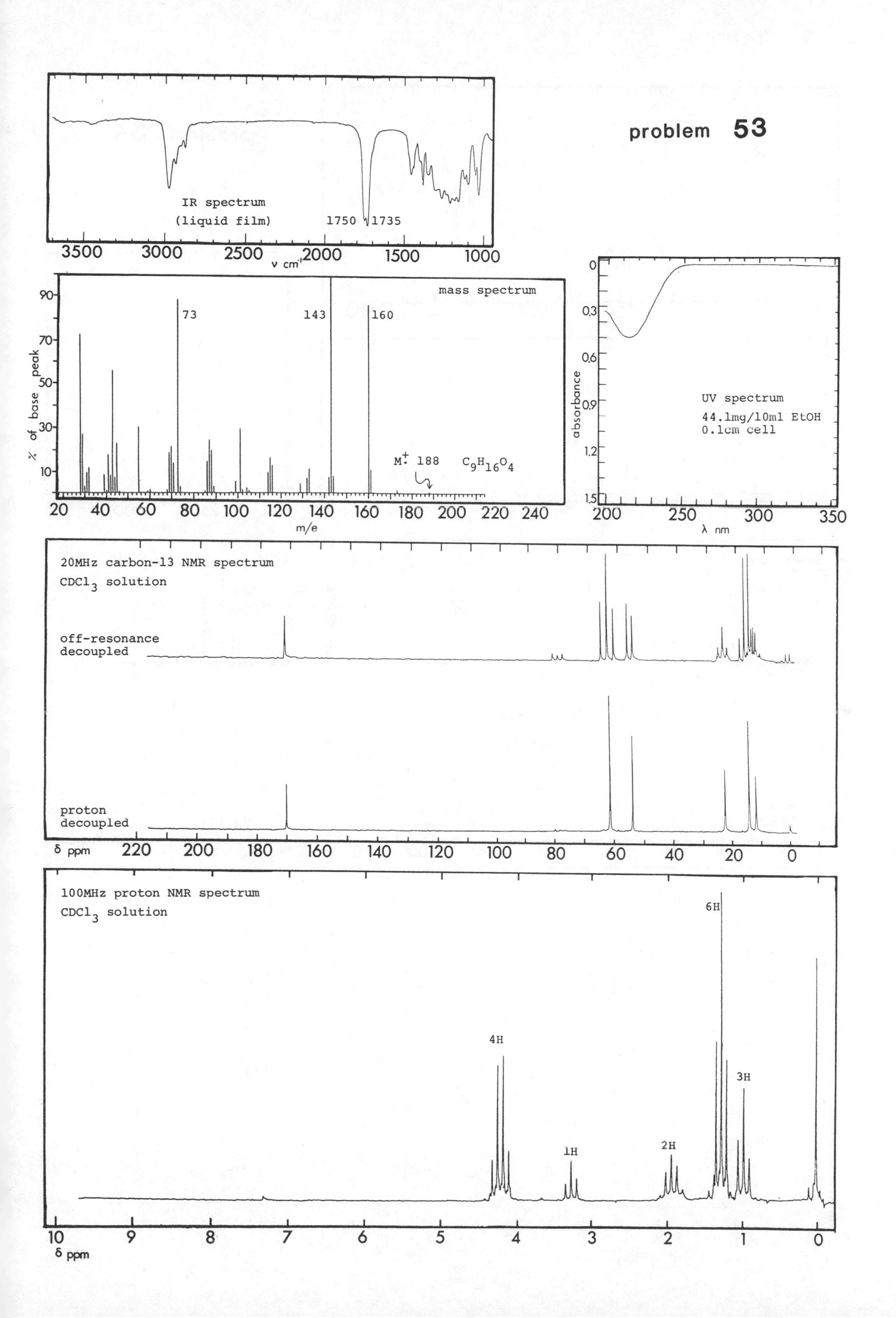

IR spectrum
(liquid film)
1750
1735
3500
3000
2500
2000
1500
1000
ν cm⁻¹
mass spectrum
73
143
160
M⁺ 188
C9H16O4
% of base peak
90
70
50
30
10
20
40
60
80
100
120
140
160
180
200
220
240
m/e
UV spectrum
44.1mg/10ml EtOH
0.1cm cell
absorbance
0
0.3
0.6
0.9
1.2
1.5
200
250
300
350
λ nm
20MHz carbon-13 NMR spectrum
CDCl3 solution
off-resonance
decoupled
proton
decoupled
δ ppm
220
200
180
160
140
120
100
80
60
40
20
0
100MHz proton NMR spectrum
CDCl3 solution
6H
4H
3H
1H
2H
10
9
8
7
6
5
4
3
2
1
0
δ ppm

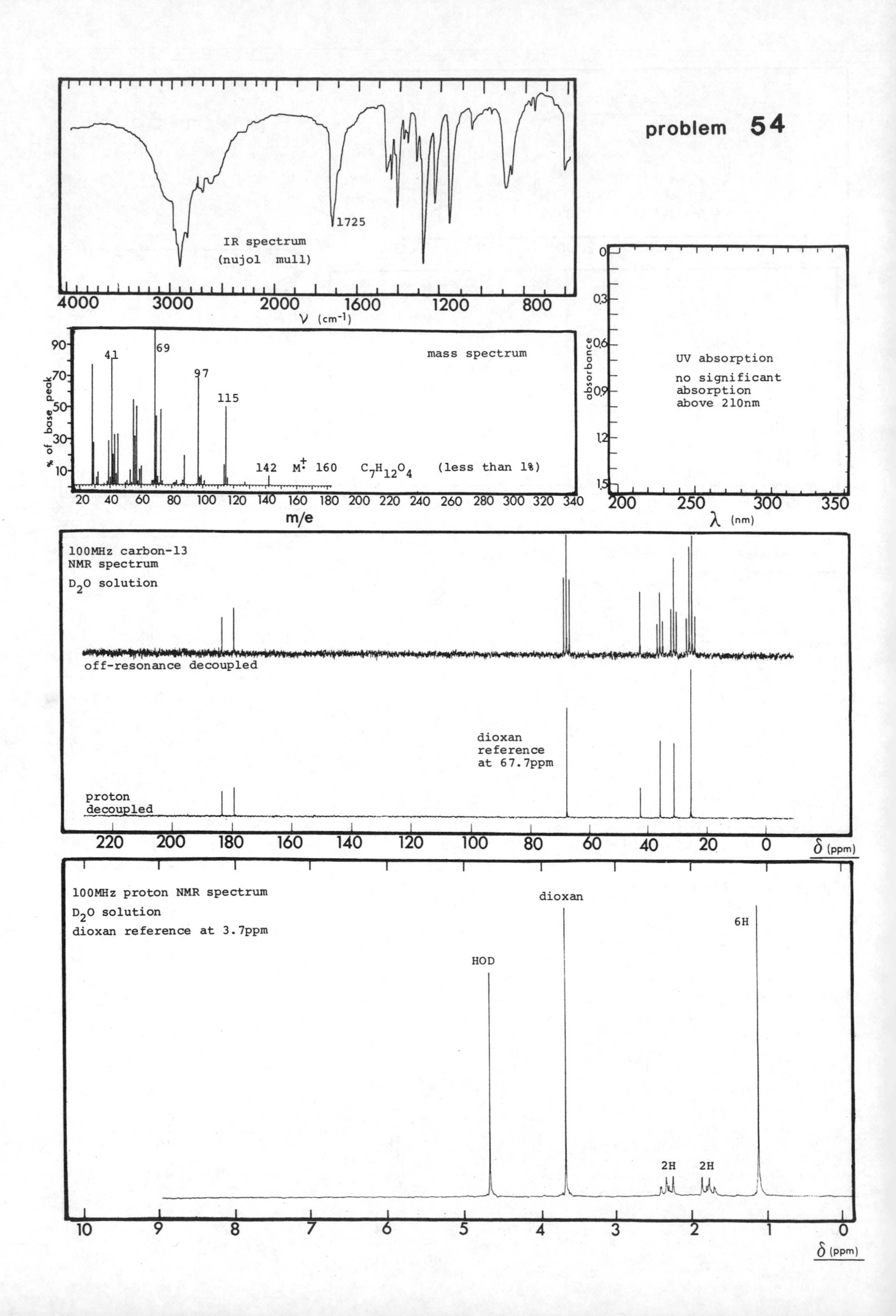

problem 54
IR spectrum
(nujol mull)
1725
4000
3000
2000
1600
1200
800
ν (cm-1)
mass spectrum
% of base peak
41
69
97
115
142
M+· 160
C7H12O4
(less than 1%)
m/e
UV absorption
no significant absorption above 210nm
absorbance
λ (nm)
100MHz carbon-13 NMR spectrum
D2O solution
off-resonance decoupled
dioxan reference at 67.7ppm
proton decoupled
δ (ppm)
100MHz proton NMR spectrum
D2O solution
dioxan reference at 3.7ppm
dioxan
HOD
6H
2H
2H
δ (ppm)

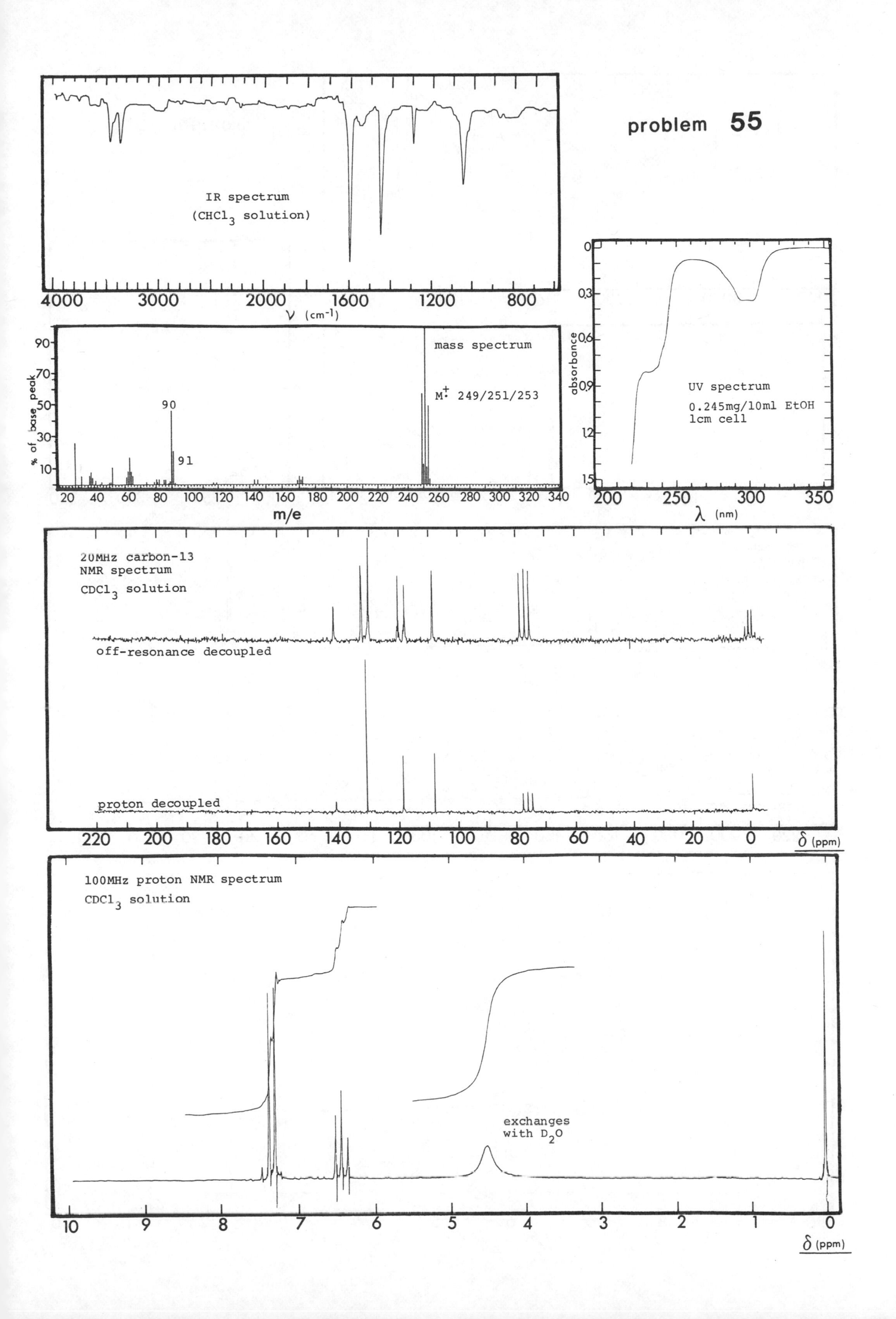

problem 55
IR spectrum
(CHCl3 solution)
4000
3000
2000
1600
1200
800
ν (cm-1)
mass spectrum
M+· 249/251/253
90
91
% of base peak
m/e
UV spectrum
0.245mg/10ml EtOH
1cm cell
absorbance
λ (nm)
20MHz carbon-13
NMR spectrum
CDCl3 solution
off-resonance decoupled
proton decoupled
δ (ppm)
100MHz proton NMR spectrum
CDCl3 solution
exchanges
with D2O
δ (ppm)

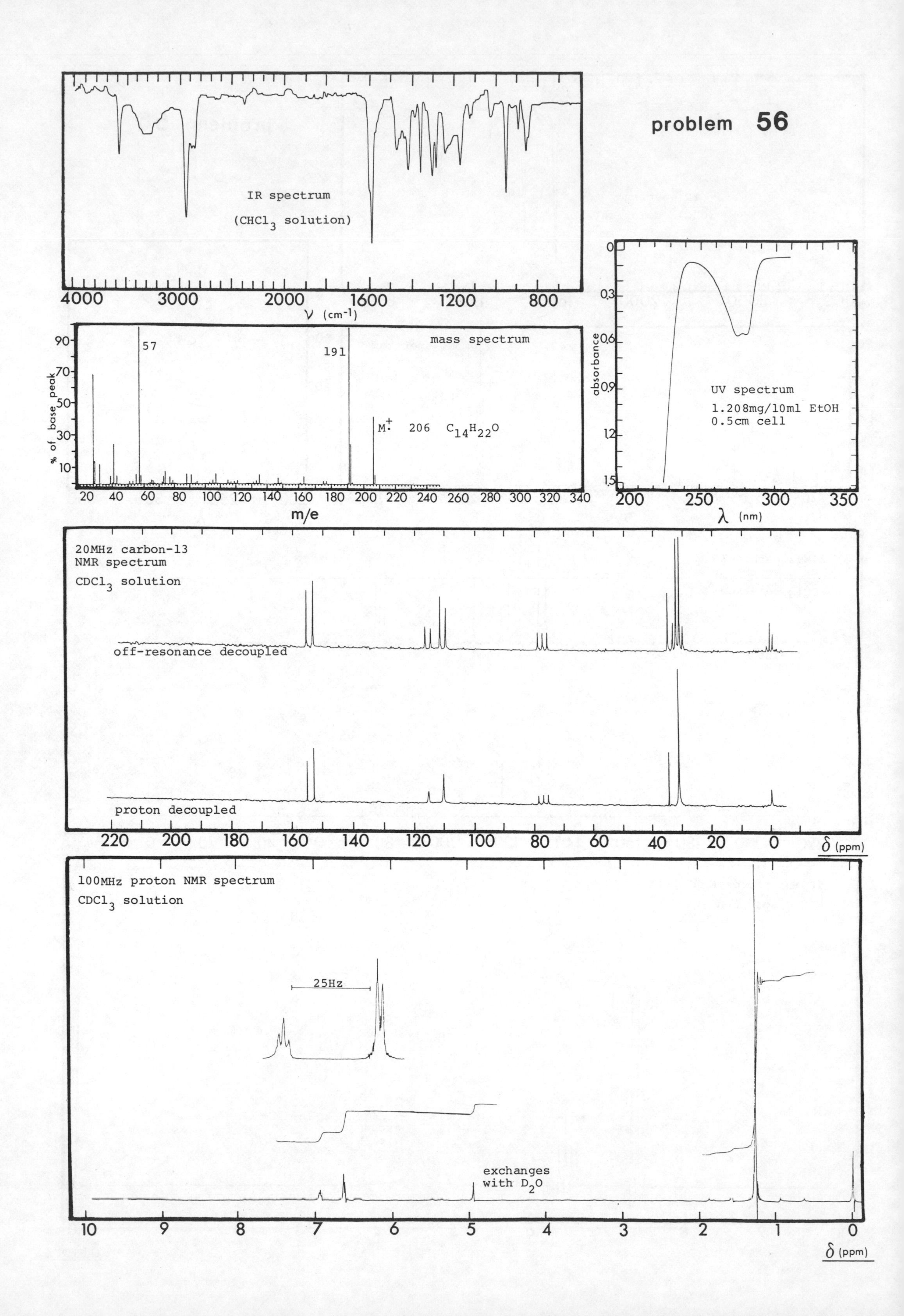

problem 56
IR spectrum
(CHCl3 solution)
4000
3000
2000
1600
1200
800
ν (cm-1)
mass spectrum
% of base peak
57
191
M+ 206 C14H22O
m/e
UV spectrum
1.208mg/10ml EtOH
0.5cm cell
absorbance
λ (nm)
20MHz carbon-13
NMR spectrum
CDCl3 solution
off-resonance decoupled
proton decoupled
δ (ppm)
100MHz proton NMR spectrum
CDCl3 solution
25Hz
exchanges
with D2O
δ (ppm)

problem **57**

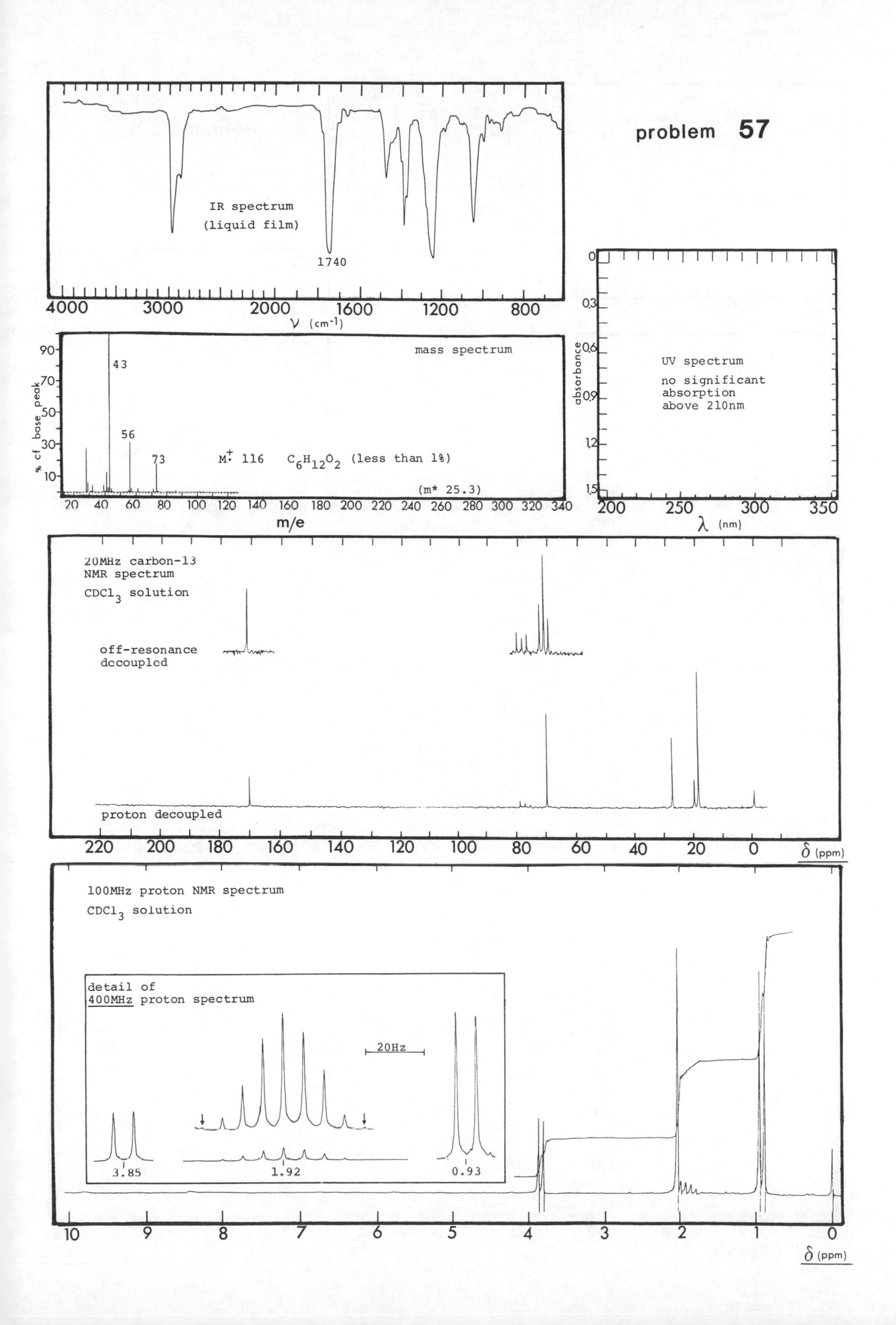

IR spectrum
(liquid film)
1740
4000
3000
2000
1600
1200
800
ν (cm⁻¹)
mass spectrum
% of base peak
43
56
73
M⁺· 116 C6H12O2 (less than 1%)
(m* 25.3)
m/e
UV spectrum
no significant absorption above 210nm
absorbance
λ (nm)
20MHz carbon-13 NMR spectrum
CDCl3 solution
off-resonance decoupled
proton decoupled
δ (ppm)
100MHz proton NMR spectrum
CDCl3 solution
detail of
400MHz proton spectrum
20Hz
3.85
1.92
0.93
δ (ppm)

problem 58

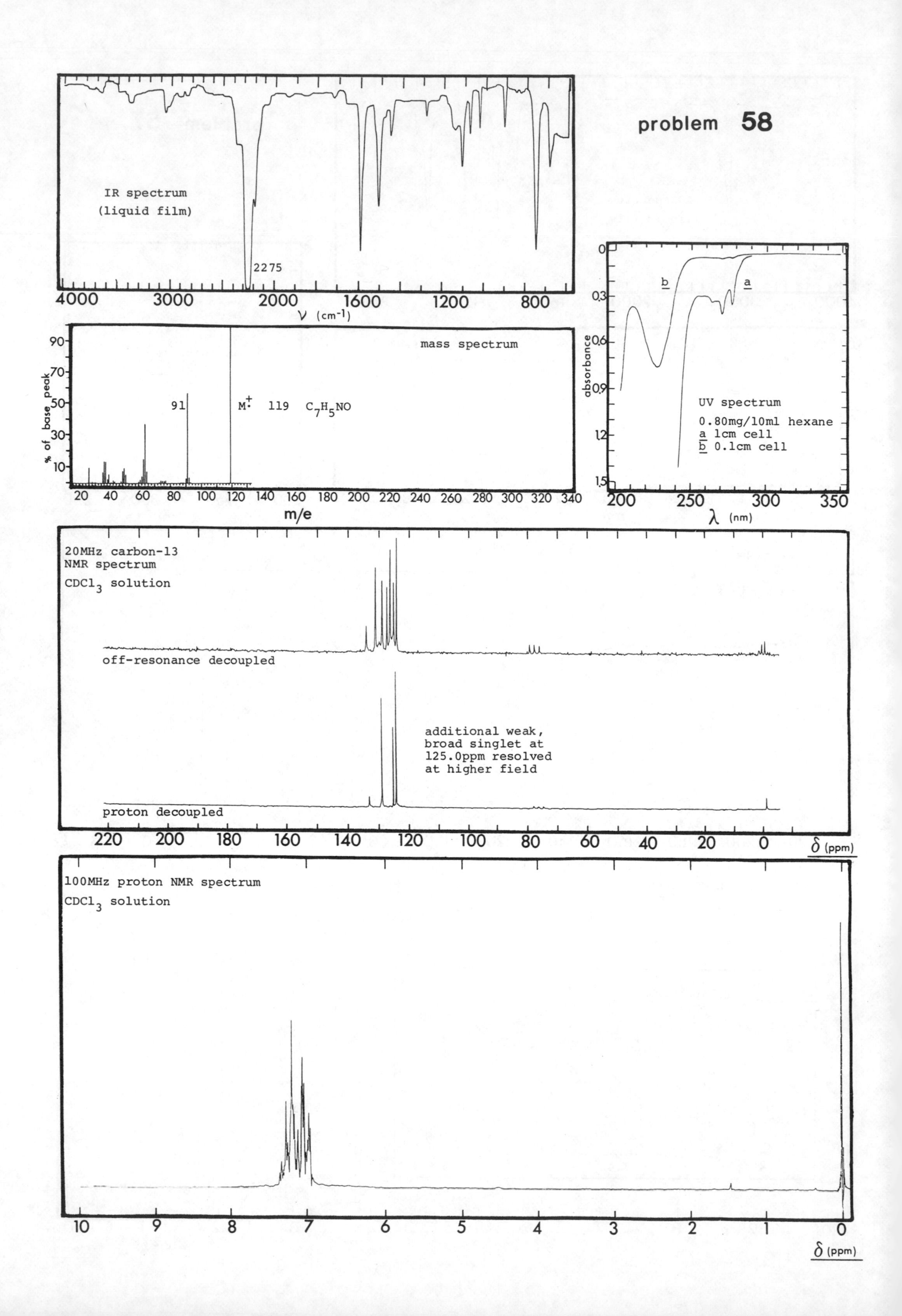

IR spectrum
(liquid film)
2275
4000
3000
2000
1600
1200
800
ν (cm-1)
mass spectrum
% of base peak
91
M+· 119 C7H5NO
m/e
UV spectrum
0.80mg/10ml hexane
a 1cm cell
b 0.1cm cell
absorbance
λ (nm)
20MHz carbon-13
NMR spectrum
CDCl3 solution
off-resonance decoupled
additional weak,
broad singlet at
125.0ppm resolved
at higher field
proton decoupled
δ (ppm)
100MHz proton NMR spectrum
CDCl3 solution
δ (ppm)

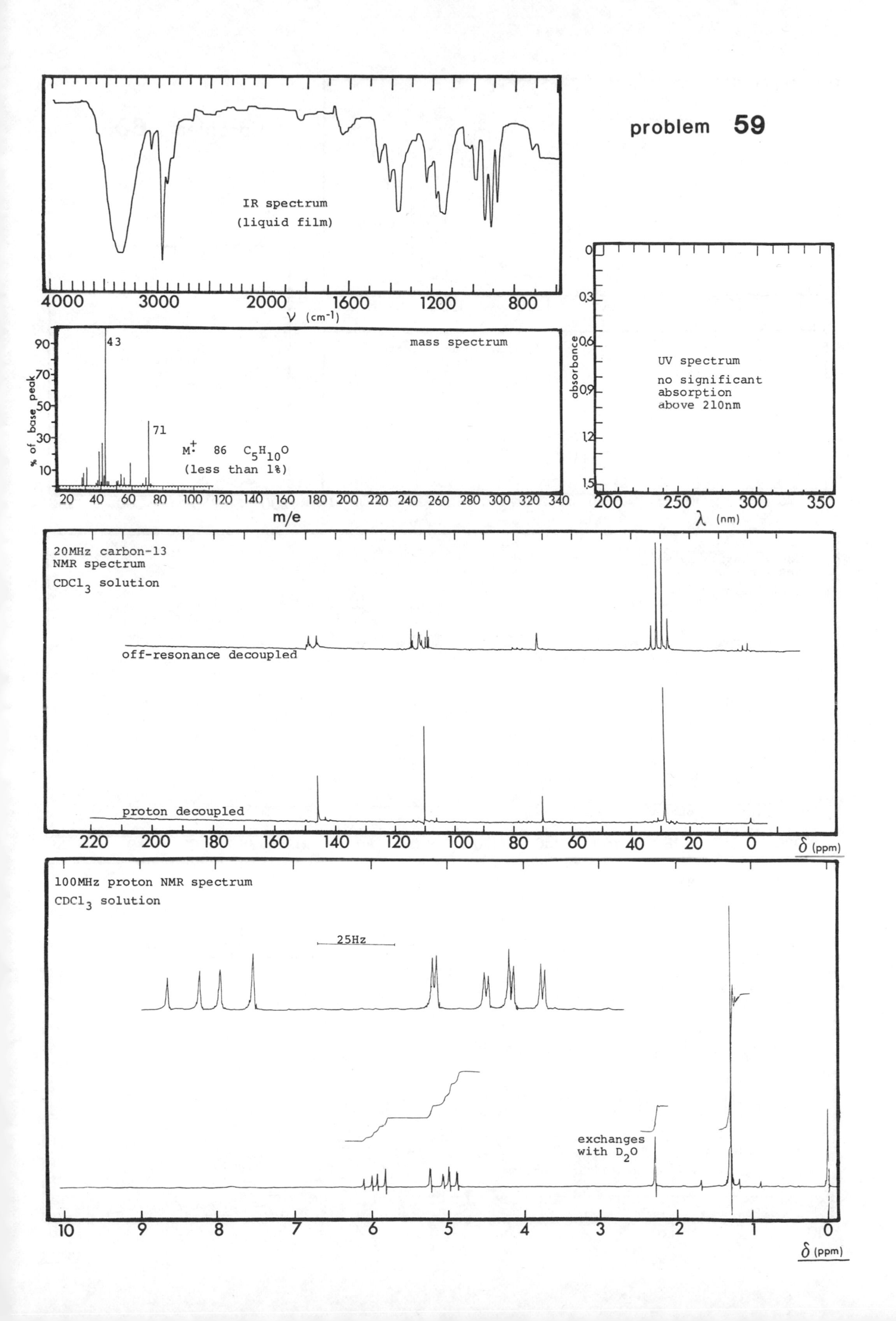

problem 59
IR spectrum
(liquid film)
4000
3000
2000
1600
1200
800
ν (cm-1)
mass spectrum
% of base peak
43
71
M+ 86 C5H10O
(less than 1%)
m/e
UV spectrum
no significant absorption above 210nm
absorbance
λ (nm)
20MHz carbon-13 NMR spectrum
CDCl3 solution
off-resonance decoupled
proton decoupled
δ (ppm)
100MHz proton NMR spectrum
CDCl3 solution
25Hz
exchanges with D2O
δ (ppm)

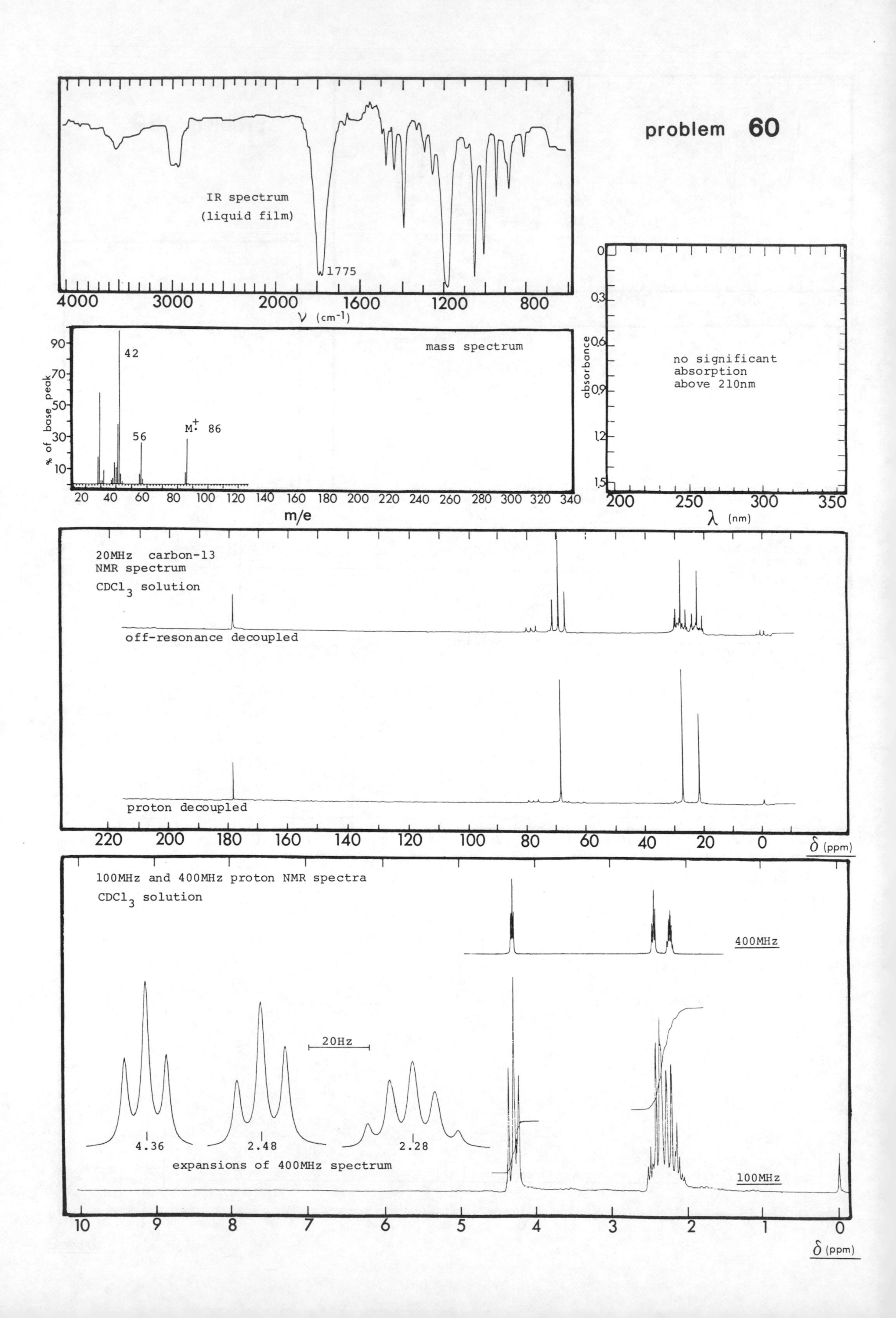
problem 60
IR spectrum
(liquid film)
1775
4000
3000
2000
1600
1200
800
ν (cm-1)
mass spectrum
% of base peak
90
70
50
30
10
42
56
M+· 86
20 40 60 80 100 120 140 160 180 200 220 240 260 280 300 320 340
m/e
absorbance
0
0.3
0.6
0.9
1.2
1.5
no significant
absorption
above 210nm
200 250 300 350
λ (nm)
20MHz carbon-13
NMR spectrum
CDCl3 solution
off-resonance decoupled
proton decoupled
220 200 180 160 140 120 100 80 60 40 20 0
δ (ppm)
100MHz and 400MHz proton NMR spectra
CDCl3 solution
400MHz
20Hz
4.36
2.48
2.28
expansions of 400MHz spectrum
100MHz
10 9 8 7 6 5 4 3 2 1 0
δ (ppm)

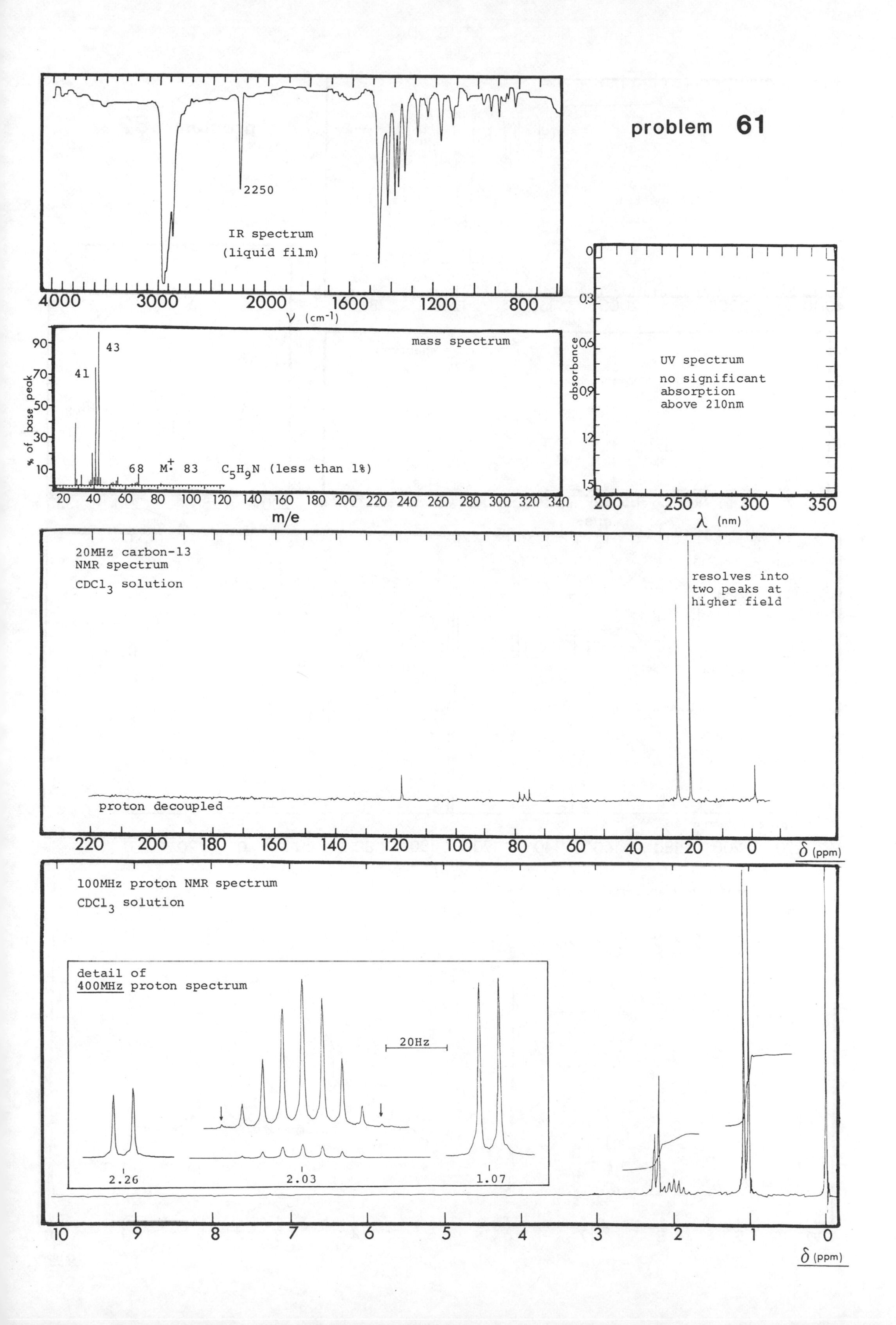

problem 61
2250
IR spectrum
(liquid film)
4000
3000
2000
1600
1200
800
ν (cm⁻¹)
mass spectrum
% of base peak
90
70
50
30
10
43
41
68
M⁺· 83
C5H9N (less than 1%)
20
40
60
80
100
120
140
160
180
200
220
240
260
280
300
320
340
m/e
UV spectrum
no significant absorption above 210nm
absorbance
0
0.3
0.6
0.9
1.2
1.5
200
250
300
350
λ (nm)
20MHz carbon-13 NMR spectrum
CDCl3 solution
resolves into two peaks at higher field
proton decoupled
220
200
180
160
140
120
100
80
60
40
20
0
δ (ppm)
100MHz proton NMR spectrum
CDCl3 solution
detail of 400MHz proton spectrum
20Hz
2.26
2.03
1.07
10
9
8
7
6
5
4
3
2
1
0
δ (ppm)

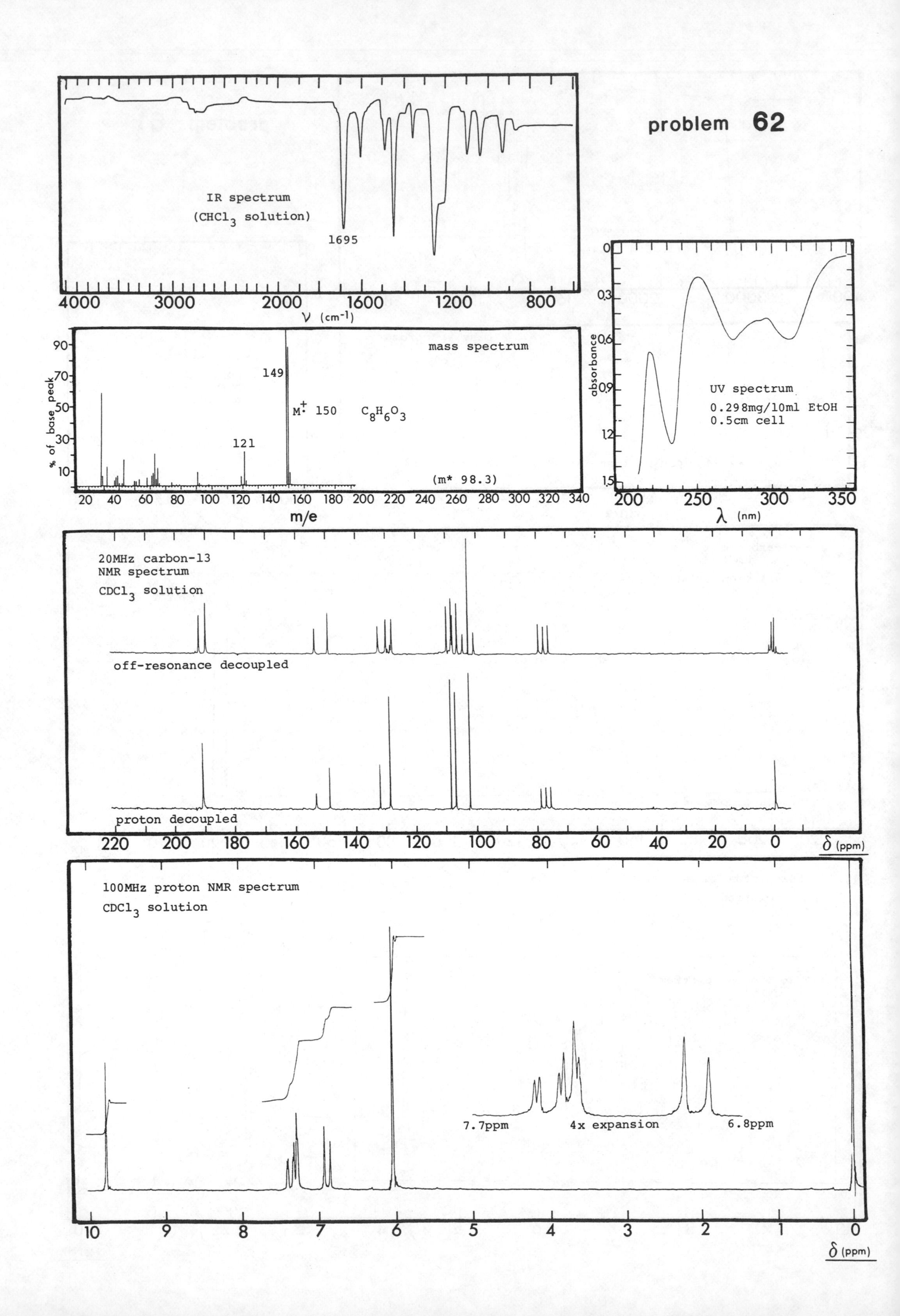
problem 62
IR spectrum
(CHCl3 solution)
1695
4000
3000
2000
1600
1200
800
ν (cm-1)
mass spectrum
% of base peak
149
M+ 150
C8H6O3
121
(m* 98.3)
m/e
UV spectrum
0.298mg/10ml EtOH
0.5cm cell
absorbance
λ (nm)
20MHz carbon-13
NMR spectrum
CDCl3 solution
off-resonance decoupled
proton decoupled
δ (ppm)
100MHz proton NMR spectrum
CDCl3 solution
7.7ppm
4x expansion
6.8ppm
δ (ppm)

problem **63**

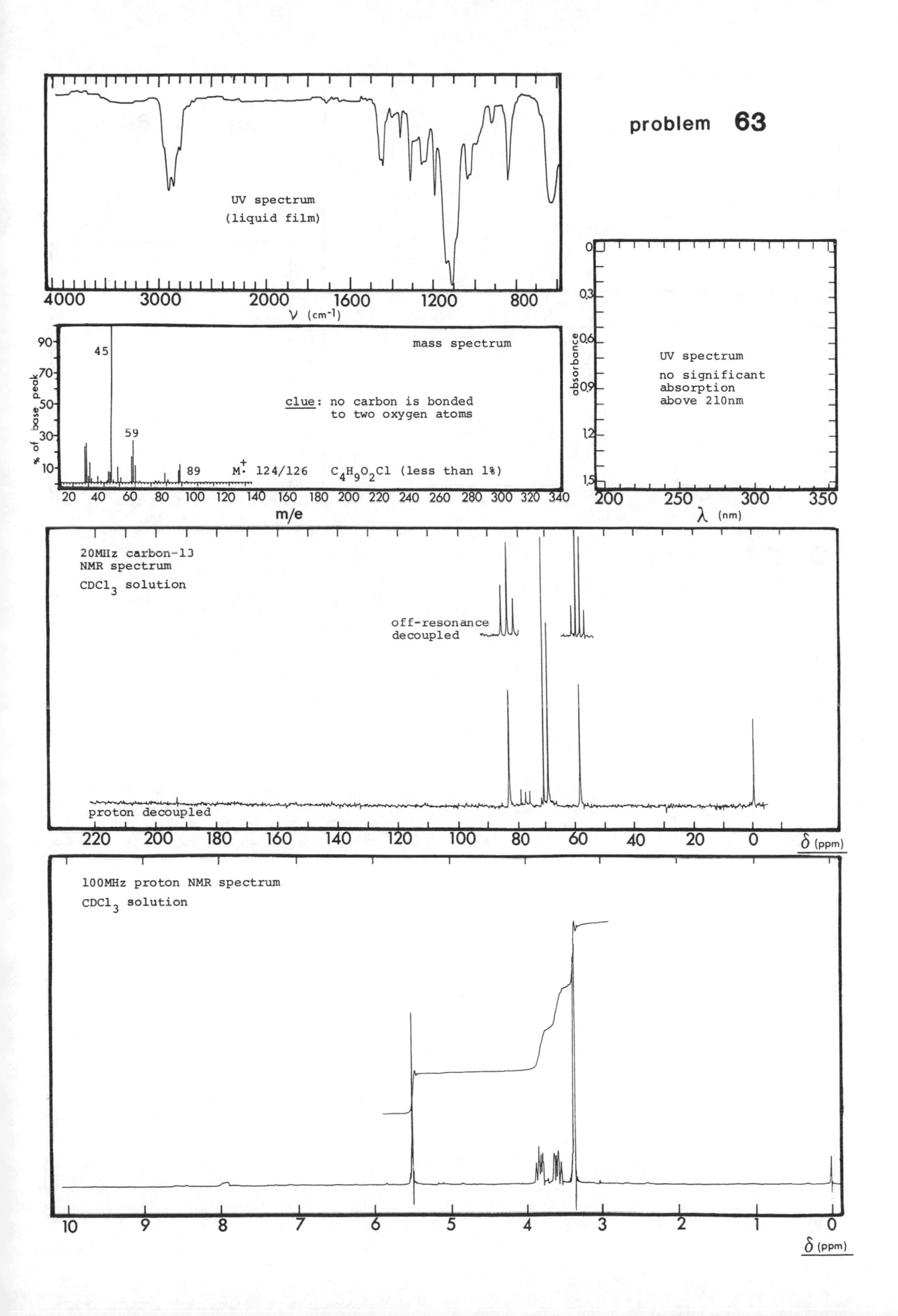

UV spectrum
(liquid film)
4000
3000
2000
1600
1200
800
ν (cm-1)
mass spectrum
45
59
89
M+ 124/126 C4H9O2Cl (less than 1%)
clue: no carbon is bonded
to two oxygen atoms
% of base peak
m/e
UV spectrum
no significant
absorption
above 210nm
absorbance
λ (nm)
20MHz carbon-13
NMR spectrum
CDCl3 solution
off-resonance
decoupled
proton decoupled
δ (ppm)
100MHz proton NMR spectrum
CDCl3 solution
δ (ppm)

problem 64

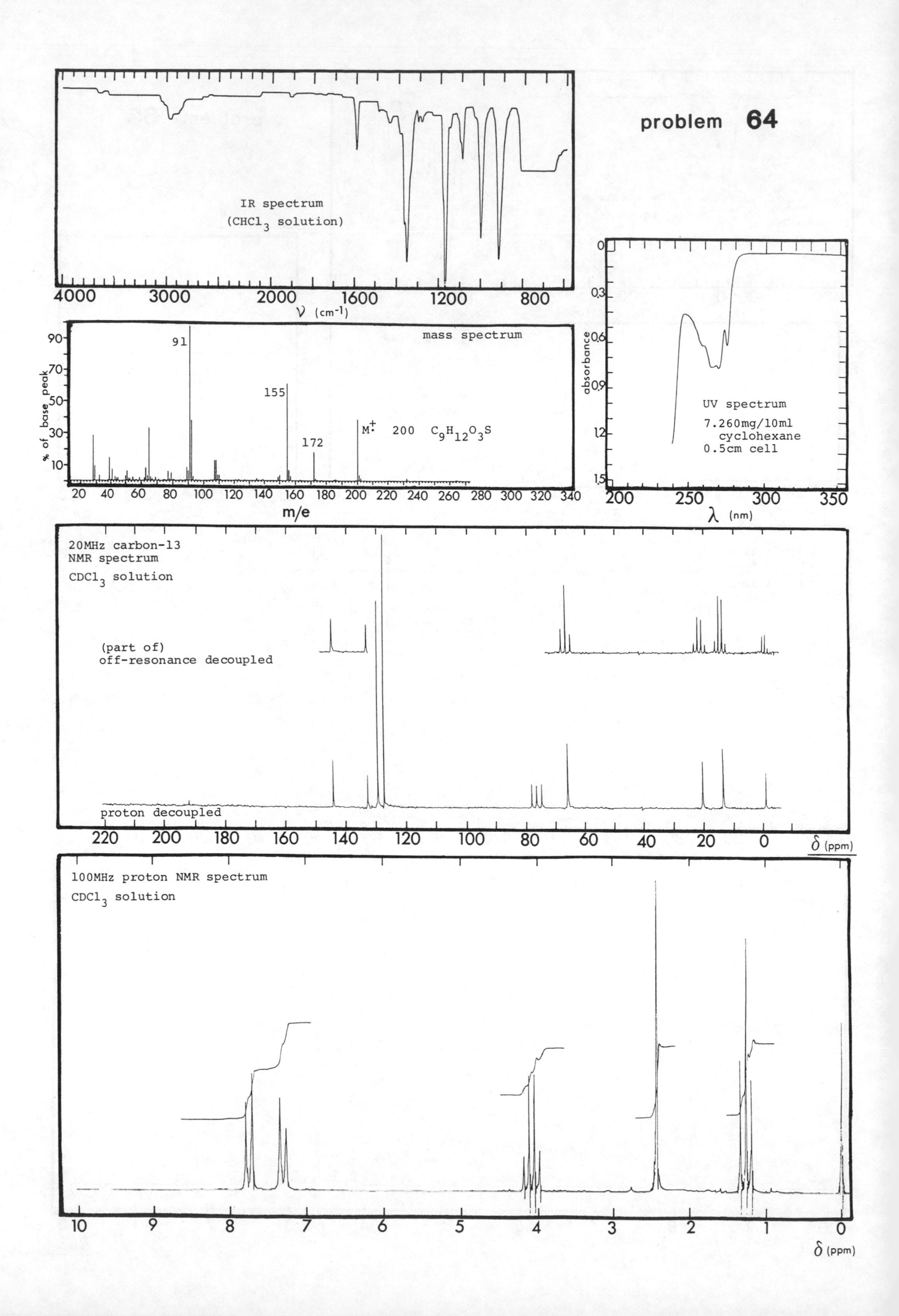

IR spectrum
(CHCl3 solution)
4000
3000
2000
1600
1200
800
ν (cm-1)
mass spectrum
% of base peak
91
155
172
M+ 200 C9H12O3S
m/e
UV spectrum
7.260mg/10ml
cyclohexane
0.5cm cell
absorbance
λ (nm)
20MHz carbon-13
NMR spectrum
CDCl3 solution
(part of)
off-resonance decoupled
proton decoupled
δ (ppm)
100MHz proton NMR spectrum
CDCl3 solution
δ (ppm)

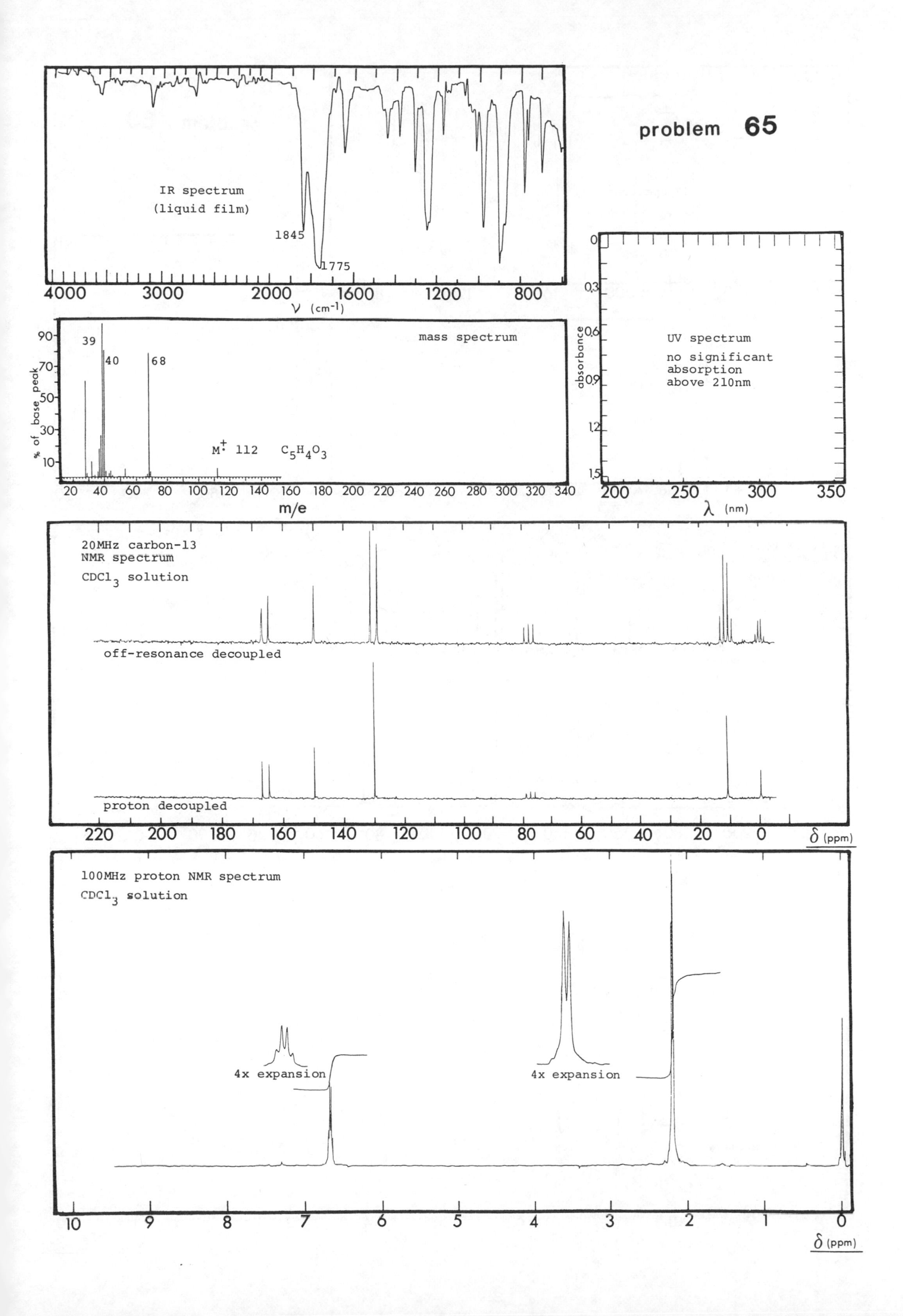
problem 65
IR spectrum
(liquid film)
1845
1775
4000
3000
2000
1600
1200
800
ν (cm⁻¹)
mass spectrum
39
40
68
M+ 112 C5H4O3
% of base peak
m/e
UV spectrum
no significant
absorption
above 210nm
absorbance
λ (nm)
20MHz carbon-13
NMR spectrum
CDCl3 solution
off-resonance decoupled
proton decoupled
δ (ppm)
100MHz proton NMR spectrum
CDCl3 solution
4x expansion
4x expansion
δ (ppm)

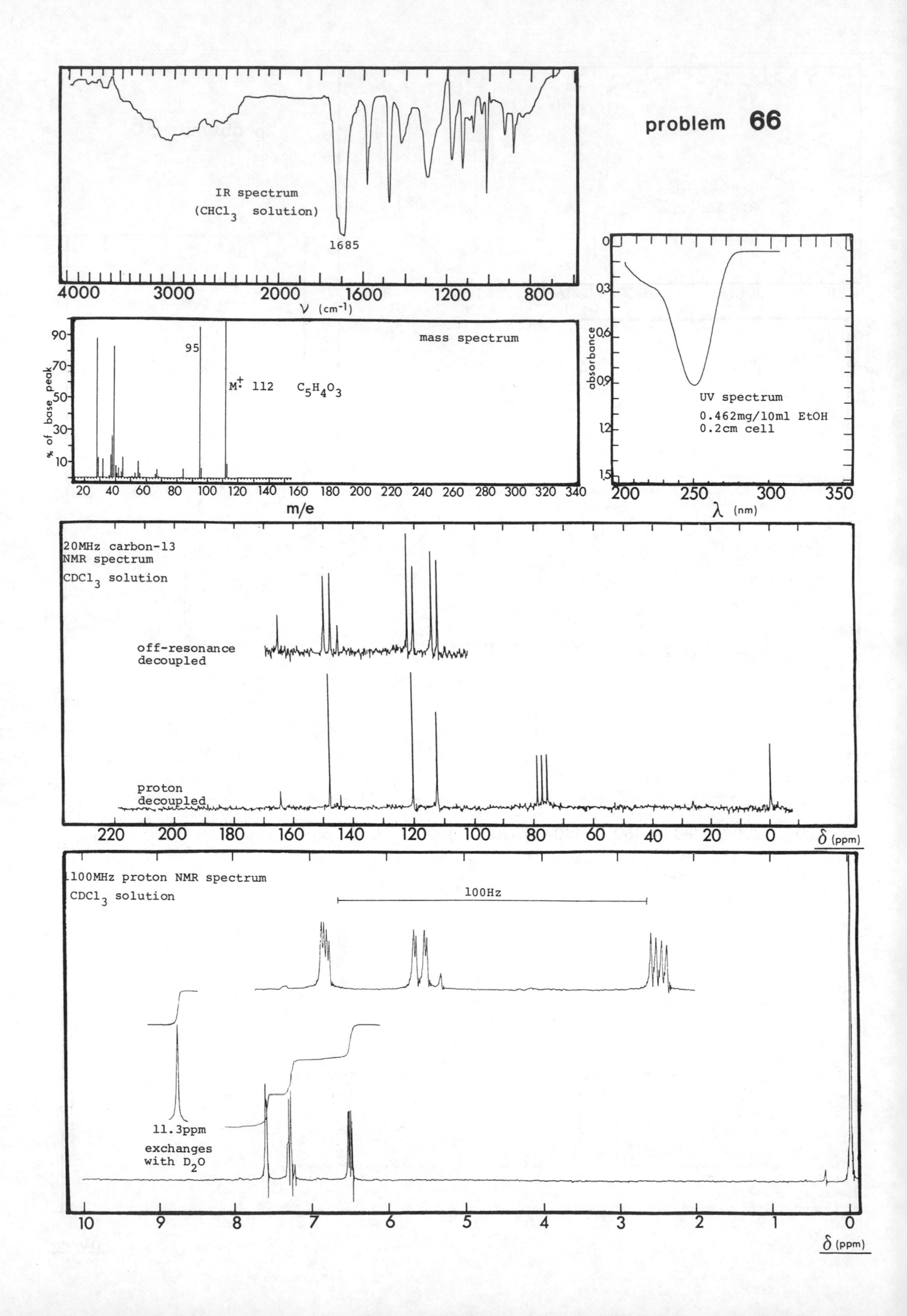

problem 66
IR spectrum
(CHCl3 solution)
1685
4000
3000
2000
1600
1200
800
ν (cm-1)
mass spectrum
95
M+ 112
C5H4O3
% of base peak
m/e
UV spectrum
0.462mg/10ml EtOH
0.2cm cell
absorbance
λ (nm)
20MHz carbon-13
NMR spectrum
CDCl3 solution
off-resonance
decoupled
proton
decoupled
δ (ppm)
100MHz proton NMR spectrum
CDCl3 solution
100Hz
11.3ppm
exchanges
with D2O
δ (ppm)

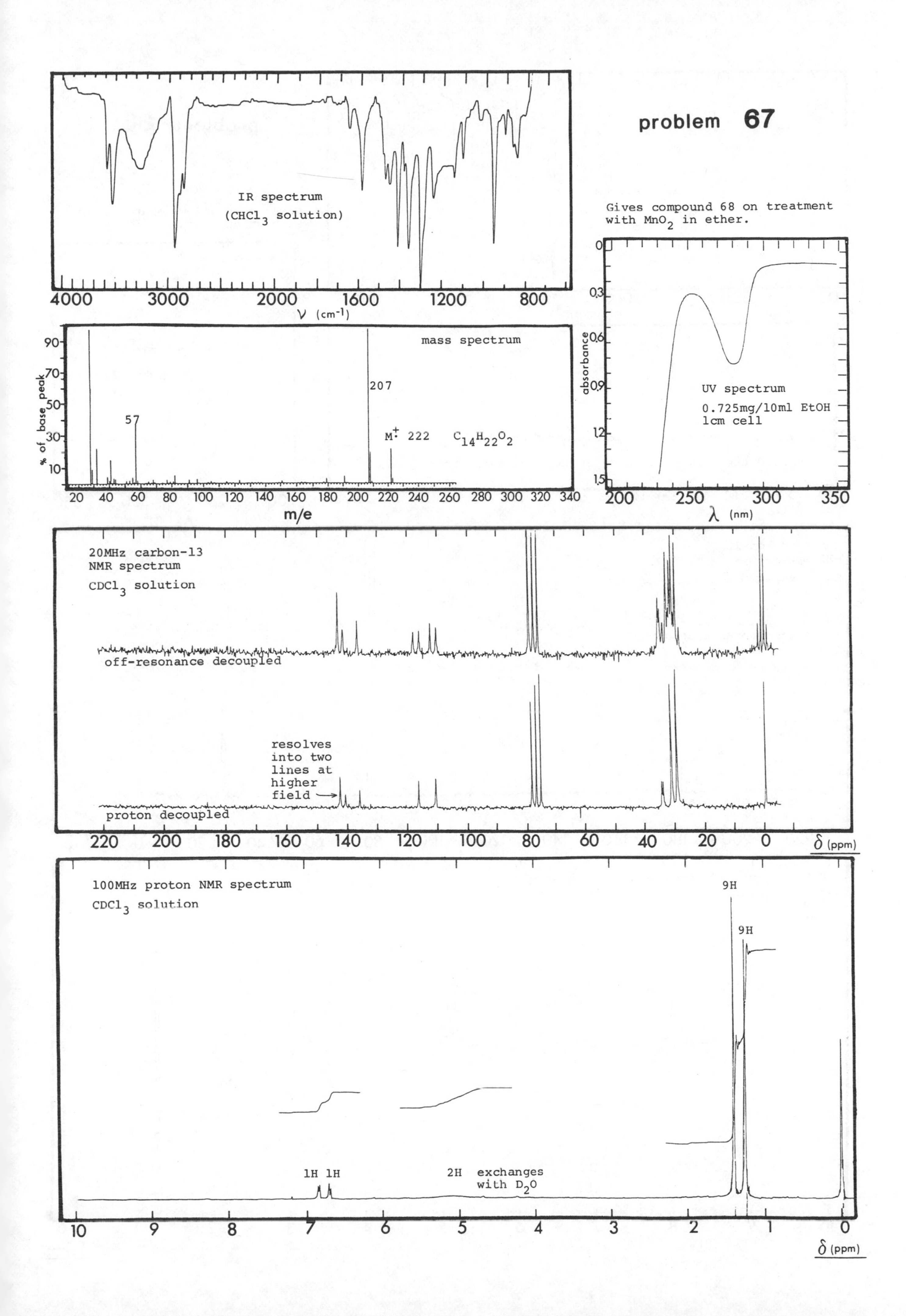
problem 67
IR spectrum
(CHCl3 solution)
4000
3000
2000
1600
1200
800
ν (cm-1)
Gives compound 68 on treatment with MnO2 in ether.
mass spectrum
% of base peak
57
207
M+ 222
C14H22O2
m/e
UV spectrum
0.725mg/10ml EtOH
1cm cell
absorbance
λ (nm)
20MHz carbon-13
NMR spectrum
CDCl3 solution
off-resonance decoupled
resolves
into two
lines at
higher
field
proton decoupled
δ (ppm)
100MHz proton NMR spectrum
CDCl3 solution
9H
9H
1H 1H
2H exchanges
with D2O
δ (ppm)

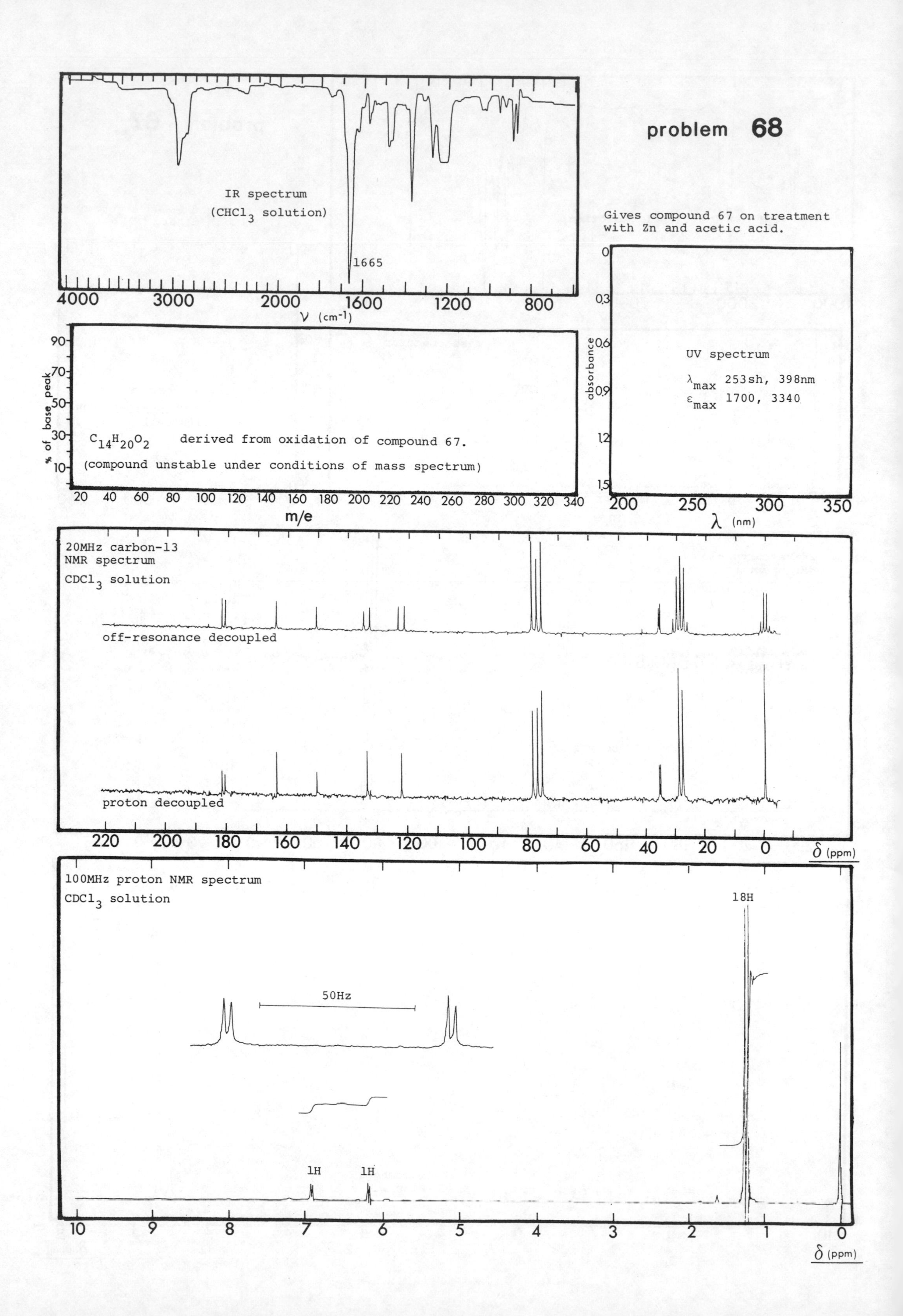
problem 68
IR spectrum
(CHCl3 solution)
1665
4000
3000
2000
1600
1200
800
ν (cm-1)
Gives compound 67 on treatment with Zn and acetic acid.
UV spectrum
λmax 253sh, 398nm
εmax 1700, 3340
absorbance
λ (nm)
% of base peak
C14H20O2 derived from oxidation of compound 67.
(compound unstable under conditions of mass spectrum)
m/e
20MHz carbon-13
NMR spectrum
CDCl3 solution
off-resonance decoupled
proton decoupled
δ (ppm)
100MHz proton NMR spectrum
CDCl3 solution
18H
50Hz
1H
1H
δ (ppm)

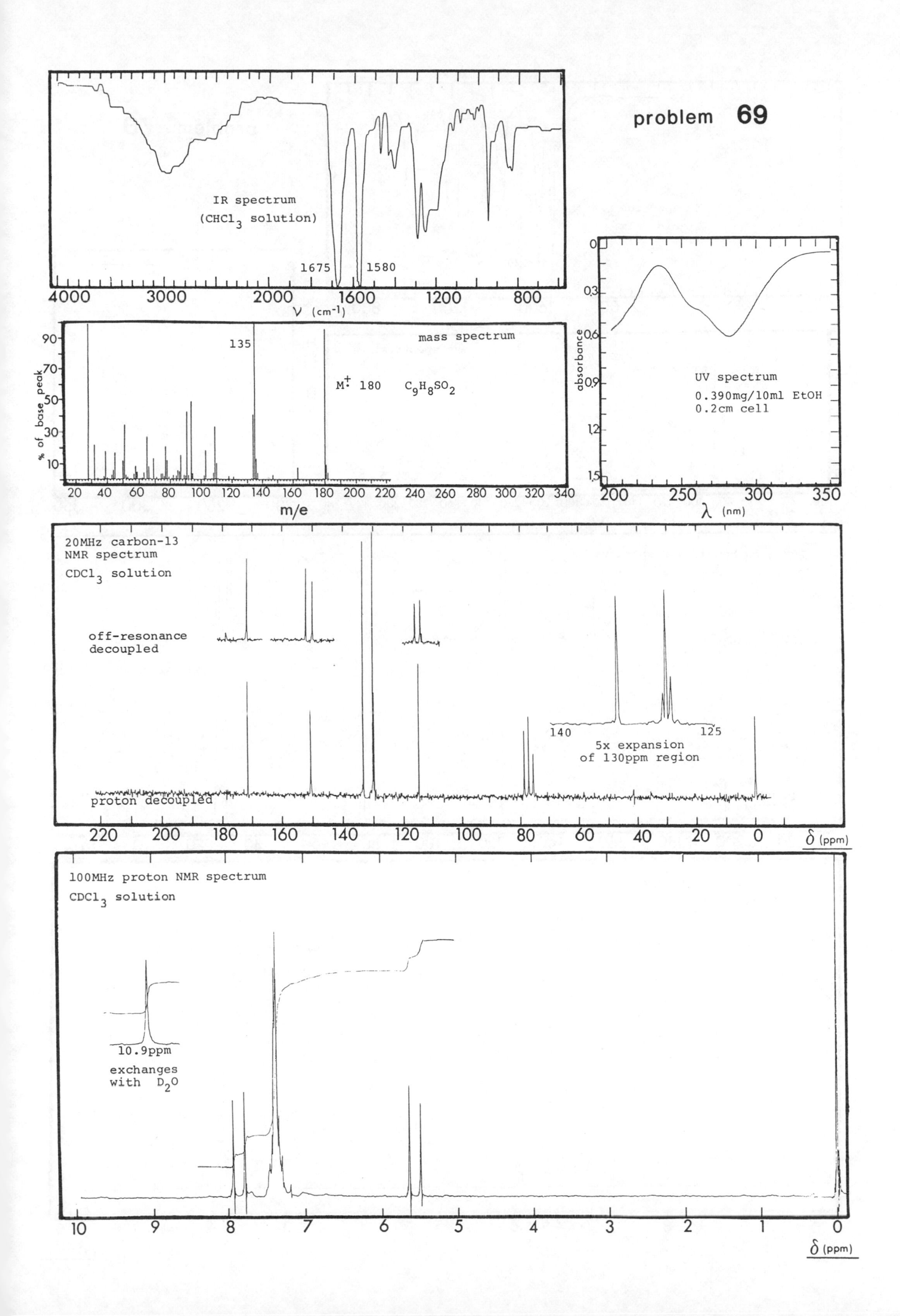
problem 69
IR spectrum
(CHCl3 solution)
1675
1580
4000
3000
2000
1600
1200
800
ν (cm-1)
mass spectrum
135
M+ 180 C9H8SO2
% of base peak
m/e
UV spectrum
0.390mg/10ml EtOH
0.2cm cell
absorbance
λ (nm)
20MHz carbon-13
NMR spectrum
CDCl3 solution
off-resonance
decoupled
5x expansion
of 130ppm region
140
125
proton decoupled
δ (ppm)
100MHz proton NMR spectrum
CDCl3 solution
10.9ppm
exchanges
with D2O
δ (ppm)

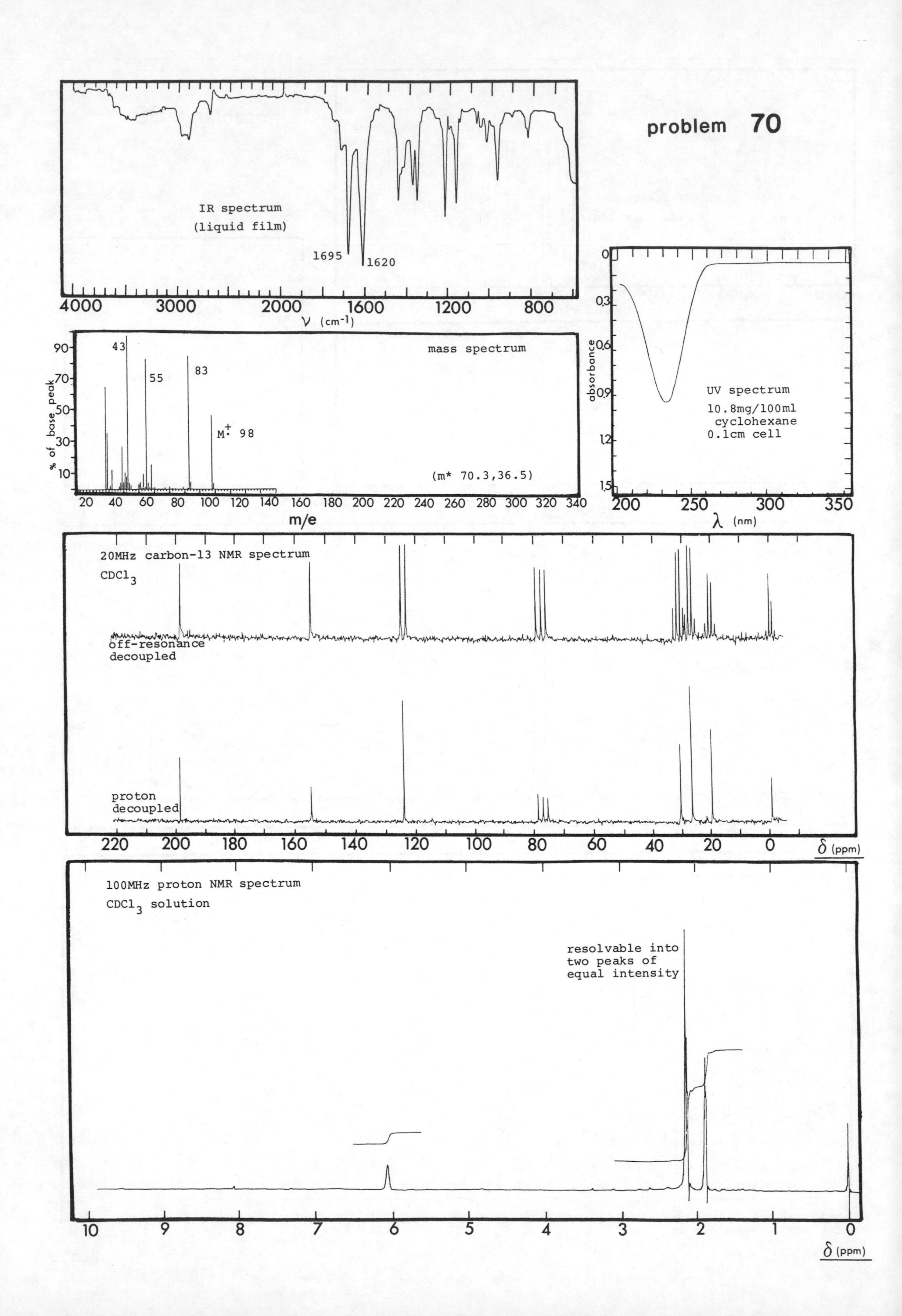
problem 70
IR spectrum
(liquid film)
1695
1620
4000
3000
2000
1600
1200
800
ν (cm-1)
mass spectrum
43
55
83
M+· 98
(m* 70.3,36.5)
% of base peak
m/e
UV spectrum
10.8mg/100ml
cyclohexane
0.1cm cell
absorbance
λ (nm)
20MHz carbon-13 NMR spectrum
CDCl3
off-resonance
decoupled
proton
decoupled
δ (ppm)
100MHz proton NMR spectrum
CDCl3 solution
resolvable into
two peaks of
equal intensity
δ (ppm)

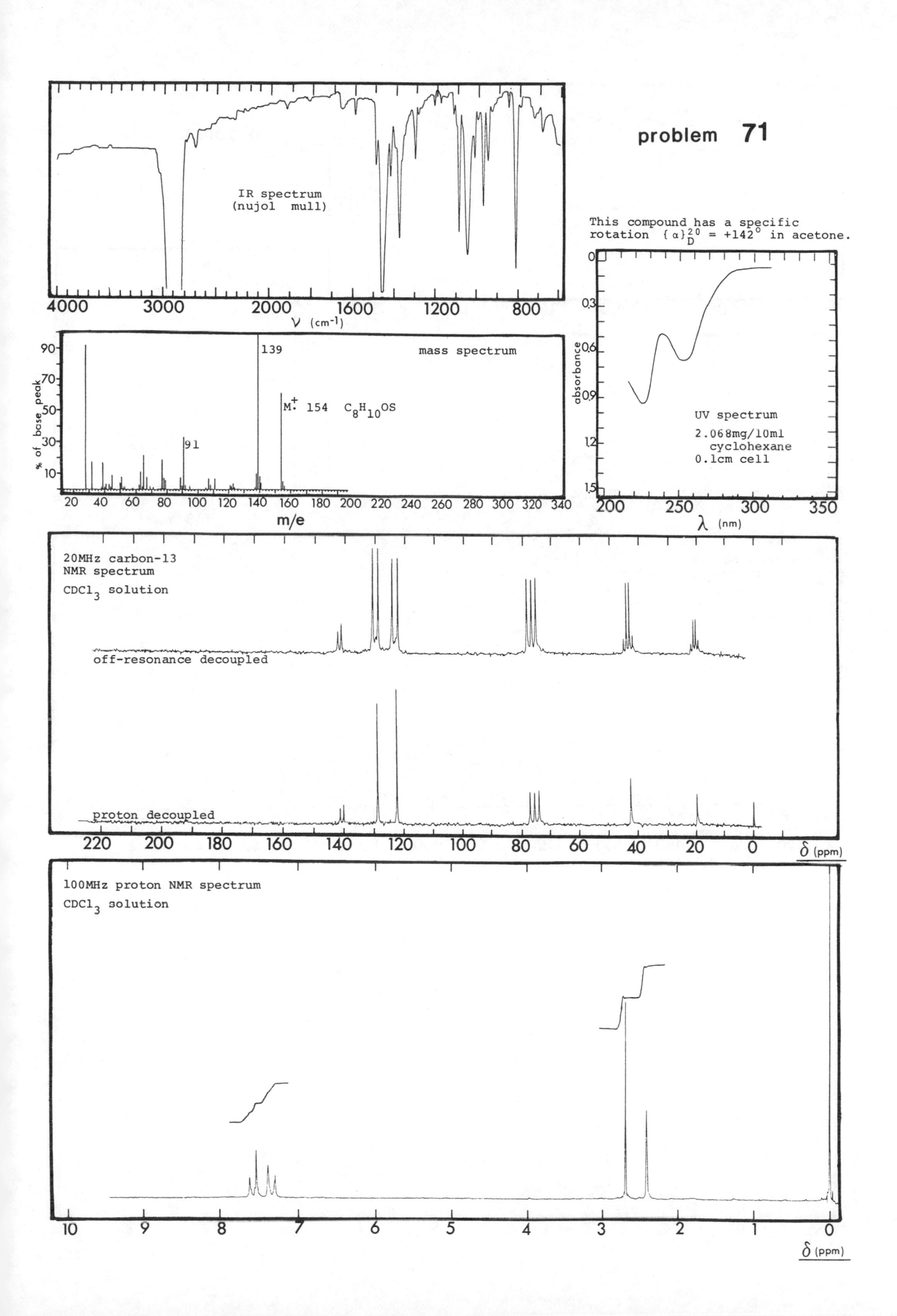

problem 71
IR spectrum
(nujol mull)
4000
3000
2000
1600
1200
800
ν (cm⁻¹)
This compound has a specific rotation {α}$^{20}_{D}$ = +142° in acetone.
mass spectrum
139
M⁺ 154 C8H10OS
91
% of base peak
m/e
UV spectrum
2.068mg/10ml
cyclohexane
0.1cm cell
absorbance
λ (nm)
20MHz carbon-13
NMR spectrum
CDCl3 solution
off-resonance decoupled
proton decoupled
δ (ppm)
100MHz proton NMR spectrum
CDCl3 solution
δ (ppm)

problem 72

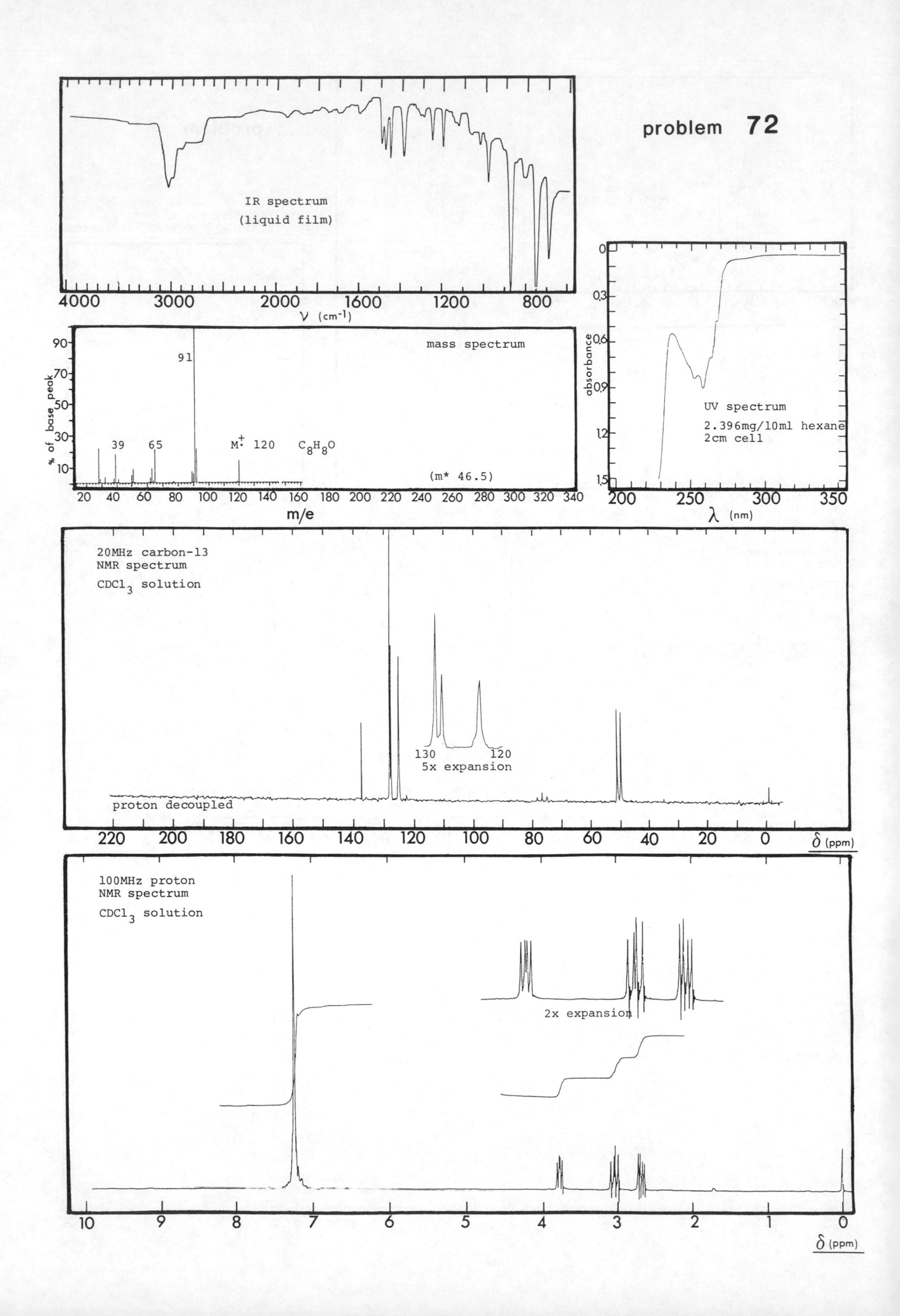

IR spectrum
(liquid film)
4000
3000
2000
1600
1200
800
ν (cm-1)
mass spectrum
90
70
50
30
10
% of base peak
91
39
65
M+ 120
C8H8O
(m* 46.5)
20 40 60 80 100 120 140 160 180 200 220 240 260 280 300 320 340
m/e
0
0.3
0.6
0.9
1.2
1.5
absorbance
UV spectrum
2.396mg/10ml hexane
2cm cell
200 250 300 350
λ (nm)
20MHz carbon-13
NMR spectrum
CDCl3 solution
130
120
5x expansion
proton decoupled
220 200 180 160 140 120 100 80 60 40 20 0
δ (ppm)
100MHz proton
NMR spectrum
CDCl3 solution
2x expansion
10 9 8 7 6 5 4 3 2 1 0
δ (ppm)

problem 73

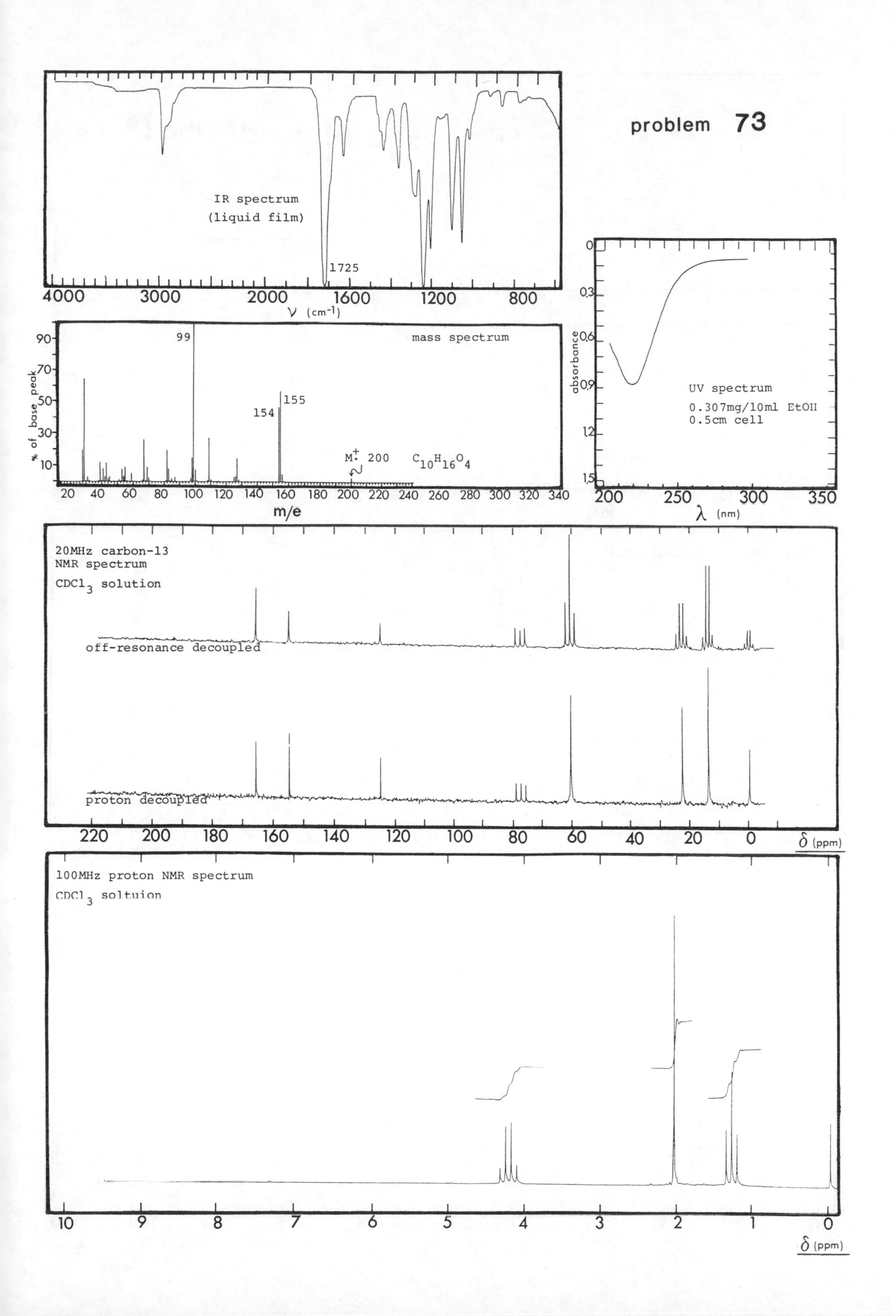

IR spectrum
(liquid film)
1725
4000
3000
2000
1600
1200
800
ν (cm-1)
mass spectrum
99
154
155
M+ 200
C10H16O4
% of base peak
m/e
UV spectrum
0.307mg/10ml EtOH
0.5cm cell
absorbance
λ (nm)
20MHz carbon-13
NMR spectrum
CDCl3 solution
off-resonance decoupled
proton decoupled
δ (ppm)
100MHz proton NMR spectrum
CDCl3 soltuion
δ (ppm)

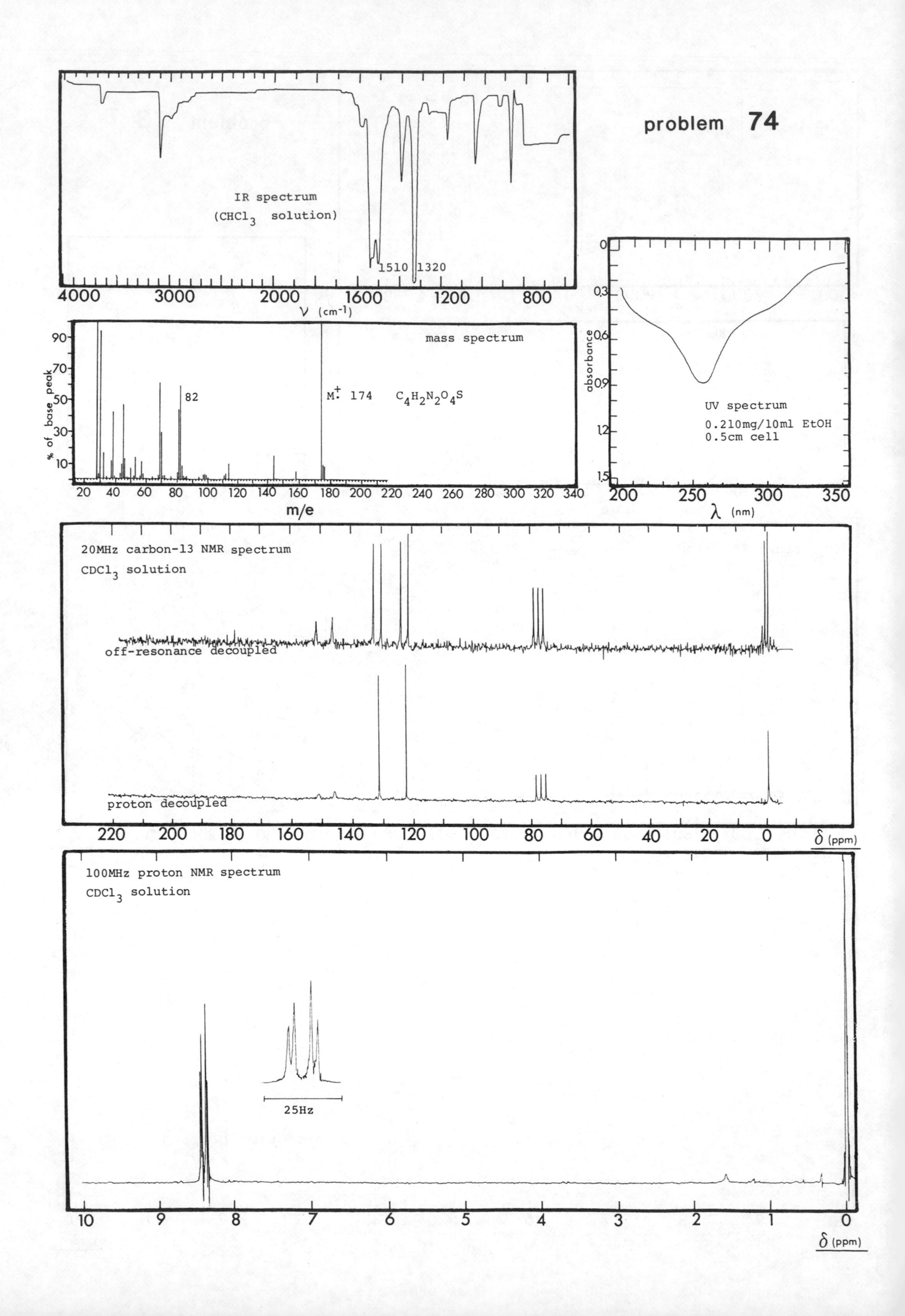
problem 74
IR spectrum
(CHCl3 solution)
1510
1320
4000
3000
2000
1600
1200
800
ν (cm-1)
mass spectrum
82
M+ 174
C4H2N2O4S
% of base peak
m/e
UV spectrum
0.210mg/10ml EtOH
0.5cm cell
absorbance
λ (nm)
20MHz carbon-13 NMR spectrum
CDCl3 solution
off-resonance decoupled
proton decoupled
δ (ppm)
100MHz proton NMR spectrum
CDCl3 solution
25Hz
δ (ppm)

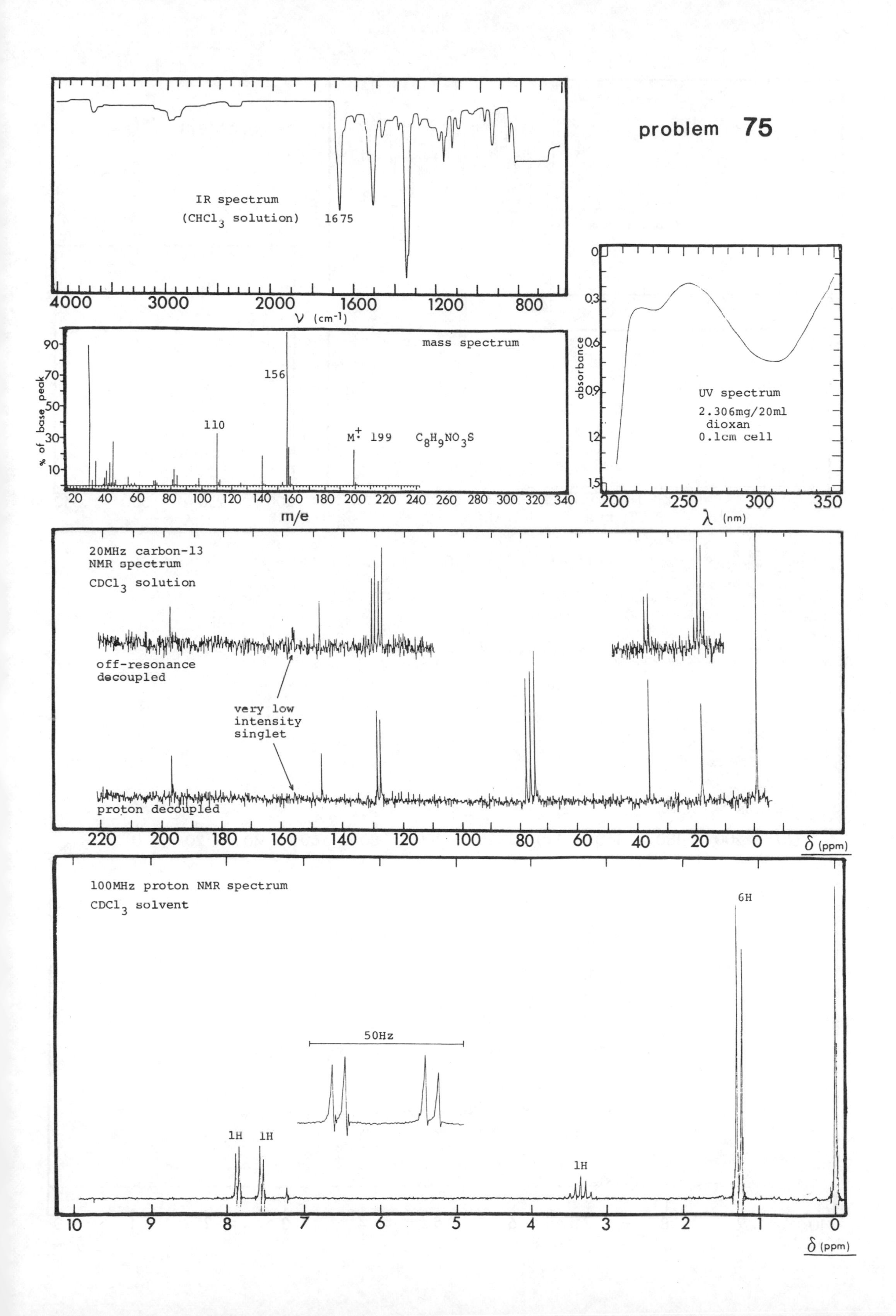

problem 75
IR spectrum
(CHCl3 solution)
1675
4000
3000
2000
1600
1200
800
ν (cm-1)
mass spectrum
% of base peak
156
110
M+ 199
C8H9NO3S
m/e
UV spectrum
2.306mg/20ml
dioxan
0.1cm cell
absorbance
λ (nm)
20MHz carbon-13
NMR spectrum
CDCl3 solution
off-resonance
decoupled
very low
intensity
singlet
proton decoupled
δ (ppm)
100MHz proton NMR spectrum
CDCl3 solvent
6H
50Hz
1H
1H
1H
δ (ppm)

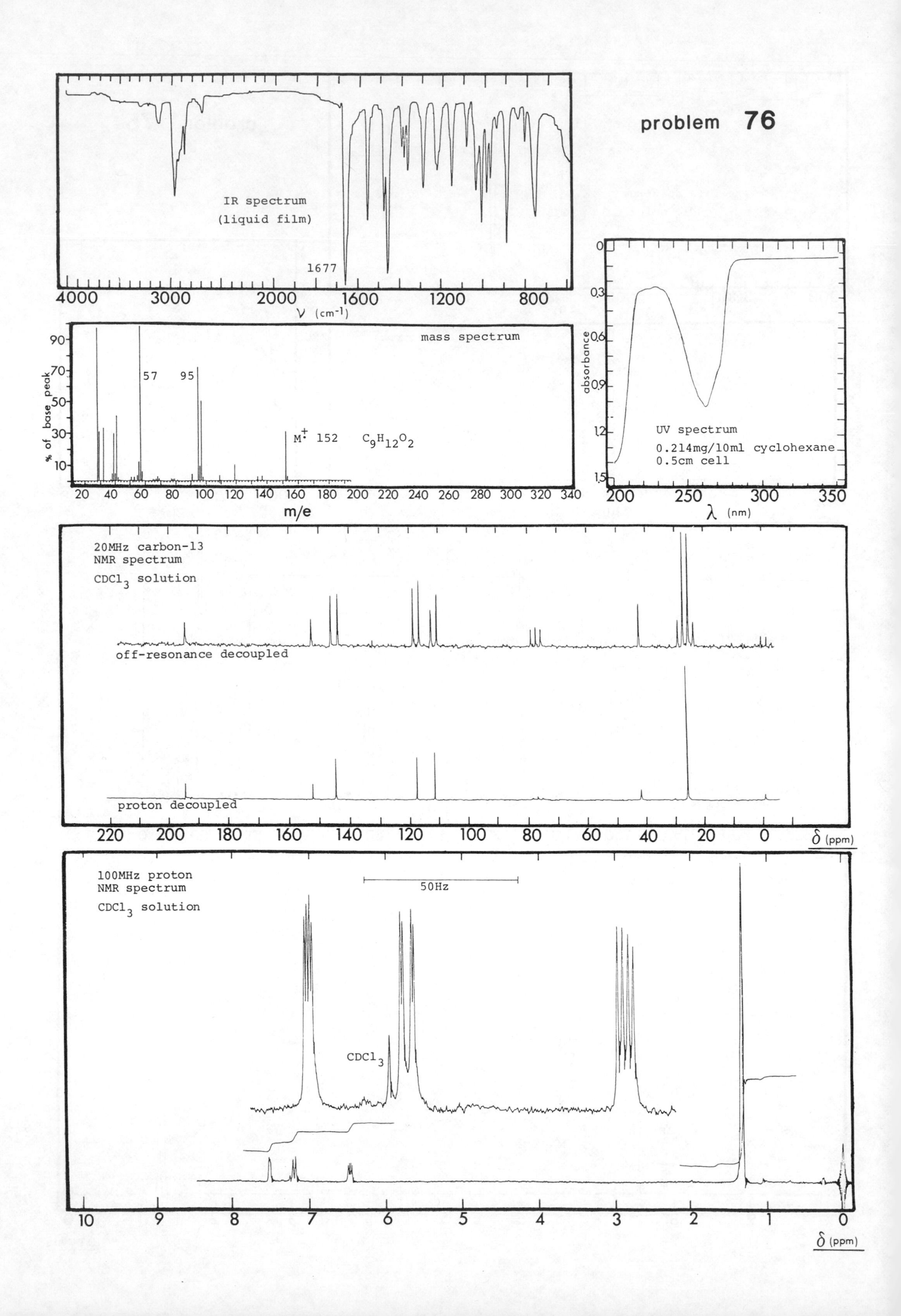

problem 76
IR spectrum
(liquid film)
1677
4000
3000
2000
1600
1200
800
ν (cm-1)
mass spectrum
% of base peak
57
95
M+ 152
C9H12O2
m/e
UV spectrum
0.214mg/10ml cyclohexane
0.5cm cell
absorbance
λ (nm)
20MHz carbon-13
NMR spectrum
CDCl3 solution
off-resonance decoupled
proton decoupled
δ (ppm)
100MHz proton
NMR spectrum
CDCl3 solution
50Hz
CDCl3
δ (ppm)

problem **77**

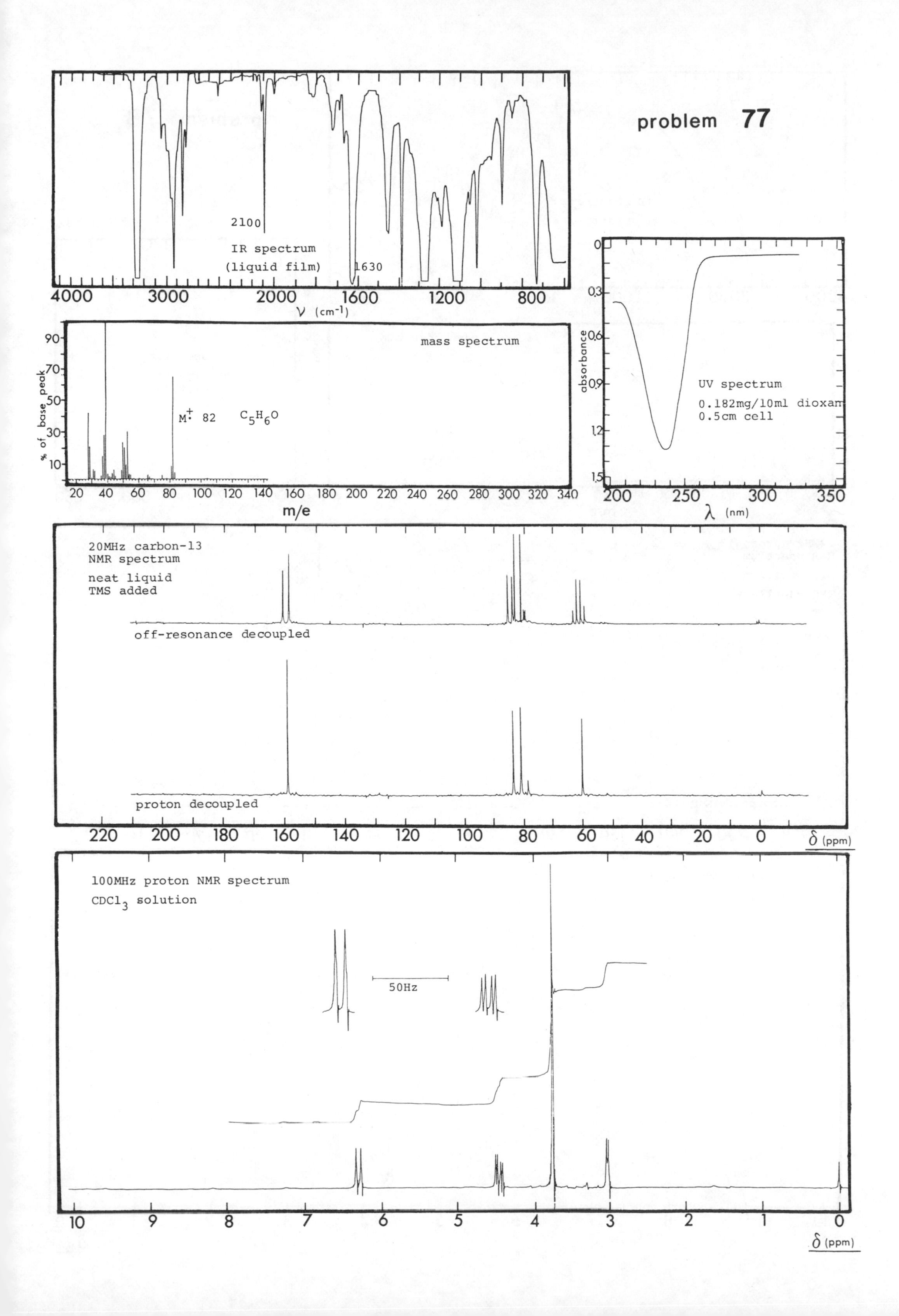

2100
IR spectrum
(liquid film)
1630
4000
3000
2000
1600
1200
800
ν (cm-1)
mass spectrum
% of base peak
M+. 82
C5H6O
m/e
UV spectrum
0.182mg/10ml dioxan
0.5cm cell
absorbance
λ (nm)
20MHz carbon-13
NMR spectrum
neat liquid
TMS added
off-resonance decoupled
proton decoupled
δ (ppm)
100MHz proton NMR spectrum
CDCl3 solution
50Hz
δ (ppm)

problem **78**

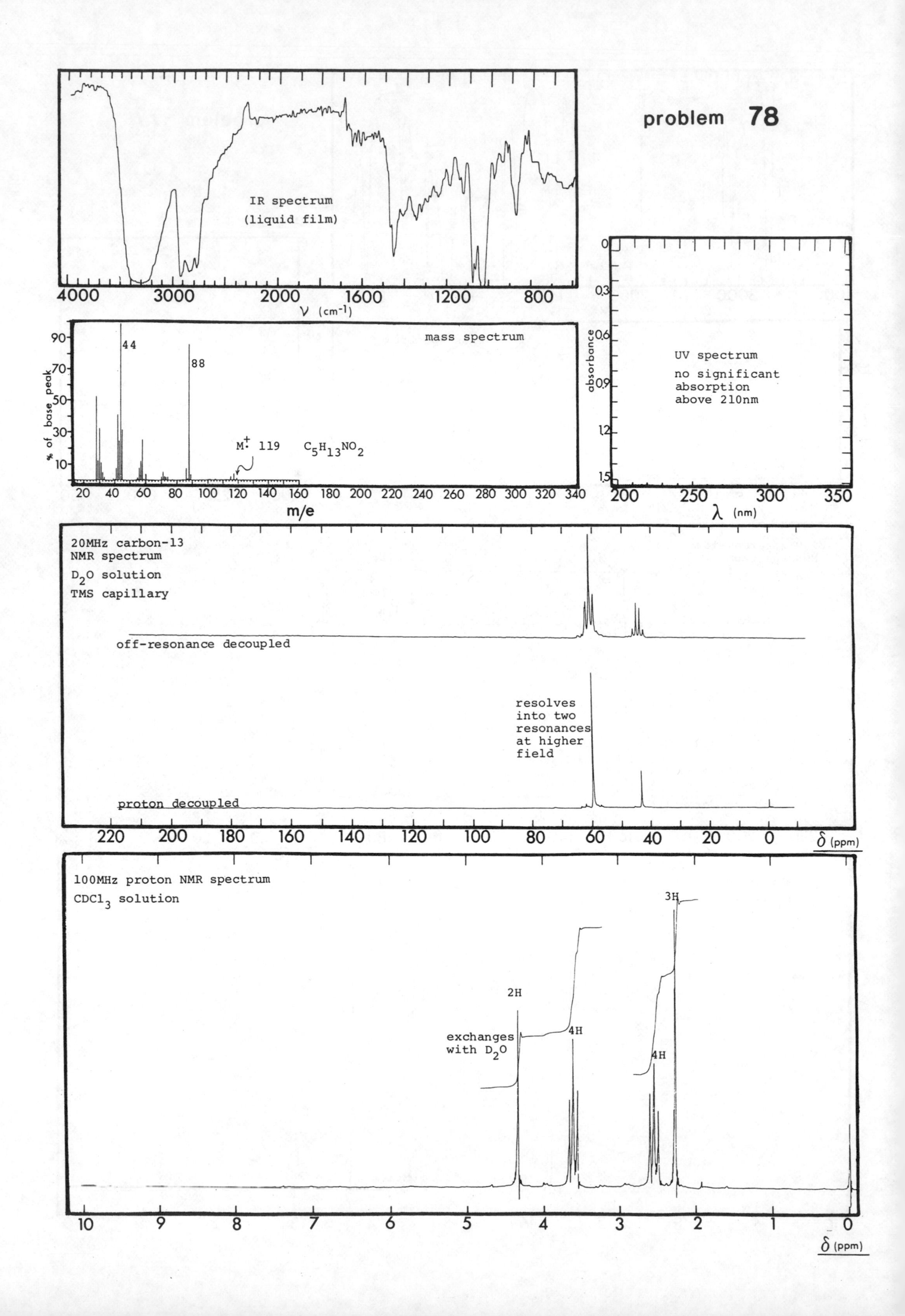

IR spectrum
(liquid film)
4000
3000
2000
1600
1200
800
ν (cm⁻¹)
mass spectrum
% of base peak
44
88
M⁺· 119
C5H13NO2
m/e
UV spectrum
no significant absorption above 210nm
absorbance
λ (nm)
20MHz carbon-13 NMR spectrum
D2O solution
TMS capillary
off-resonance decoupled
resolves into two resonances at higher field
proton decoupled
δ (ppm)
100MHz proton NMR spectrum
CDCl3 solution
2H
exchanges with D2O
4H
4H
3H
δ (ppm)

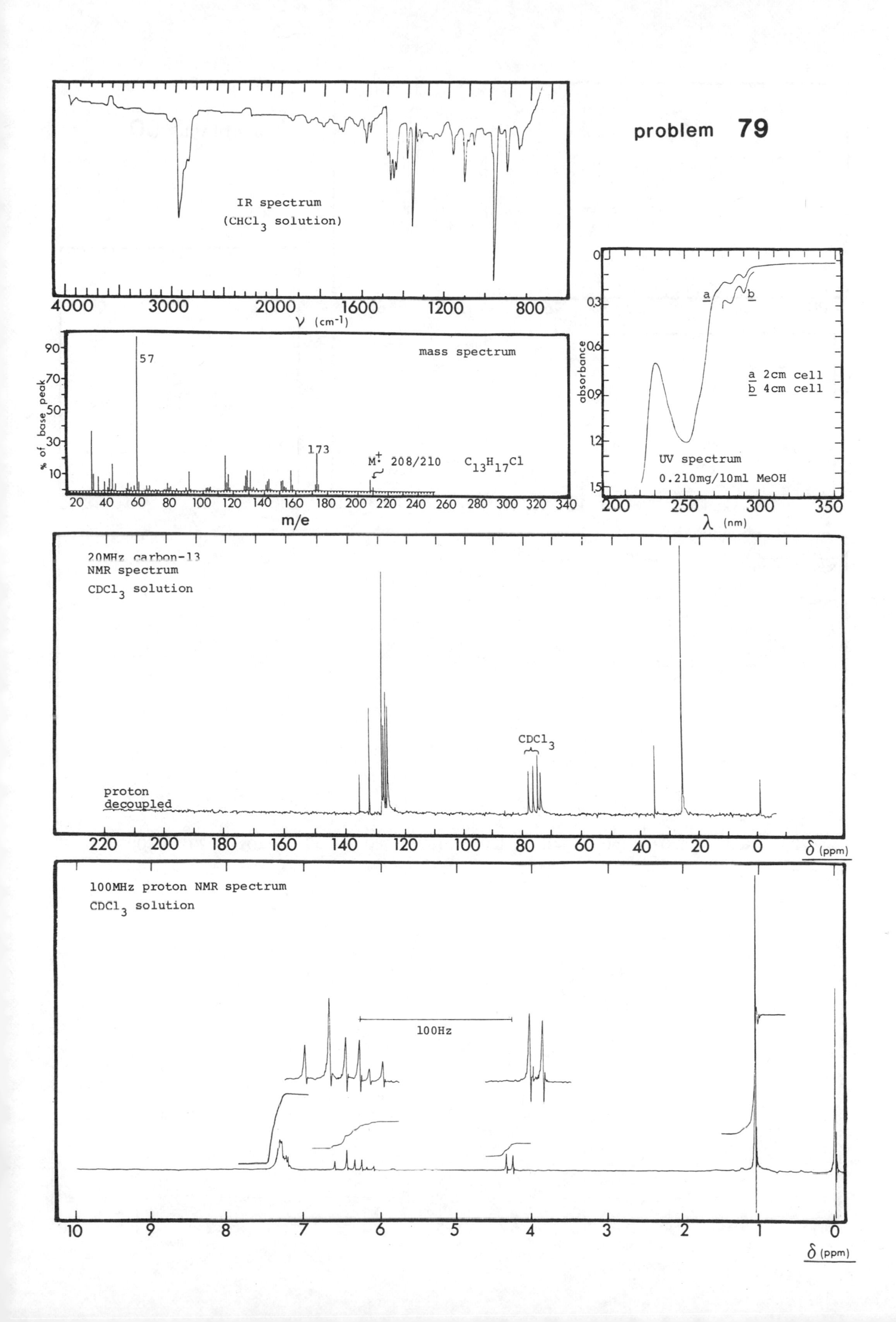

problem 79
IR spectrum
(CHCl3 solution)
4000 3000 2000 1600 1200 800
ν (cm-1)
mass spectrum
57
173
M+· 208/210
C13H17Cl
% of base peak
m/e
UV spectrum
0.210mg/10ml MeOH
a 2cm cell
b 4cm cell
absorbance
λ (nm)
20MHz carbon-13
NMR spectrum
CDCl3 solution
CDCl3
proton
decoupled
δ (ppm)
100MHz proton NMR spectrum
CDCl3 solution
100Hz
δ (ppm)

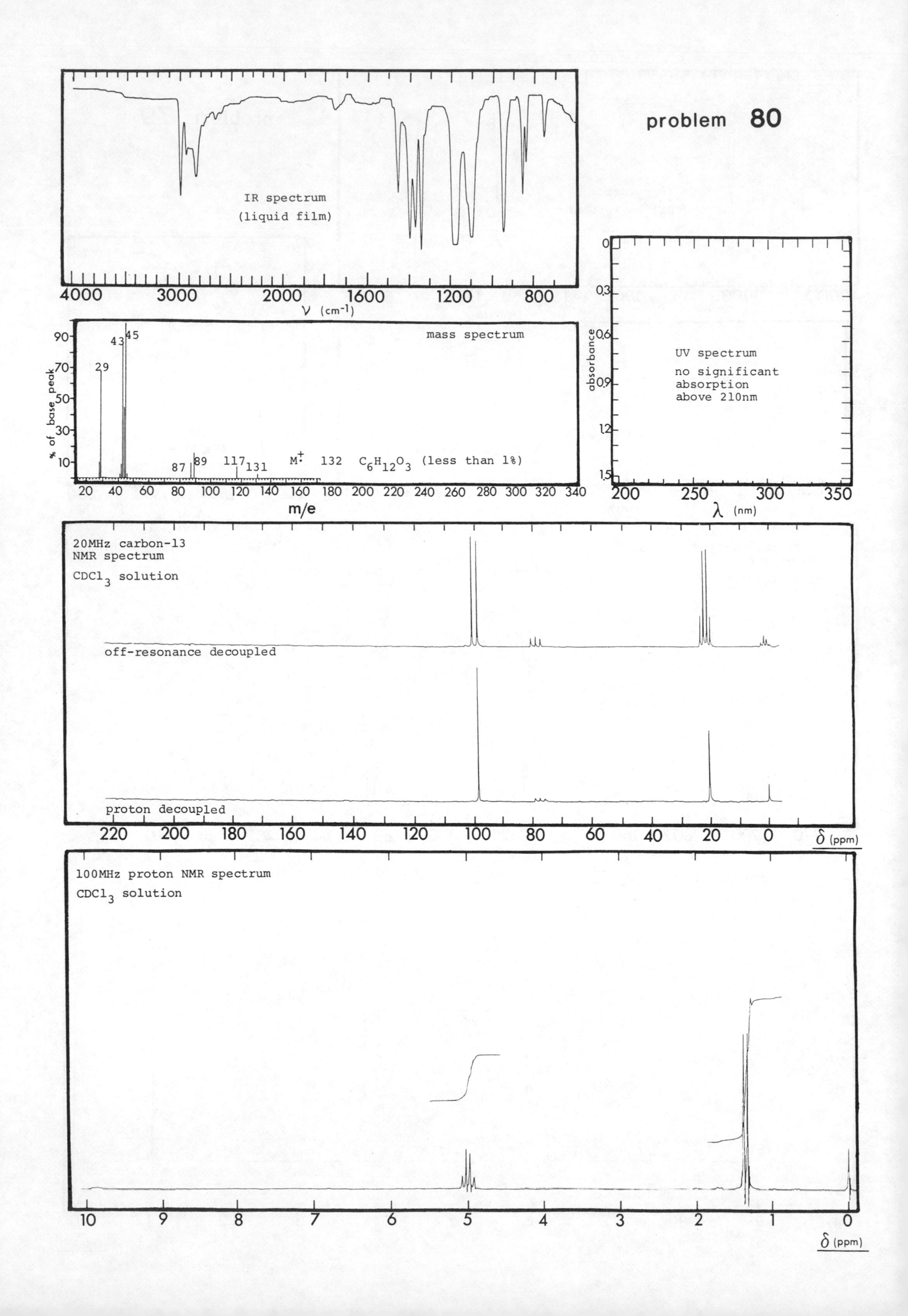
problem 80
IR spectrum
(liquid film)
4000
3000
2000
1600
1200
800
ν (cm-1)
mass spectrum
% of base peak
90
70
50
30
10
29
43
45
87
89
117
131
M+ 132 C6H12O3 (less than 1%)
20 40 60 80 100 120 140 160 180 200 220 240 260 280 300 320 340
m/e
UV spectrum
no significant absorption above 210nm
absorbance
0
0.3
0.6
0.9
1.2
1.5
200 250 300 350
λ (nm)
20MHz carbon-13 NMR spectrum
CDCl3 solution
off-resonance decoupled
proton decoupled
220 200 180 160 140 120 100 80 60 40 20 0
δ (ppm)
100MHz proton NMR spectrum
CDCl3 solution
10 9 8 7 6 5 4 3 2 1 0
δ (ppm)

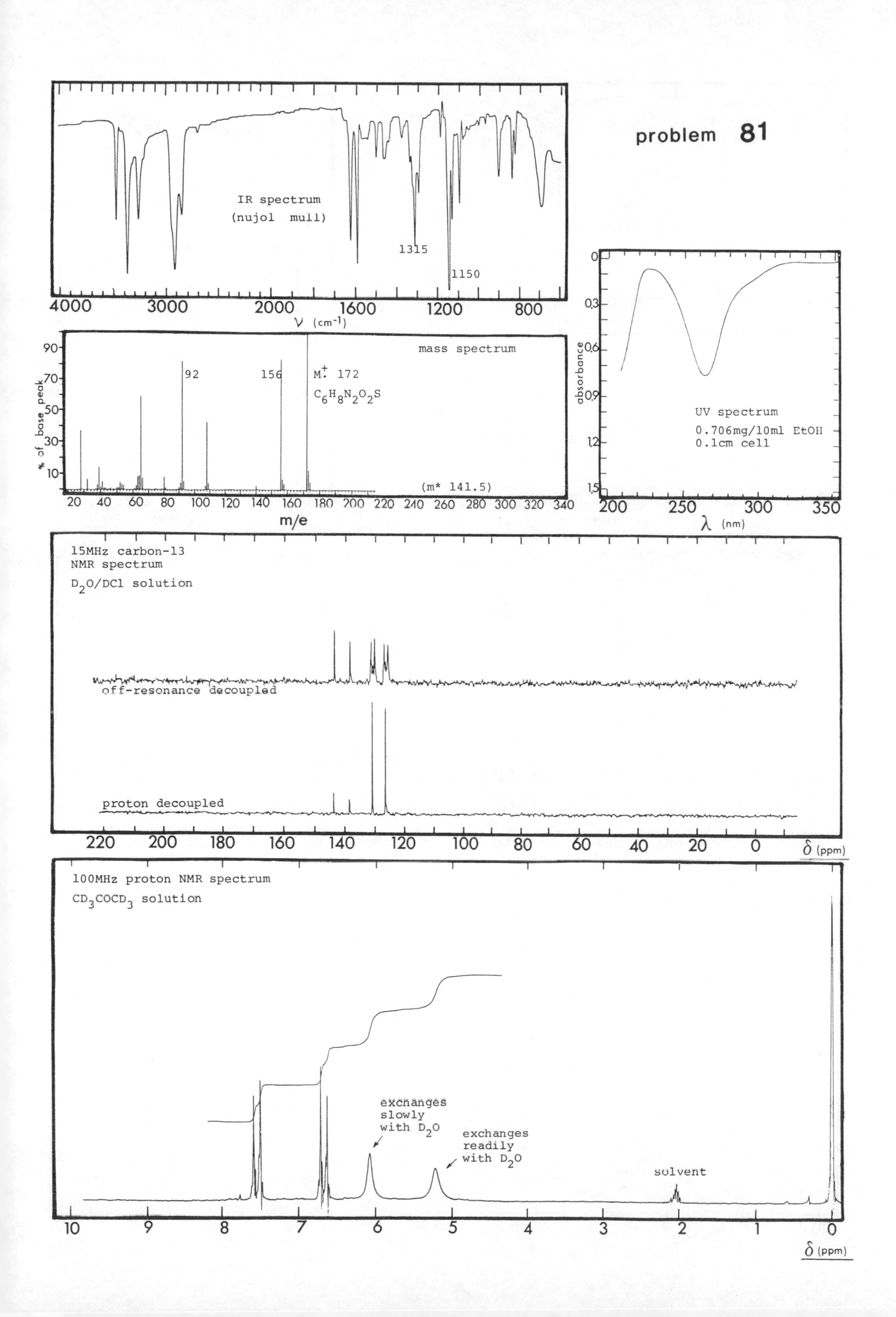

problem 81
IR spectrum
(nujol mull)
1315
1150
4000
3000
2000
1600
1200
800
ν (cm-1)
mass spectrum
92
156
M+ 172
C6H8N2O2S
(m* 141.5)
% of base peak
m/e
UV spectrum
0.706mg/10ml EtOH
0.1cm cell
absorbance
λ (nm)
15MHz carbon-13
NMR spectrum
D2O/DCl solution
off-resonance decoupled
proton decoupled
δ (ppm)
100MHz proton NMR spectrum
CD3COCD3 solution
exchanges
slowly
with D2O
exchanges
readily
with D2O
solvent
δ (ppm)

problem 82

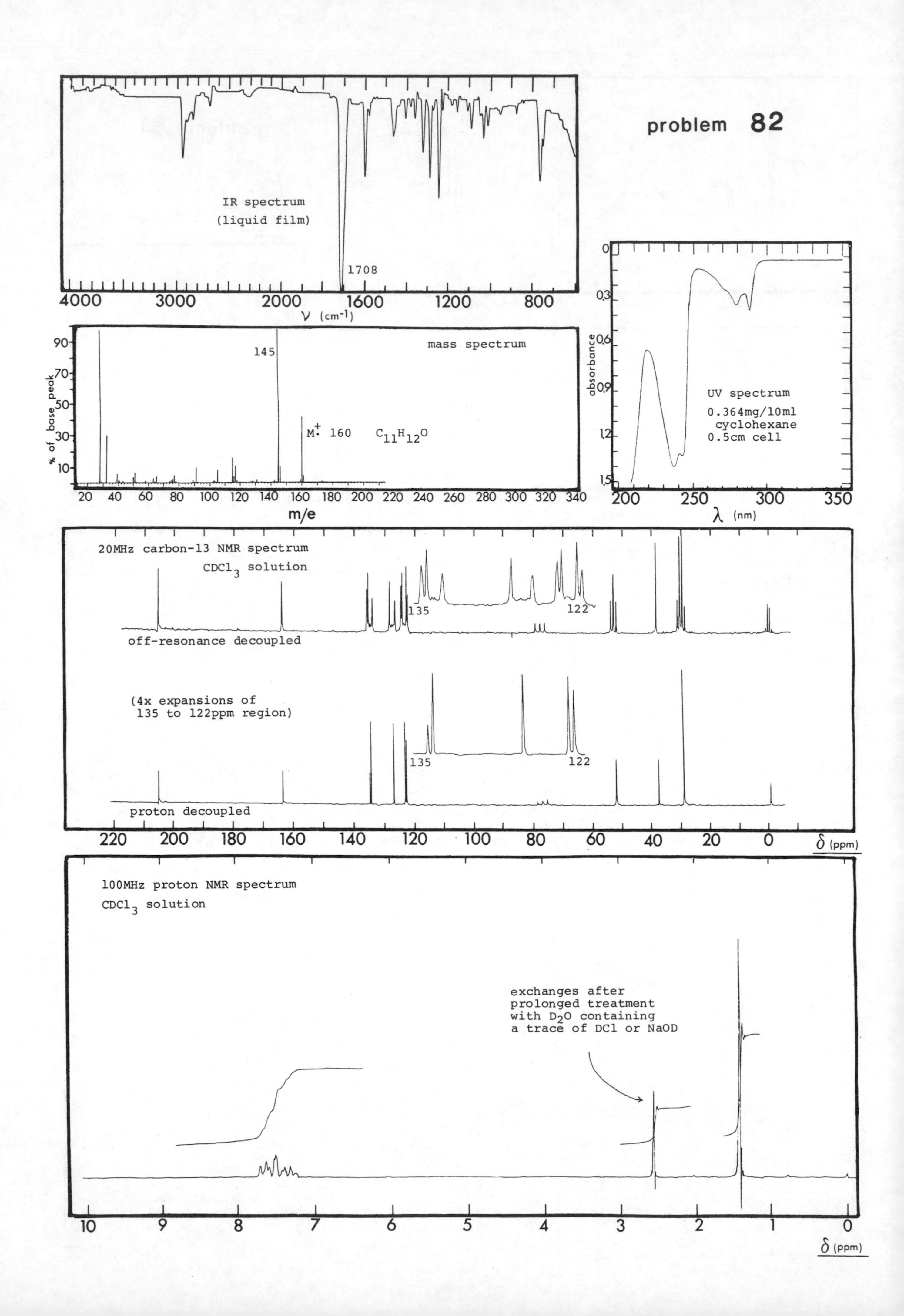

IR spectrum
(liquid film)
1708
4000
3000
2000
1600
1200
800
ν (cm-1)
mass spectrum
145
M+ 160
C11H12O
% of base peak
m/e
UV spectrum
0.364mg/10ml
cyclohexane
0.5cm cell
absorbance
λ (nm)
20MHz carbon-13 NMR spectrum
CDCl3 solution
135
122
off-resonance decoupled
(4x expansions of
135 to 122ppm region)
proton decoupled
δ (ppm)
100MHz proton NMR spectrum
CDCl3 solution
exchanges after
prolonged treatment
with D2O containing
a trace of DCl or NaOD
δ (ppm)

problem **83**

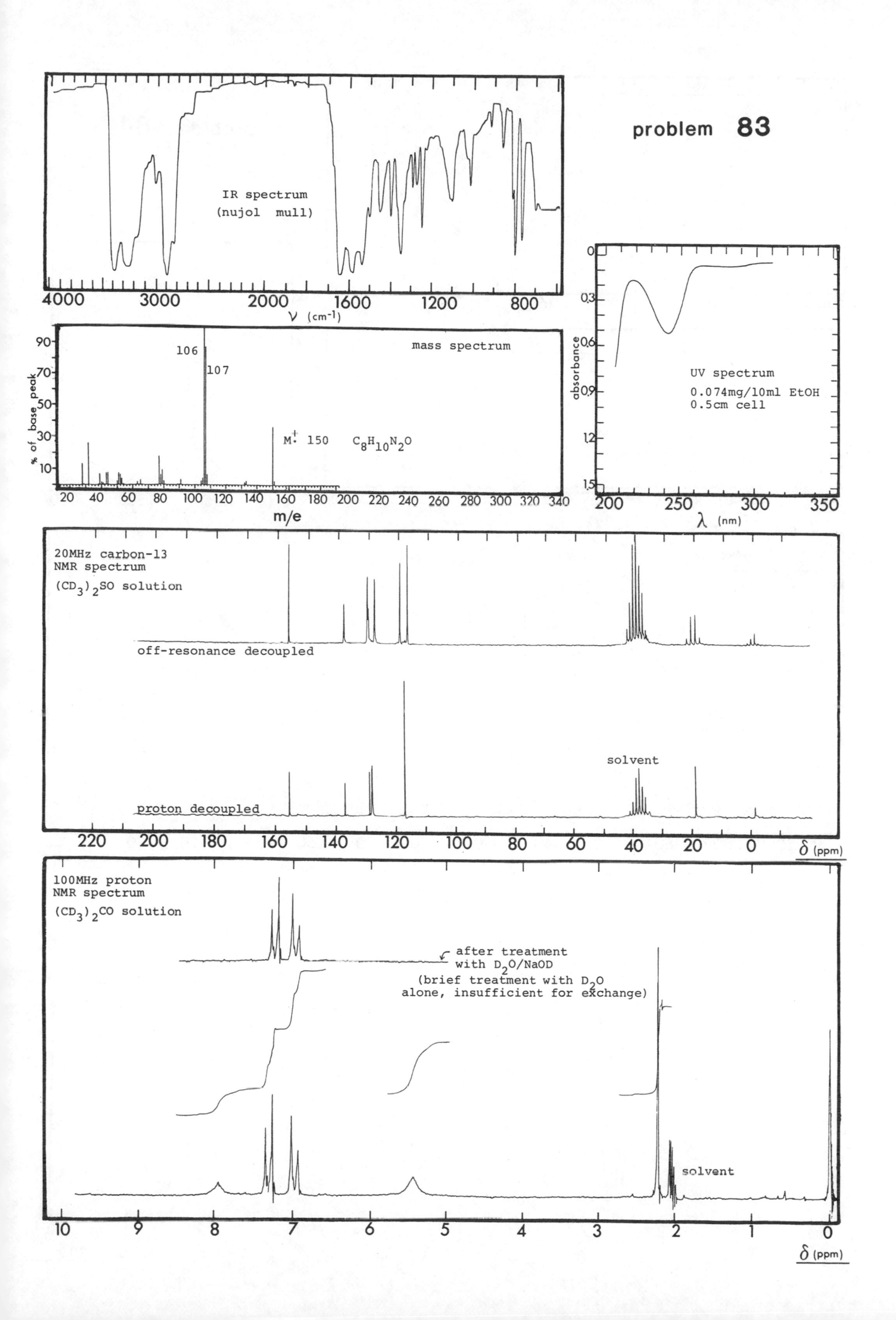

IR spectrum
(nujol mull)
4000
3000
2000
1600
1200
800
ν (cm-1)
mass spectrum
106
107
M+ 150
C8H10N2O
% of base peak
m/e
UV spectrum
0.074mg/10ml EtOH
0.5cm cell
absorbance
λ (nm)
20MHz carbon-13
NMR spectrum
(CD3)2SO solution
off-resonance decoupled
proton decoupled
solvent
δ (ppm)
100MHz proton
NMR spectrum
(CD3)2CO solution
after treatment
with D2O/NaOD
(brief treatment with D2O
alone, insufficient for exchange)
solvent
δ (ppm)

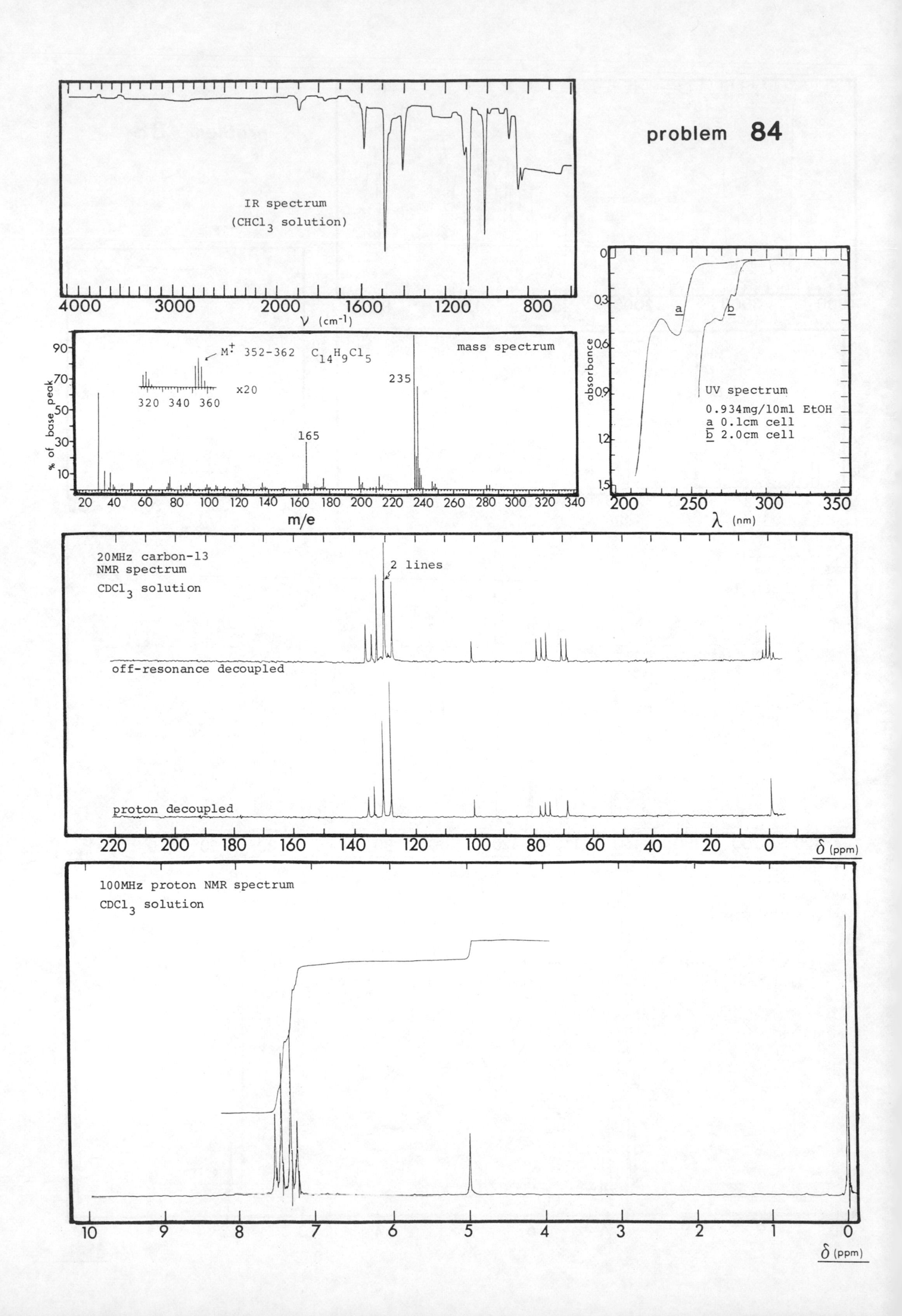

problem 84
IR spectrum
(CHCl3 solution)
4000
3000
2000
1600
1200
800
ν (cm-1)
mass spectrum
M+ 352-362 C14H9Cl5
x20
320 340 360
235
165
% of base peak
m/e
UV spectrum
0.934mg/10ml EtOH
a 0.1cm cell
b 2.0cm cell
absorbance
λ (nm)
20MHz carbon-13
NMR spectrum
CDCl3 solution
2 lines
off-resonance decoupled
proton decoupled
δ (ppm)
100MHz proton NMR spectrum
CDCl3 solution
δ (ppm)

problem 85

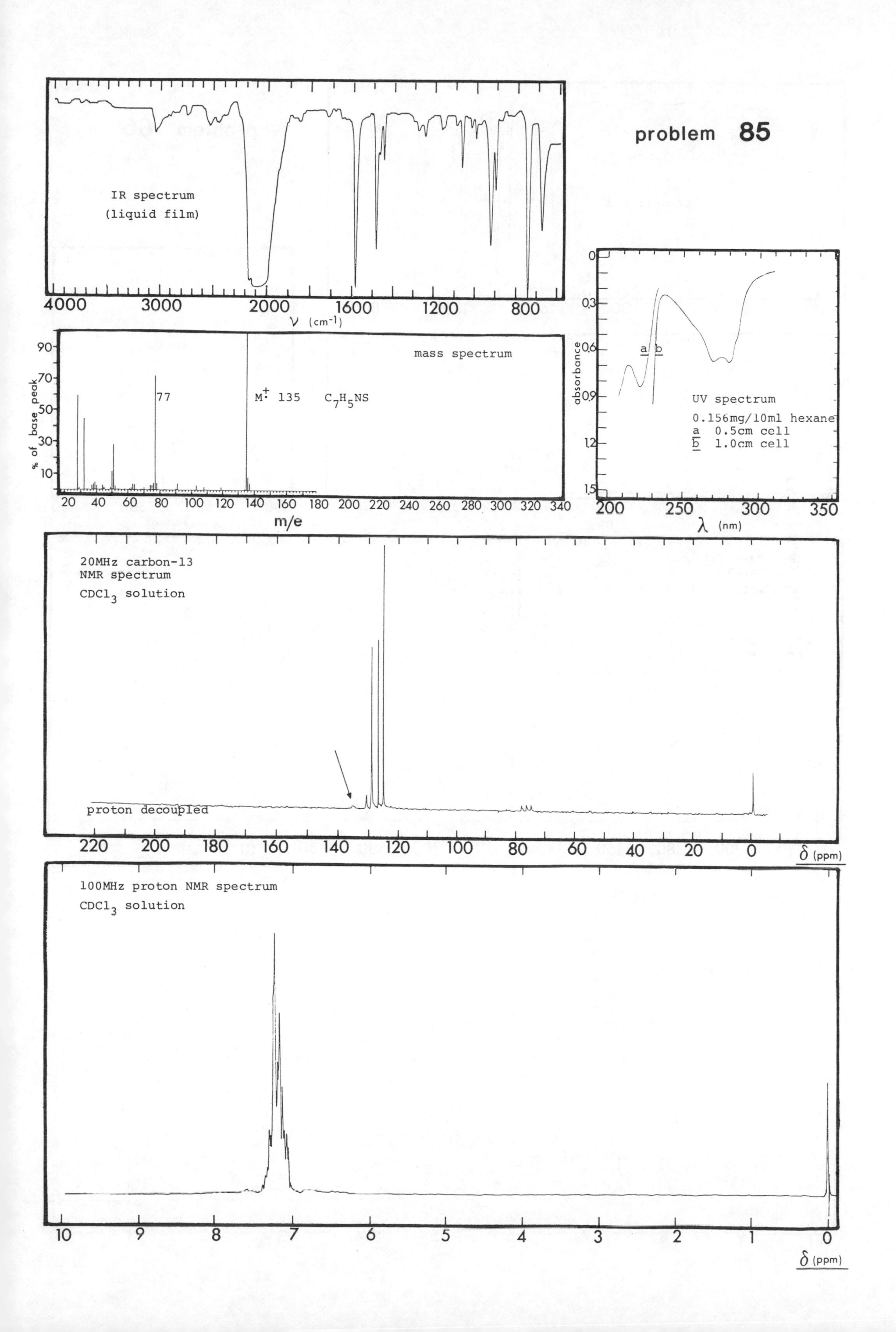
IR spectrum
(liquid film)
4000
3000
2000
1600
1200
800
ν (cm-1)
mass spectrum
77
M+· 135
C7H5NS
% of base peak
m/e
UV spectrum
0.156mg/10ml hexane
a 0.5cm cell
b 1.0cm cell
absorbance
λ (nm)
20MHz carbon-13
NMR spectrum
CDCl3 solution
proton decoupled
δ (ppm)
100MHz proton NMR spectrum
CDCl3 solution
δ (ppm)

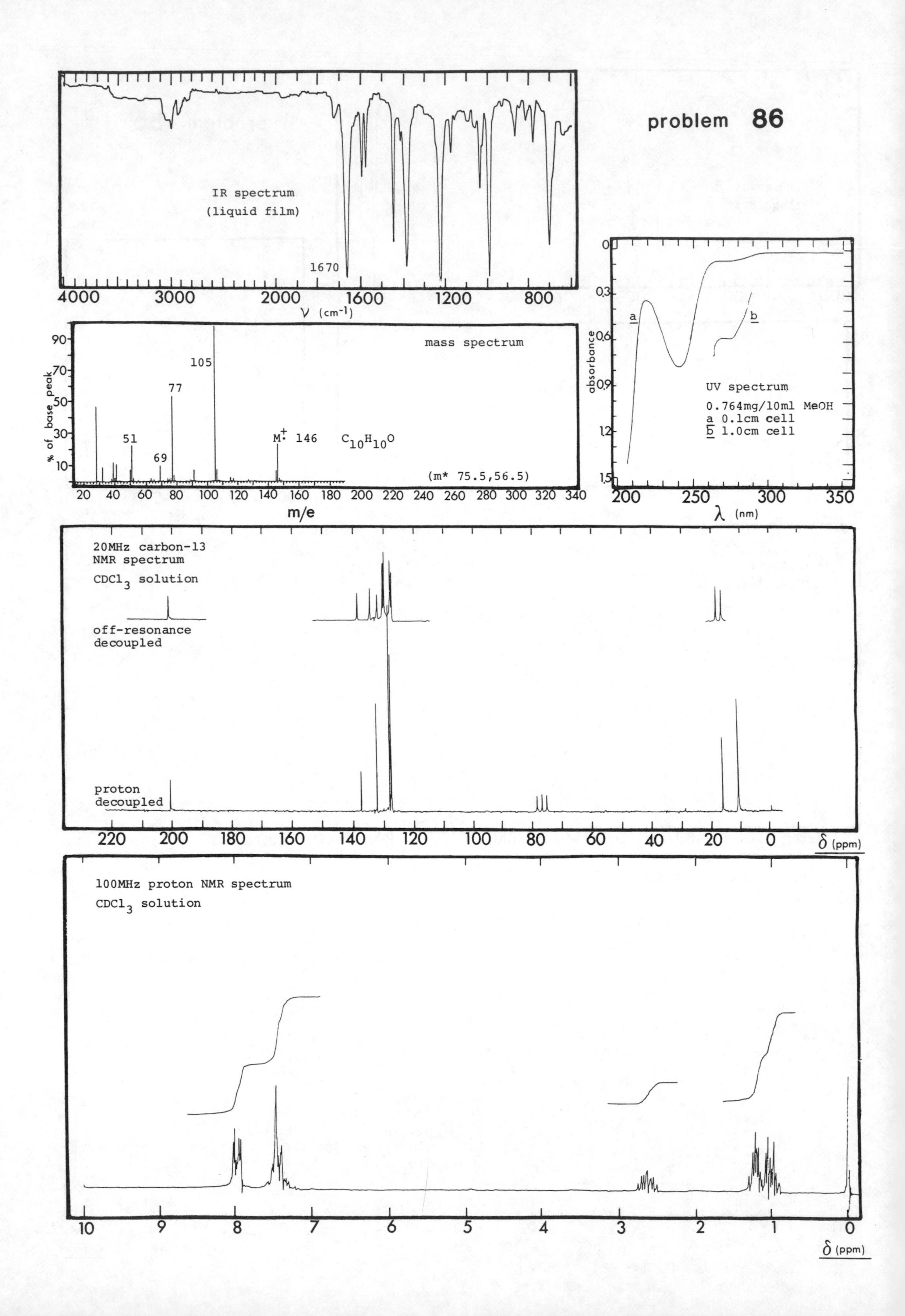
problem 86
IR spectrum
(liquid film)
1670
4000
3000
2000
1600
1200
800
ν (cm-1)
mass spectrum
105
77
51
69
M+ 146
C10H10O
(m* 75.5,56.5)
% of base peak
m/e
UV spectrum
0.764mg/10ml MeOH
a 0.1cm cell
b 1.0cm cell
absorbance
λ (nm)
20MHz carbon-13
NMR spectrum
CDCl3 solution
off-resonance
decoupled
proton
decoupled
δ (ppm)
100MHz proton NMR spectrum
CDCl3 solution
δ (ppm)

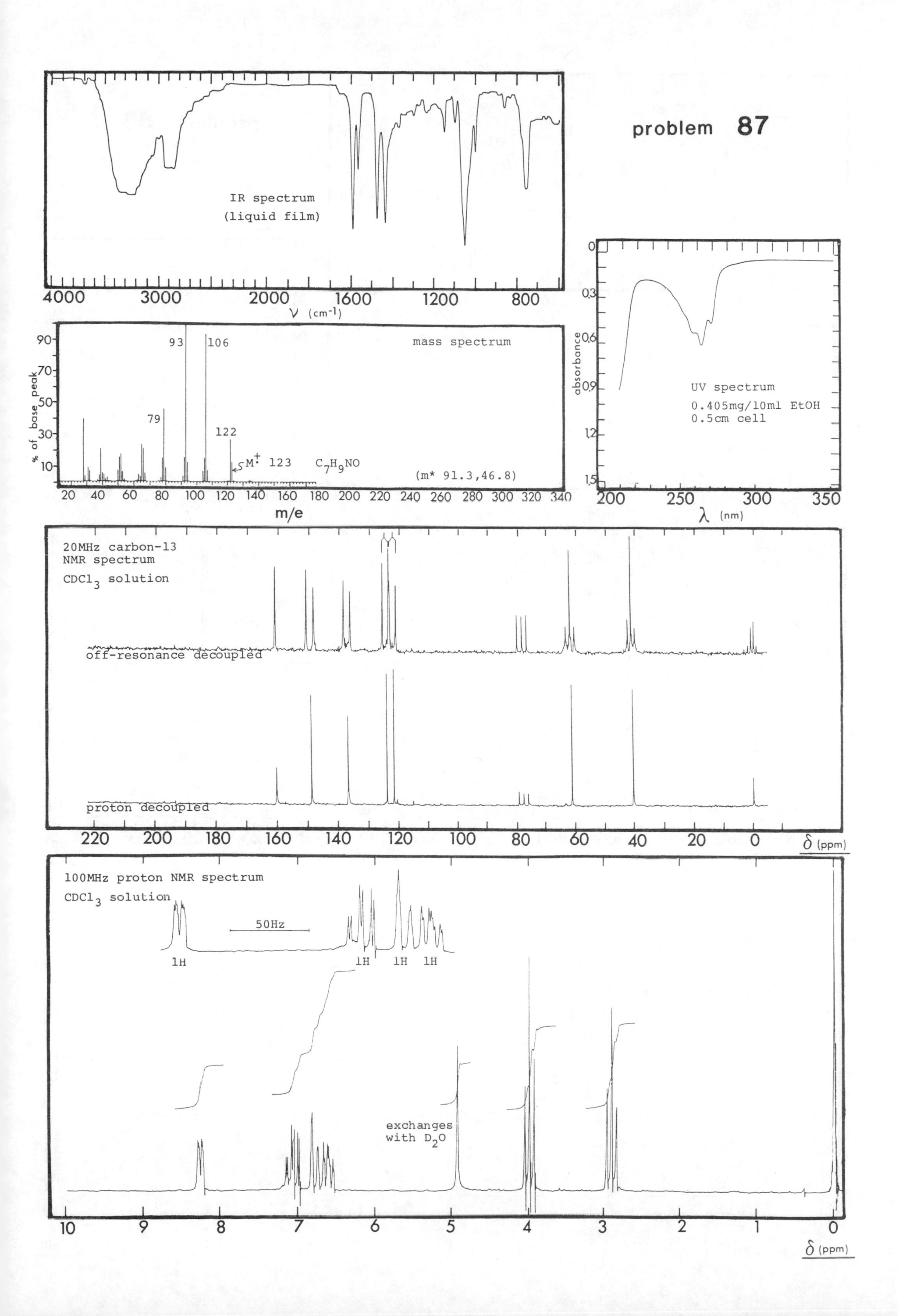
problem 87
IR spectrum
(liquid film)
4000
3000
2000
1600
1200
800
ν (cm-1)
mass spectrum
% of base peak
93
106
79
122
M+· 123
C7H9NO
(m* 91.3,46.8)
m/e
UV spectrum
0.405mg/10ml EtOH
0.5cm cell
absorbance
λ (nm)
20MHz carbon-13
NMR spectrum
CDCl3 solution
off-resonance decoupled
proton decoupled
δ (ppm)
100MHz proton NMR spectrum
CDCl3 solution
50Hz
1H
1H
1H
1H
exchanges
with D2O
δ (ppm)

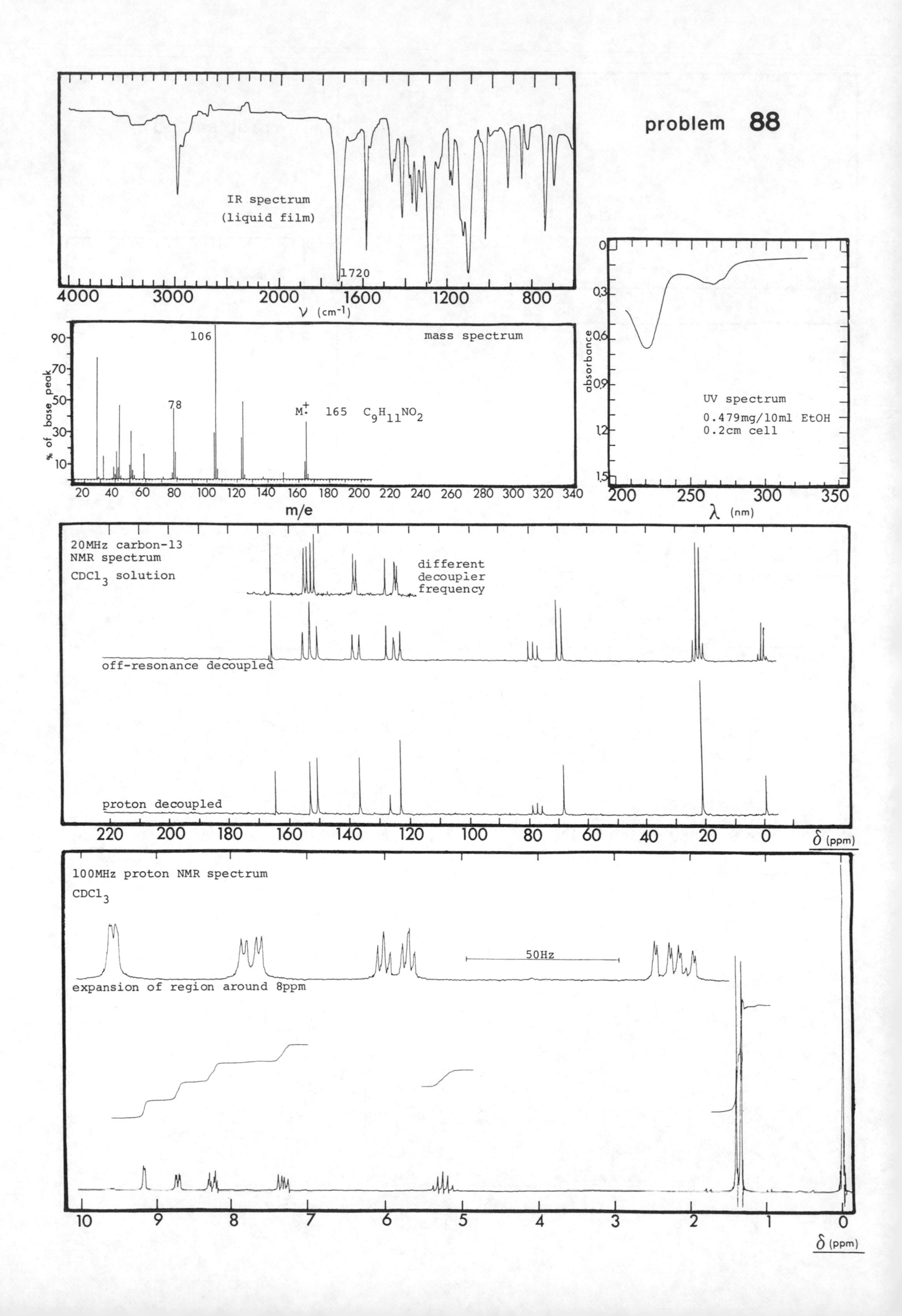

problem 88
IR spectrum
(liquid film)
1720
4000
3000
2000
1600
1200
800
ν (cm⁻¹)
mass spectrum
106
78
M⁺ 165 C9H11NO2
% of base peak
m/e
UV spectrum
0.479mg/10ml EtOH
0.2cm cell
absorbance
λ (nm)
20MHz carbon-13
NMR spectrum
CDCl3 solution
different
decoupler
frequency
off-resonance decoupled
proton decoupled
δ (ppm)
100MHz proton NMR spectrum
CDCl3
50Hz
expansion of region around 8ppm
δ (ppm)

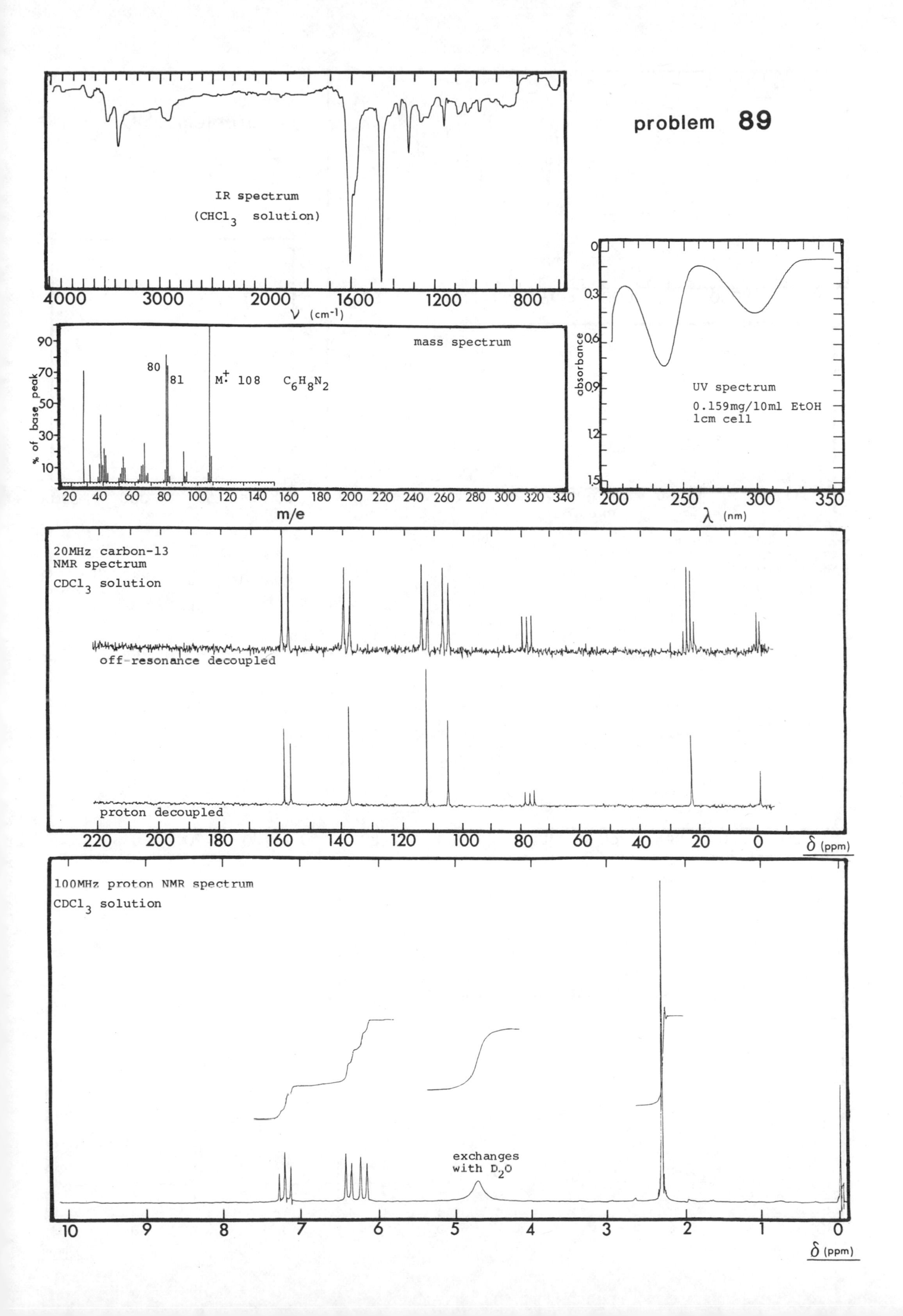
problem 89
IR spectrum
(CHCl3 solution)
4000
3000
2000
1600
1200
800
ν (cm-1)
mass spectrum
80
81
M+ 108
C6H8N2
% of base peak
m/e
UV spectrum
0.159mg/10ml EtOH
1cm cell
absorbance
λ (nm)
20MHz carbon-13
NMR spectrum
CDCl3 solution
off-resonance decoupled
proton decoupled
δ (ppm)
100MHz proton NMR spectrum
CDCl3 solution
exchanges
with D2O
δ (ppm)

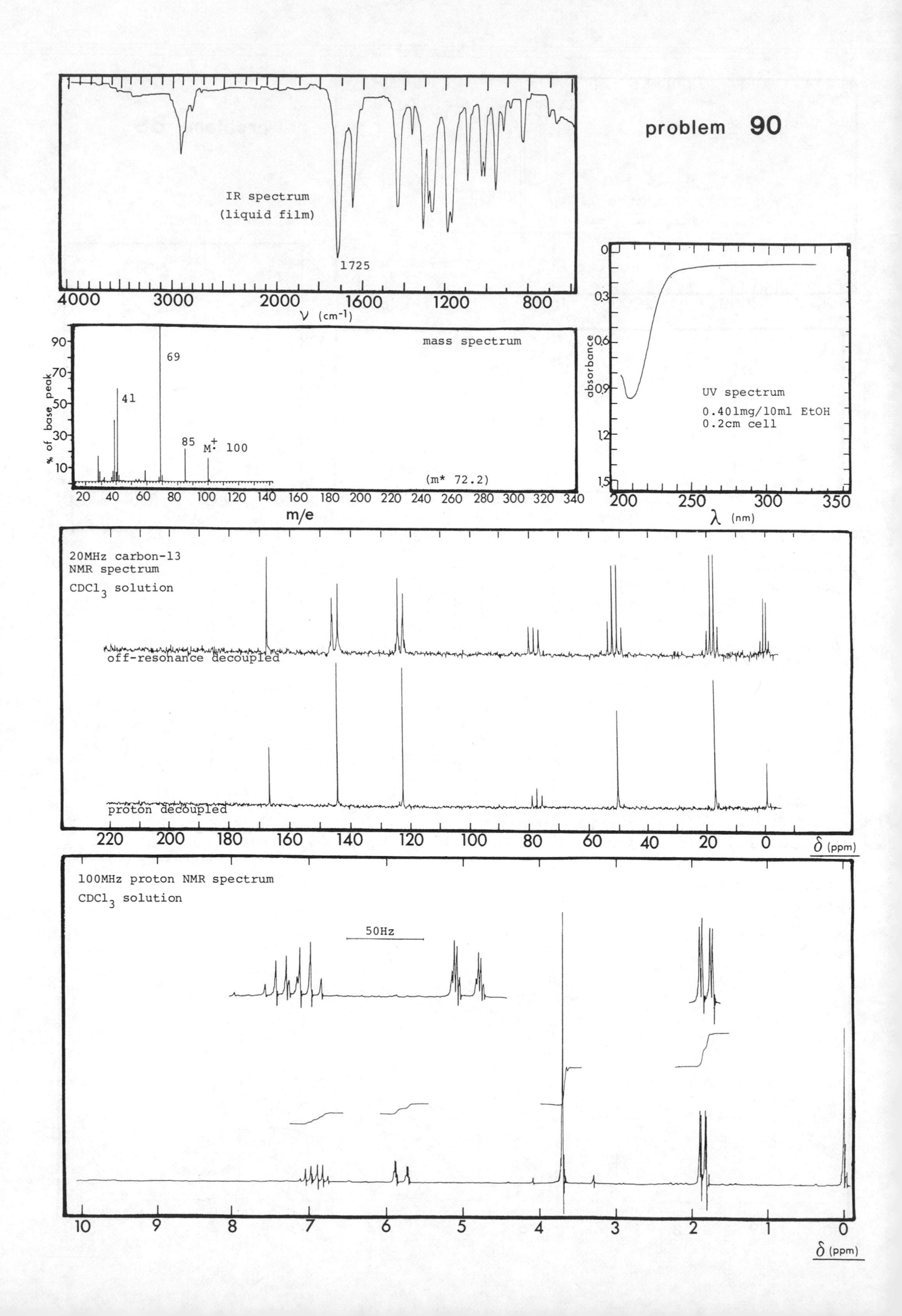

problem 90
IR spectrum
(liquid film)
1725
4000
3000
2000
1600
1200
800
ν (cm-1)
mass spectrum
69
41
85
M+· 100
(m* 72.2)
% of base peak
m/e
UV spectrum
0.401mg/10ml EtOH
0.2cm cell
absorbance
λ (nm)
20MHz carbon-13
NMR spectrum
CDCl3 solution
off-resonance decoupled
proton decoupled
δ (ppm)
100MHz proton NMR spectrum
CDCl3 solution
50Hz
δ (ppm)

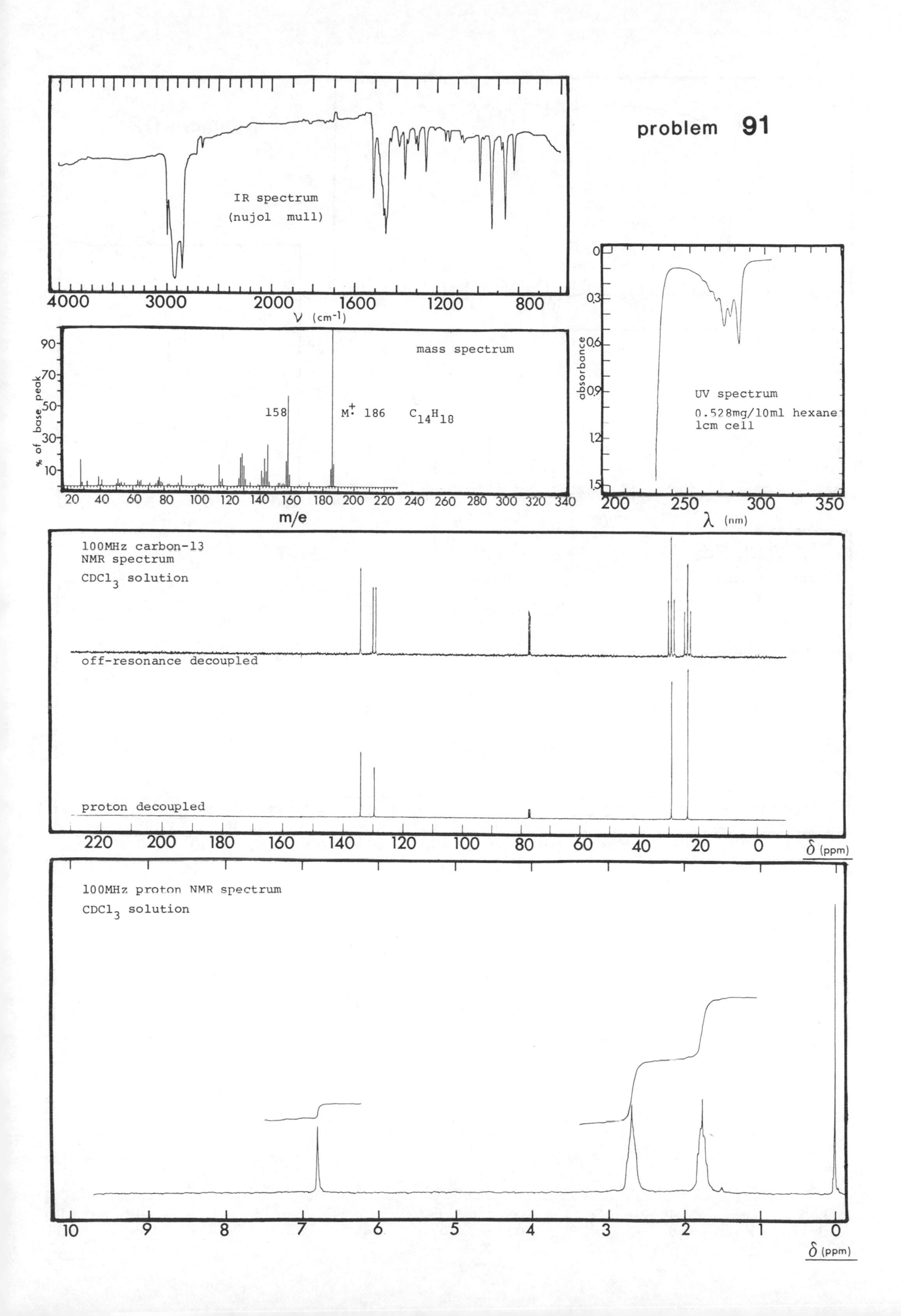

problem 91
IR spectrum
(nujol mull)
4000
3000
2000
1600
1200
800
ν (cm-1)
mass spectrum
158
M+ 186
C14H18
% of base peak
m/e
UV spectrum
0.528mg/10ml hexane
1cm cell
absorbance
λ (nm)
100MHz carbon-13
NMR spectrum
CDCl3 solution
off-resonance decoupled
proton decoupled
δ (ppm)
100MHz proton NMR spectrum
CDCl3 solution
δ (ppm)

problem 92

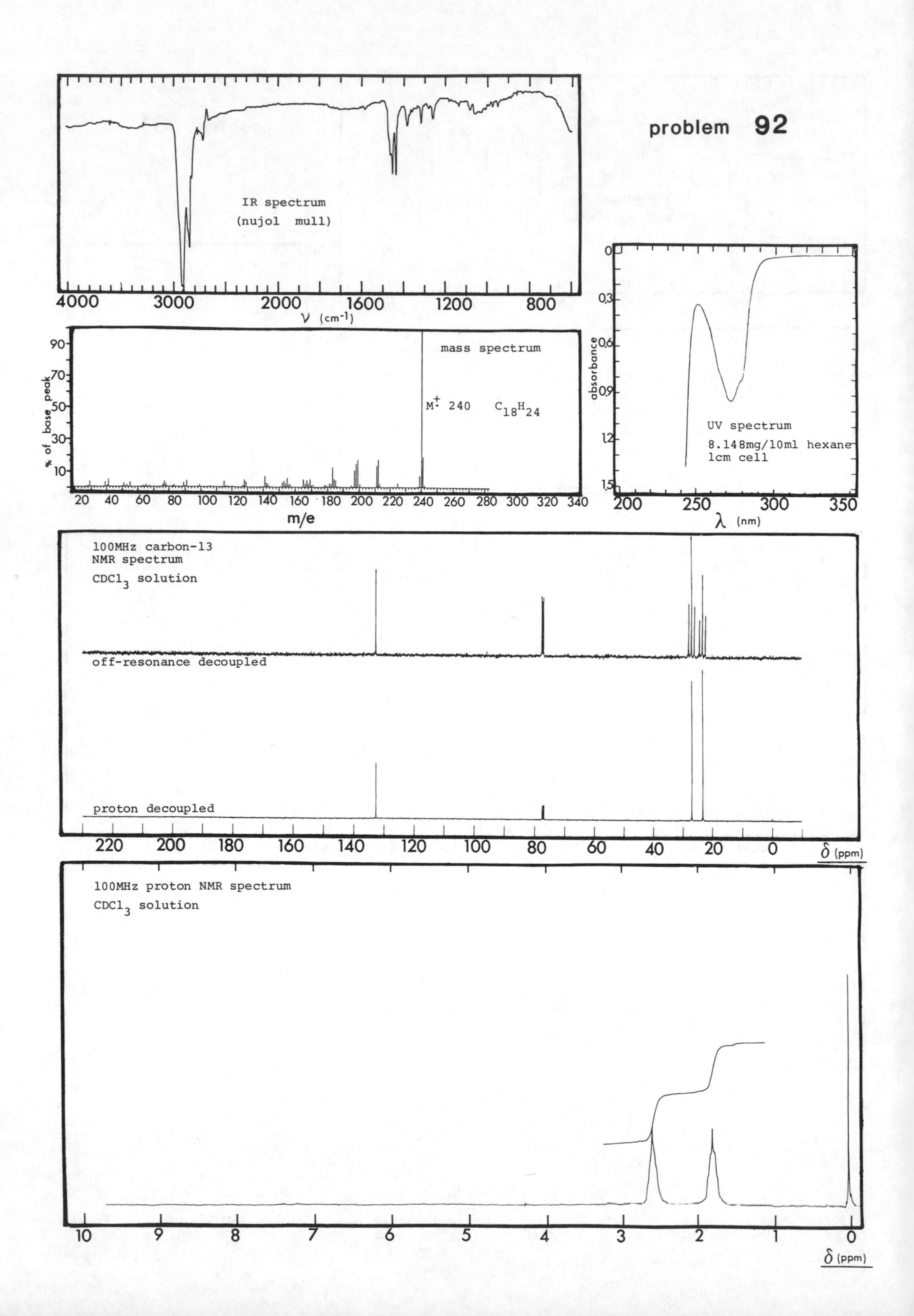

IR spectrum
(nujol mull)
4000
3000
2000
1600
1200
800
ν (cm-1)
mass spectrum
M+ 240 C18H24
% of base peak
90
70
50
30
10
20 40 60 80 100 120 140 160 180 200 220 240 260 280 300 320 340
m/e
absorbance
0
0.3
0.6
0.9
1.2
1.5
UV spectrum
8.148mg/10ml hexane
1cm cell
200 250 300 350
λ (nm)
100MHz carbon-13
NMR spectrum
CDCl3 solution
off-resonance decoupled
proton decoupled
220 200 180 160 140 120 100 80 60 40 20 0
δ (ppm)
100MHz proton NMR spectrum
CDCl3 solution
10 9 8 7 6 5 4 3 2 1 0
δ (ppm)

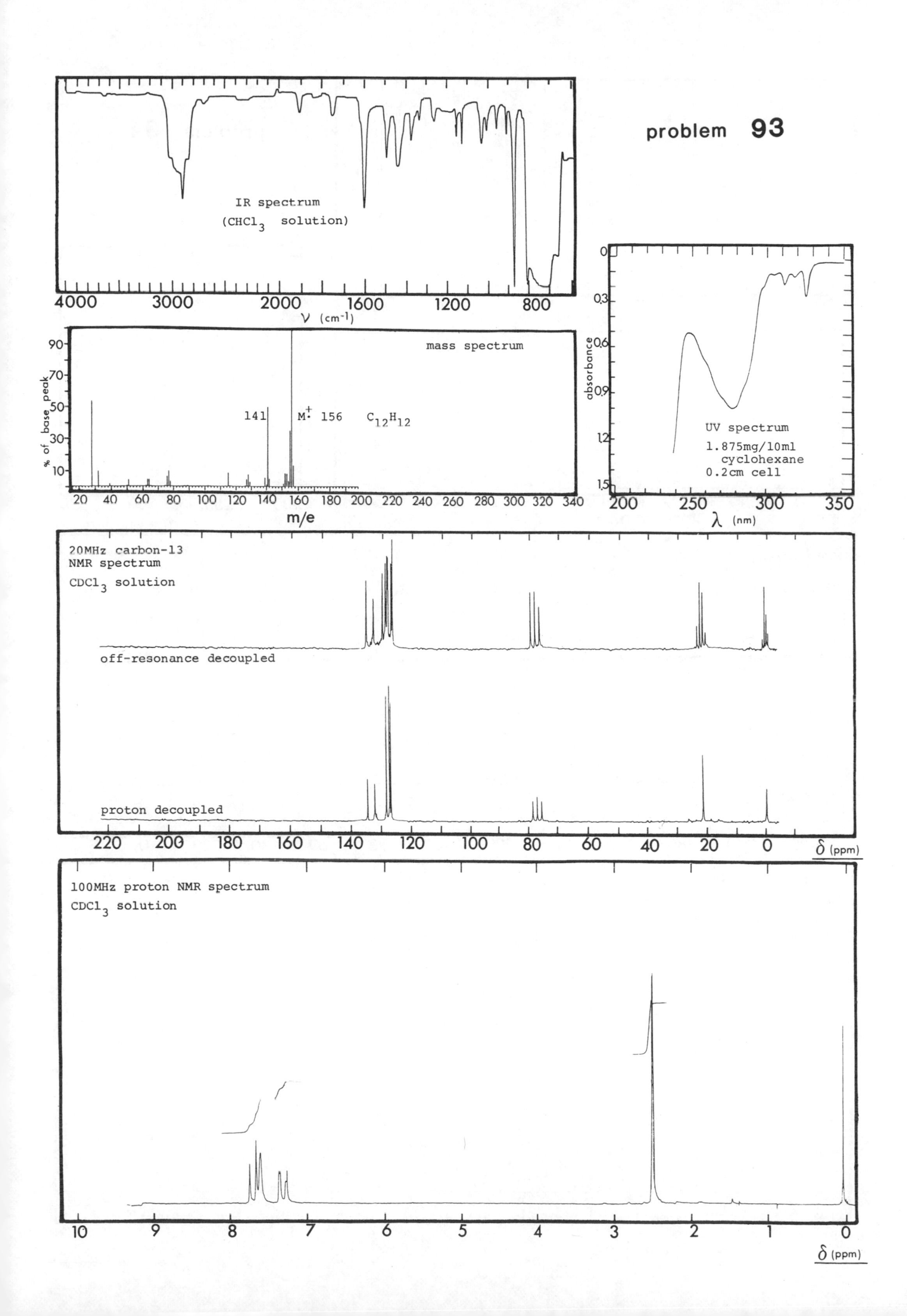

problem 93
IR spectrum
(CHCl3 solution)
4000
3000
2000
1600
1200
800
ν (cm-1)
mass spectrum
% of base peak
141
M+· 156
C12H12
m/e
UV spectrum
1.875mg/10ml
cyclohexane
0.2cm cell
absorbance
λ (nm)
20MHz carbon-13
NMR spectrum
CDCl3 solution
off-resonance decoupled
proton decoupled
δ (ppm)
100MHz proton NMR spectrum
CDCl3 solution
δ (ppm)

problem 94

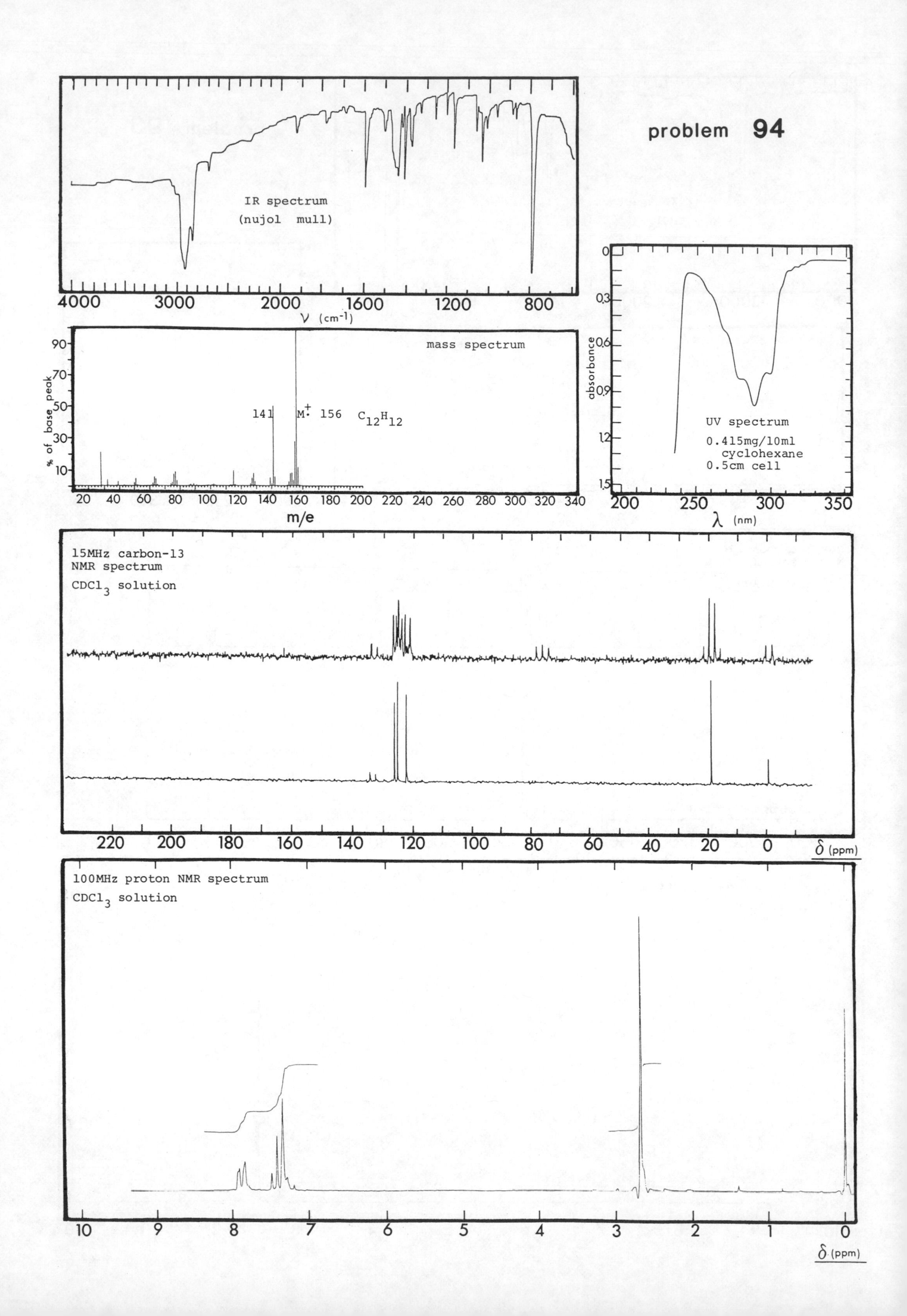

IR spectrum
(nujol mull)
4000
3000
2000
1600
1200
800
ν (cm-1)
mass spectrum
141
M+ 156
C12H12
% of base peak
m/e
UV spectrum
0.415mg/10ml
cyclohexane
0.5cm cell
absorbance
λ (nm)
15MHz carbon-13
NMR spectrum
CDCl3 solution
δ (ppm)
100MHz proton NMR spectrum
CDCl3 solution
δ (ppm)

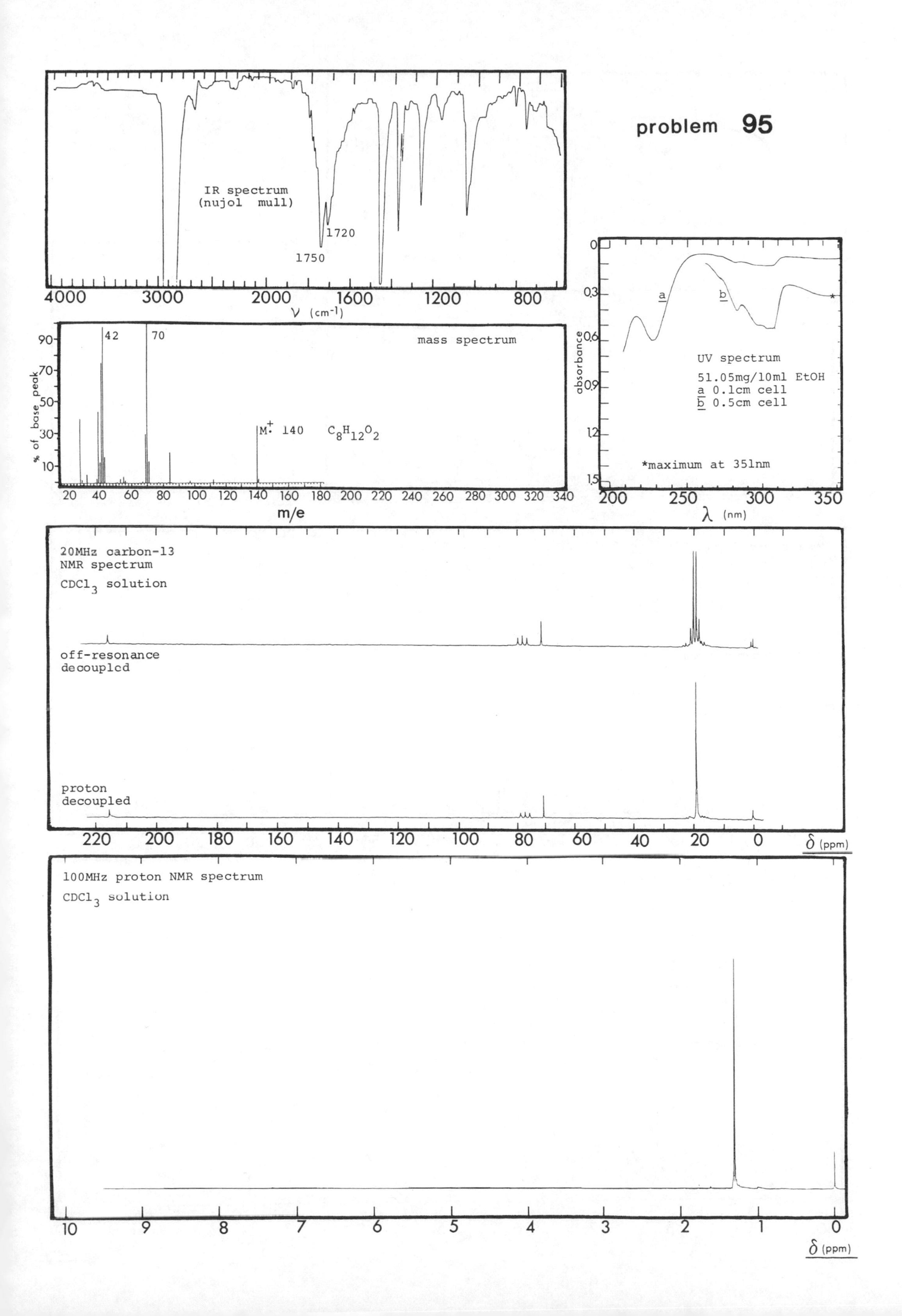
problem 95
IR spectrum
(nujol mull)
1720
1750
4000
3000
2000
1600
1200
800
ν (cm-1)
mass spectrum
42
70
M+. 140
C8H12O2
% of base peak
m/e
UV spectrum
51.05mg/10ml EtOH
a 0.1cm cell
b 0.5cm cell
*maximum at 351nm
absorbance
λ (nm)
20MHz carbon-13
NMR spectrum
CDCl3 solution
off-resonance
decoupled
proton
decoupled
δ (ppm)
100MHz proton NMR spectrum
CDCl3 solution
δ (ppm)

problem 96

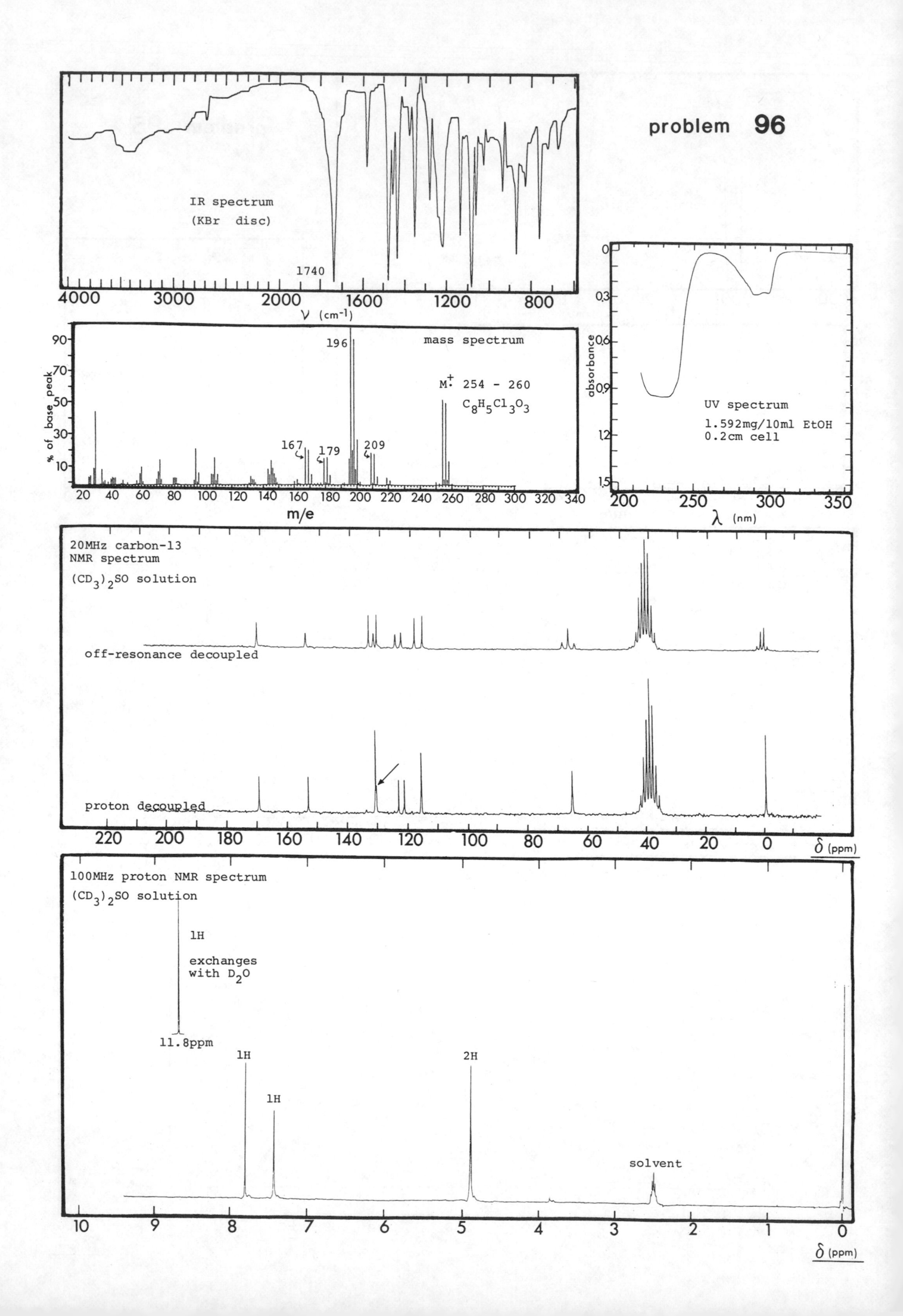

IR spectrum
(KBr disc)
1740
4000
3000
2000
1600
1200
800
ν (cm-1)
mass spectrum
196
M+· 254 - 260
C8H5Cl3O3
167
179
209
% of base peak
m/e
UV spectrum
1.592mg/10ml EtOH
0.2cm cell
absorbance
λ (nm)
20MHz carbon-13
NMR spectrum
(CD3)2SO solution
off-resonance decoupled
proton decoupled
δ (ppm)
100MHz proton NMR spectrum
(CD3)2SO solution
1H
exchanges
with D2O
11.8ppm
1H
1H
2H
solvent
δ (ppm)

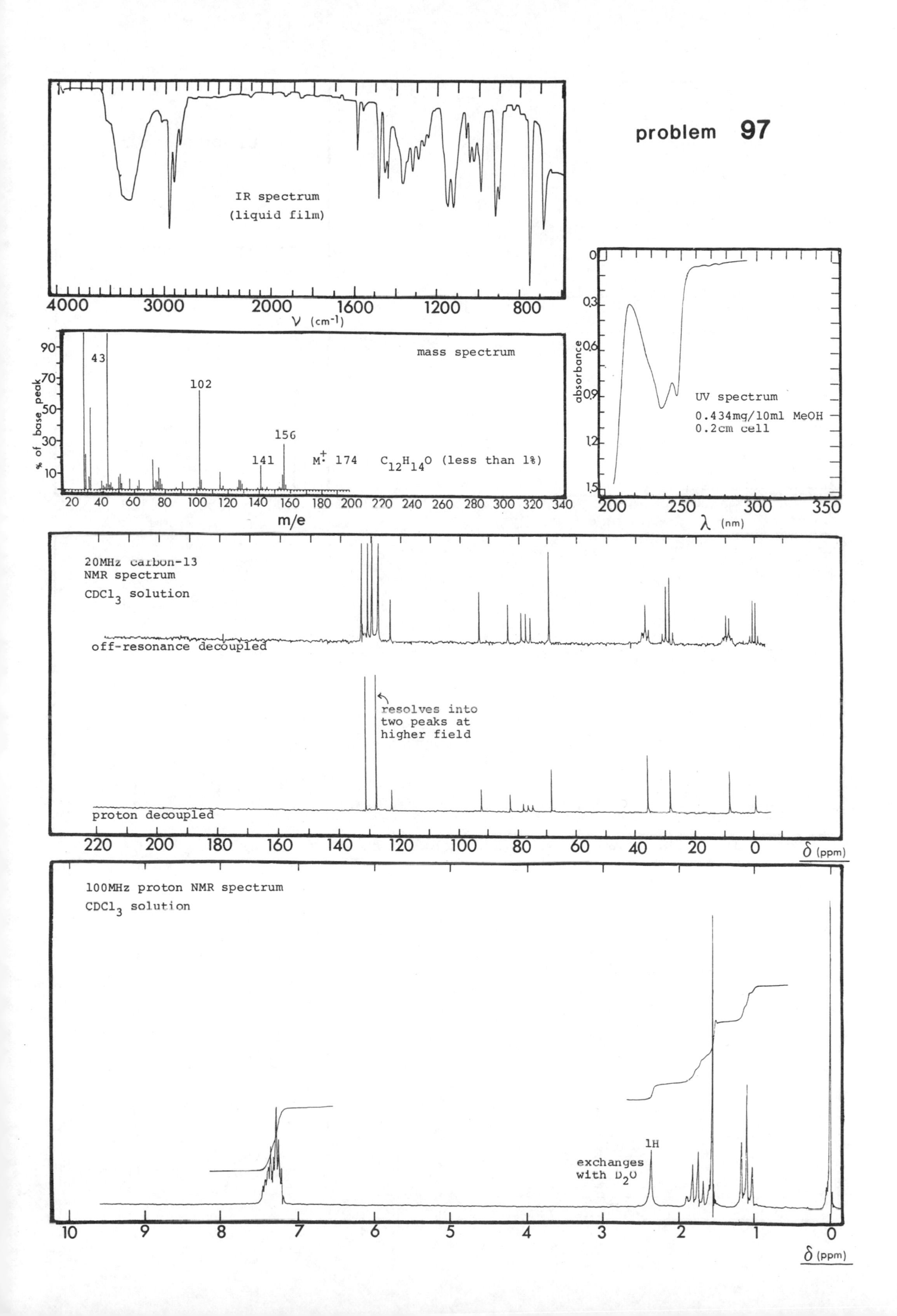

problem 97
IR spectrum
(liquid film)
4000
3000
2000
1600
1200
800
ν (cm-1)
mass spectrum
% of base peak
43
102
156
141
M+· 174
C12H14O (less than 1%)
m/e
UV spectrum
0.434mg/10ml MeOH
0.2cm cell
absorbance
λ (nm)
20MHz carbon-13
NMR spectrum
CDCl3 solution
off-resonance decoupled
resolves into
two peaks at
higher field
proton decoupled
δ (ppm)
100MHz proton NMR spectrum
CDCl3 solution
1H
exchanges
with D2O
δ (ppm)

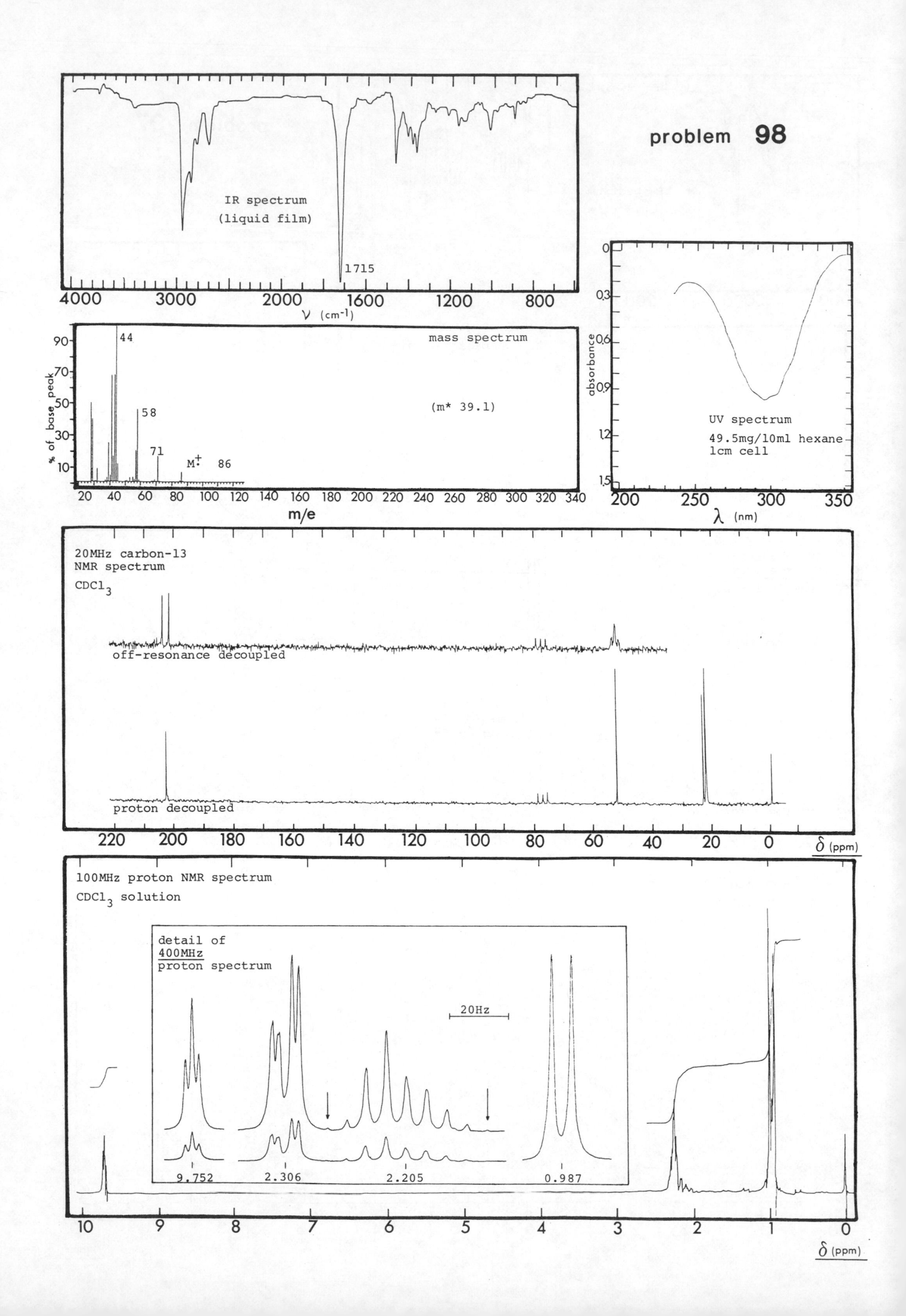

problem 98
IR spectrum
(liquid film)
1715
4000
3000
2000
1600
1200
800
ν (cm-1)
mass spectrum
% of base peak
44
58
71
M+ 86
(m* 39.1)
m/e
UV spectrum
49.5mg/10ml hexane
1cm cell
absorbance
λ (nm)
20MHz carbon-13
NMR spectrum
CDCl3
off-resonance decoupled
proton decoupled
δ (ppm)
100MHz proton NMR spectrum
CDCl3 solution
detail of
400MHz
proton spectrum
20Hz
9.752
2.306
2.205
0.987
δ (ppm)

problem 99

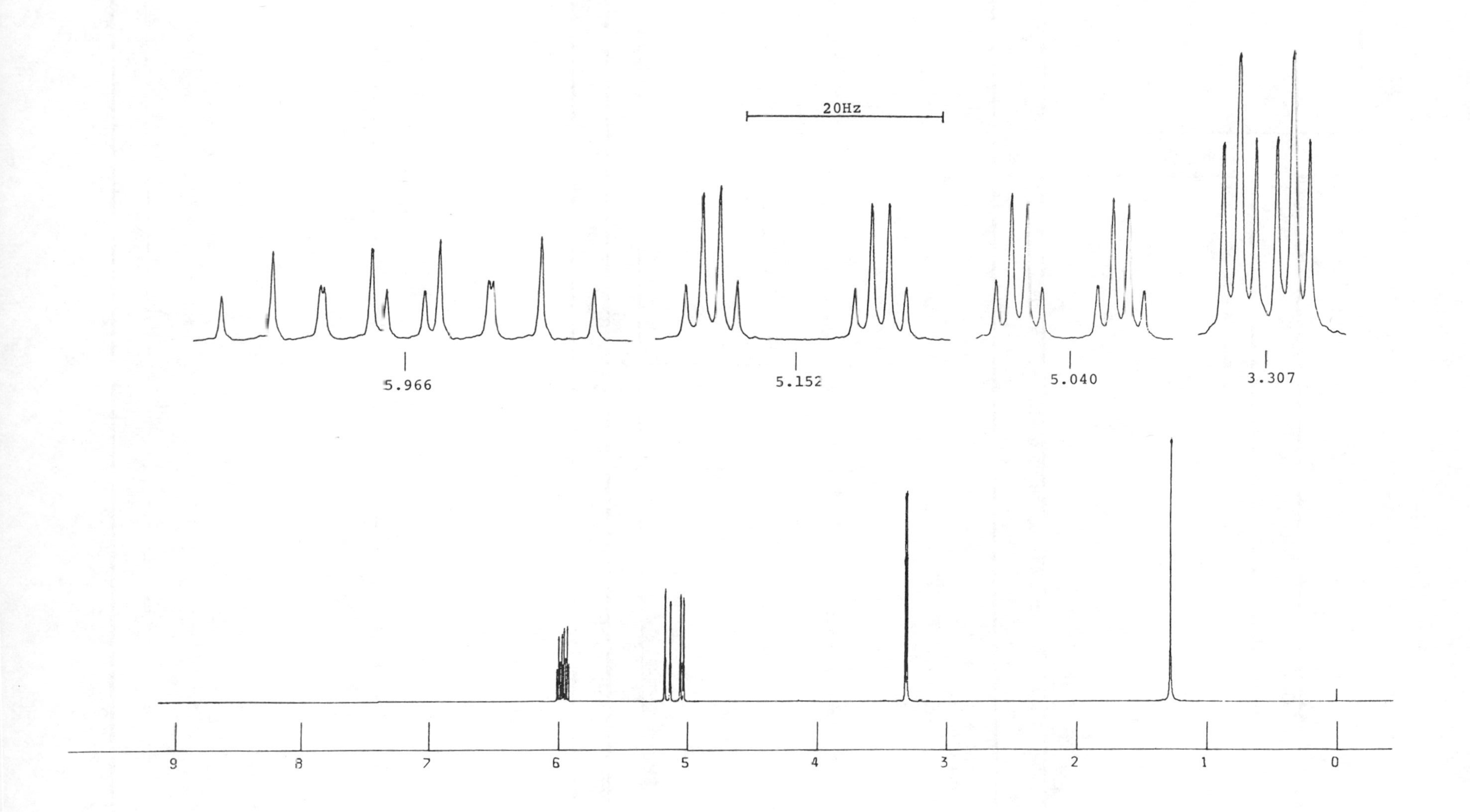
400MHz proton NMR spectrum
CDCl3 solution
20Hz
5.966
5.152
5.040
3.307
9
8
7
6
5
4
3
2
1
0

problem 99

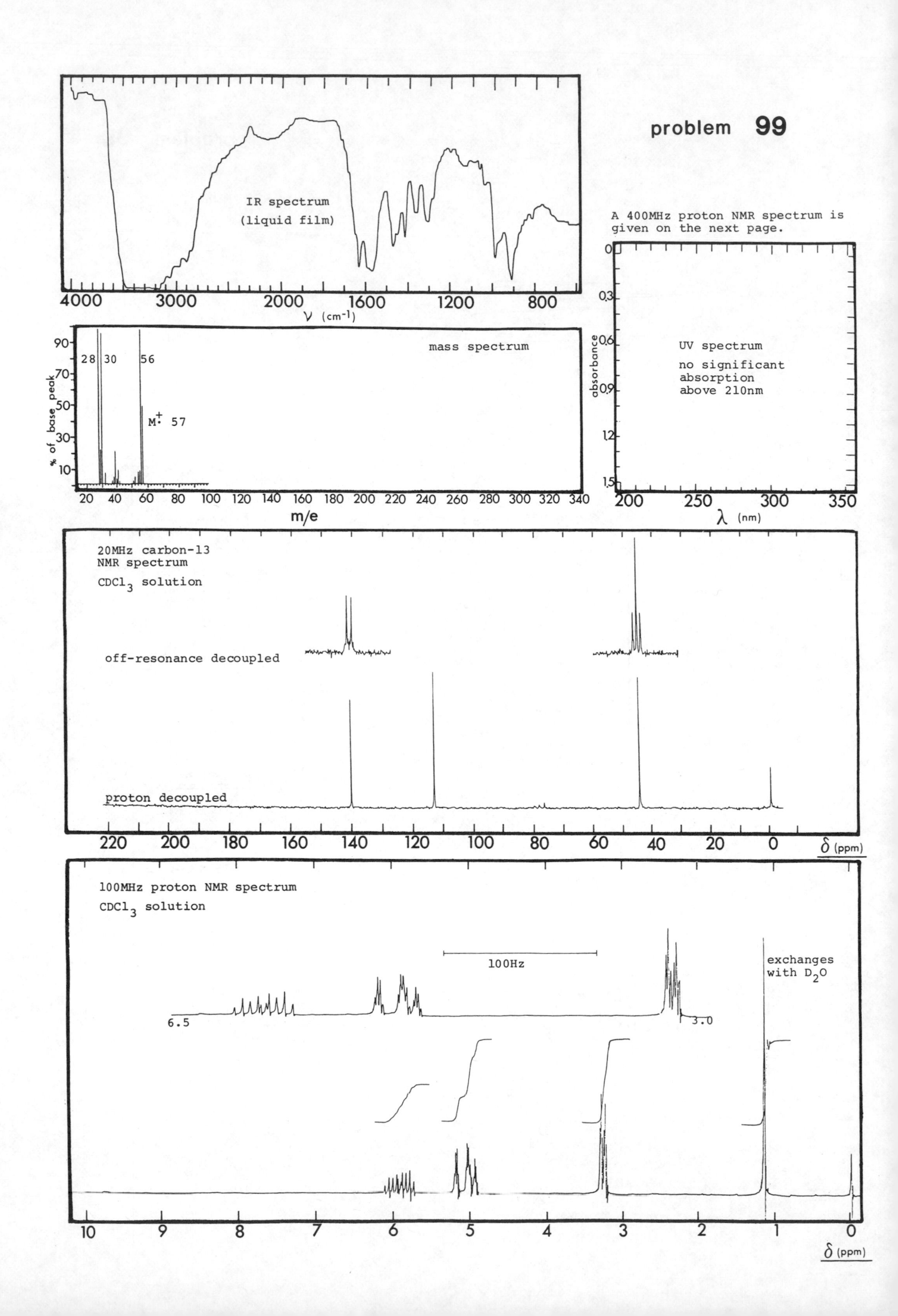

IR spectrum
(liquid film)
4000
3000
2000
1600
1200
800
ν (cm-1)
A 400MHz proton NMR spectrum is given on the next page.
mass spectrum
% of base peak
28
30
56
M+ 57
m/e
UV spectrum
no significant absorption above 210nm
absorbance
λ (nm)
20MHz carbon-13 NMR spectrum
CDCl3 solution
off-resonance decoupled
proton decoupled
δ (ppm)
100MHz proton NMR spectrum
CDCl3 solution
100Hz
6.5
3.0
exchanges with D2O
δ (ppm)

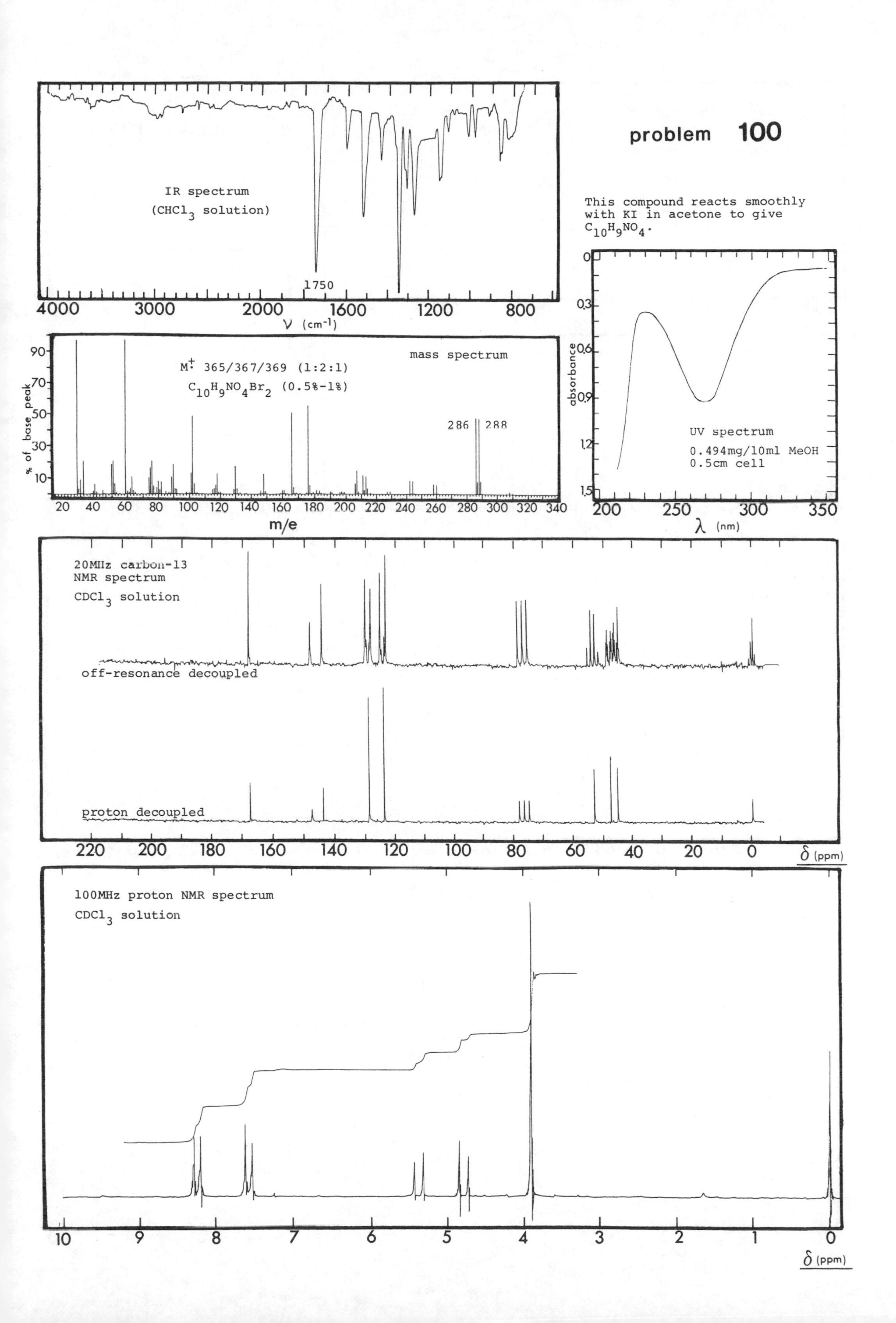
problem 100
IR spectrum
(CHCl3 solution)
1750
4000
3000
2000
1600
1200
800
ν (cm-1)
This compound reacts smoothly with KI in acetone to give C10H9NO4.
mass spectrum
M+ 365/367/369 (1:2:1)
C10H9NO4Br2 (0.5%-1%)
286
288
% of base peak
m/e
absorbance
UV spectrum
0.494mg/10ml MeOH
0.5cm cell
λ (nm)
20MHz carbon-13
NMR spectrum
CDCl3 solution
off-resonance decoupled
proton decoupled
δ (ppm)
100MHz proton NMR spectrum
CDCl3 solution
δ (ppm)

problem 101

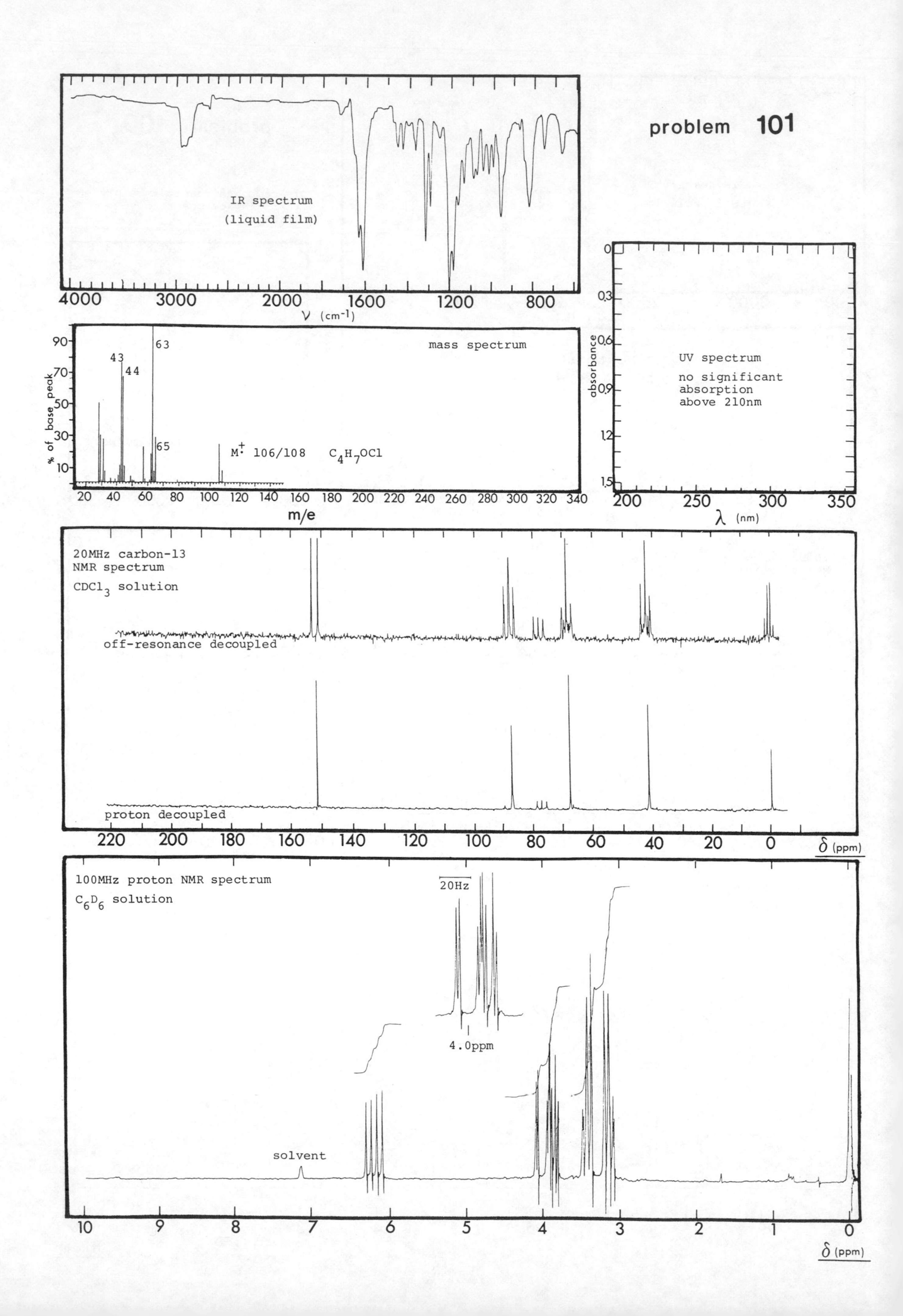

IR spectrum
(liquid film)
4000
3000
2000
1600
1200
800
ν (cm⁻¹)
mass spectrum
% of base peak
90
70
50
30
10
43
44
63
65
M⁺· 106/108 C4H7OCl
20 40 60 80 100 120 140 160 180 200 220 240 260 280 300 320 340
m/e
UV spectrum
no significant
absorption
above 210nm
absorbance
0
0.3
0.6
0.9
1.2
1.5
200 250 300 350
λ (nm)
20MHz carbon-13
NMR spectrum
CDCl3 solution
off-resonance decoupled
proton decoupled
220 200 180 160 140 120 100 80 60 40 20 0
δ (ppm)
100MHz proton NMR spectrum
C6D6 solution
20Hz
4.0ppm
solvent
10 9 8 7 6 5 4 3 2 1 0
δ (ppm)

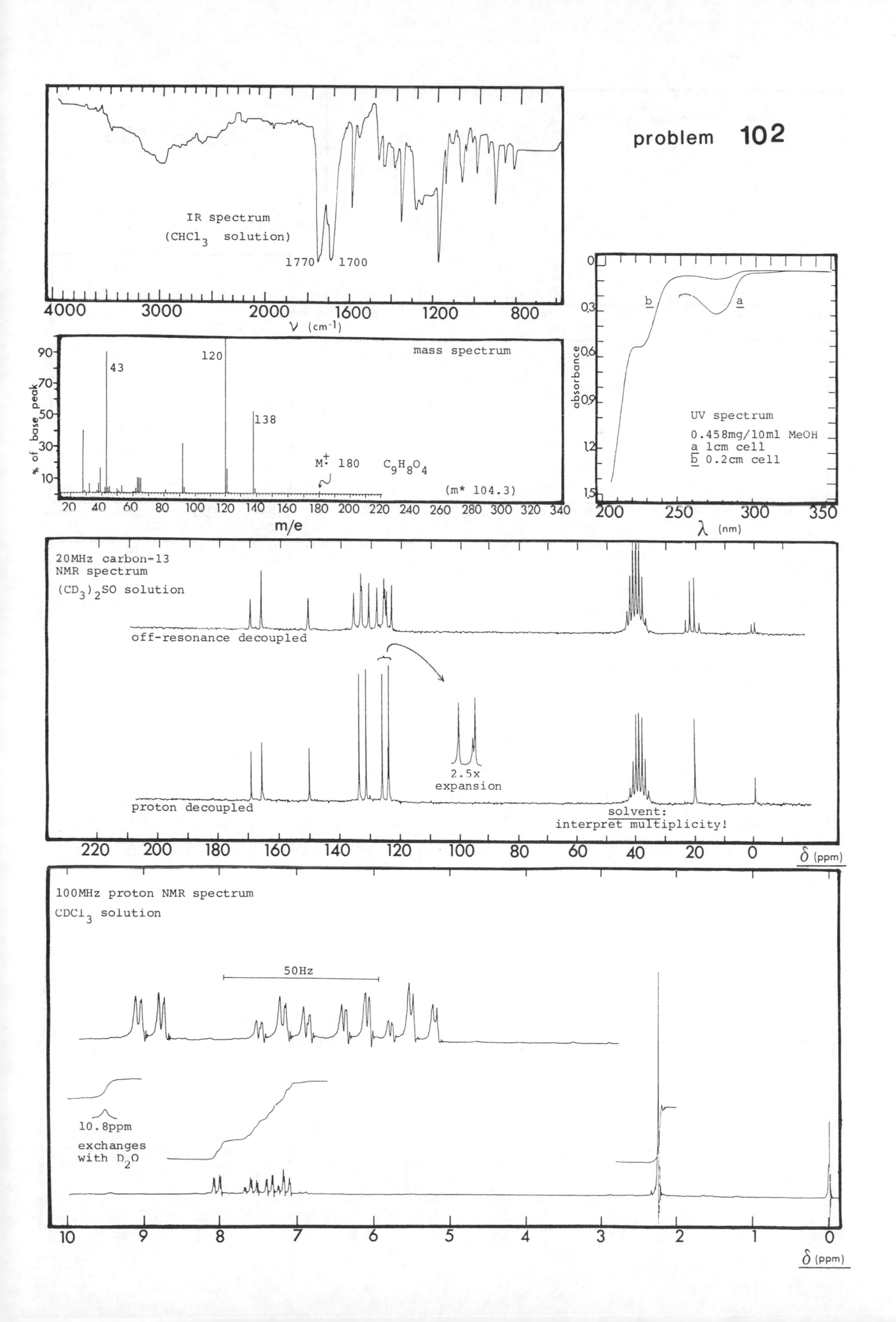

problem 102
IR spectrum
(CHCl3 solution)
1770
1700
4000
3000
2000
1600
1200
800
ν (cm-1)
mass spectrum
% of base peak
43
120
138
M+ 180
C9H8O4
(m* 104.3)
m/e
UV spectrum
0.458mg/10ml MeOH
a 1cm cell
b 0.2cm cell
absorbance
λ (nm)
20MHz carbon-13
NMR spectrum
(CD3)2SO solution
off-resonance decoupled
2.5x
expansion
proton decoupled
solvent:
interpret multiplicity!
δ (ppm)
100MHz proton NMR spectrum
CDCl3 solution
50Hz
10.8ppm
exchanges
with D2O
δ (ppm)

problem **103**

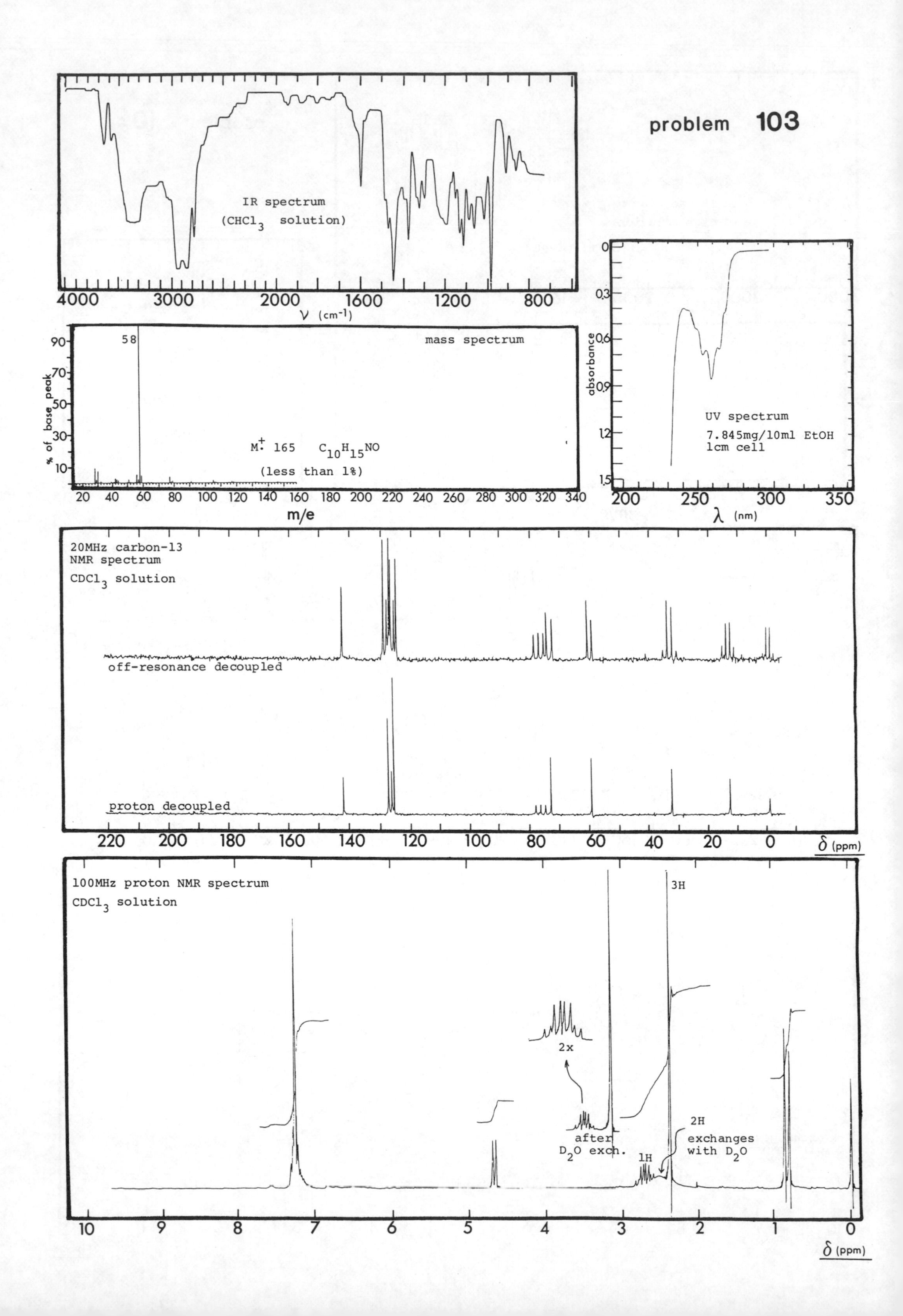

IR spectrum
(CHCl3 solution)
4000
3000
2000
1600
1200
800
ν (cm-1)
mass spectrum
58
M+. 165 C10H15NO
(less than 1%)
% of base peak
m/e
UV spectrum
7.845mg/10ml EtOH
1cm cell
absorbance
λ (nm)
20MHz carbon-13
NMR spectrum
CDCl3 solution
off-resonance decoupled
proton decoupled
δ (ppm)
100MHz proton NMR spectrum
CDCl3 solution
3H
2x
after
D2O exch.
1H
2H
exchanges
with D2O
δ (ppm)

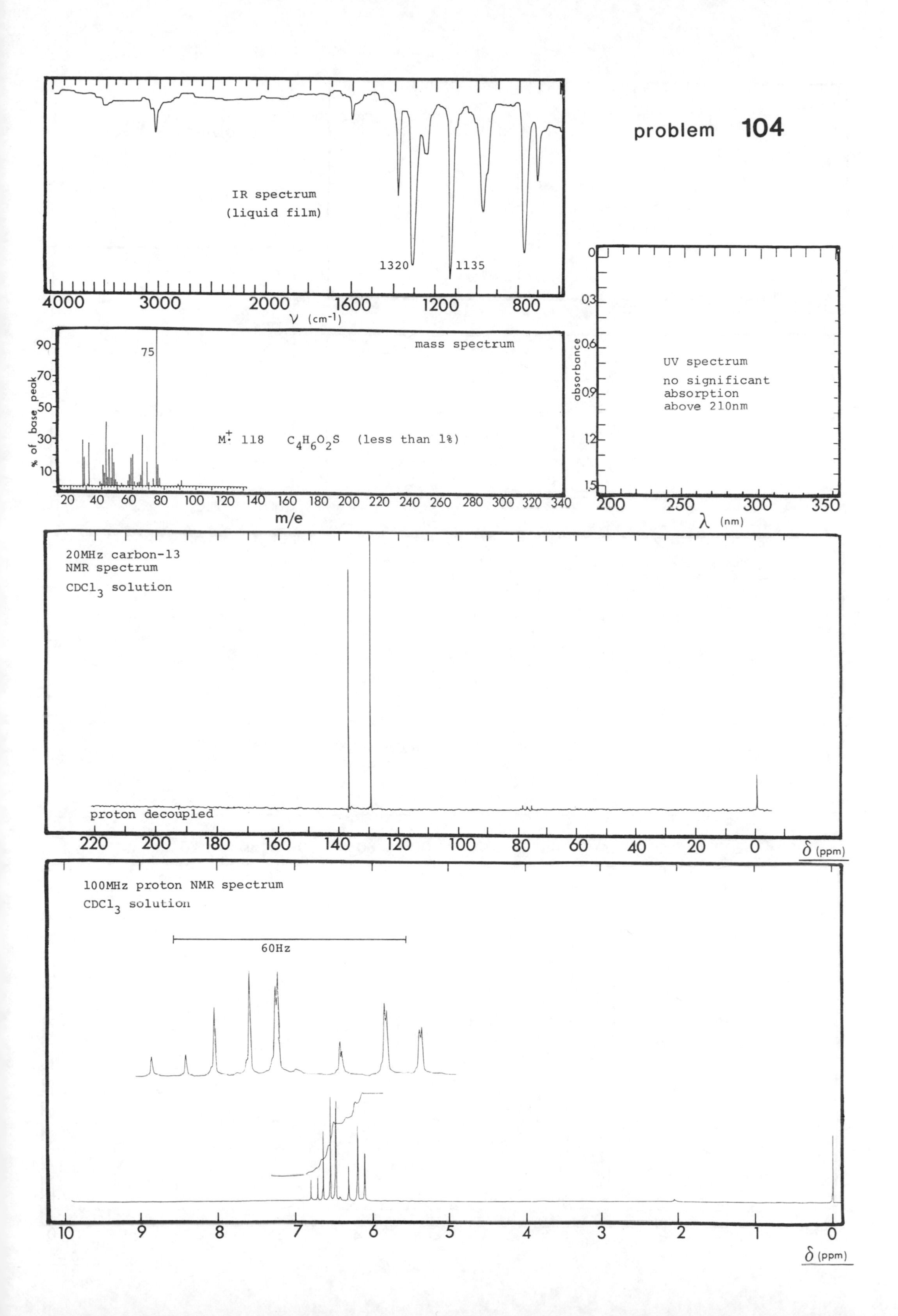

IR spectrum
(liquid film)
1320
1135
4000
3000
2000
1600
1200
800
ν (cm-1)
problem 104
mass spectrum
75
M⁺ 118 $C_4H_6O_2S$ (less than 1%)
% of base peak
m/e
UV spectrum
no significant absorption above 210nm
absorbance
λ (nm)
20MHz carbon-13 NMR spectrum
$CDCl_3$ solution
proton decoupled
δ (ppm)
100MHz proton NMR spectrum
$CDCl_3$ solution
60Hz
δ (ppm)

problem 105

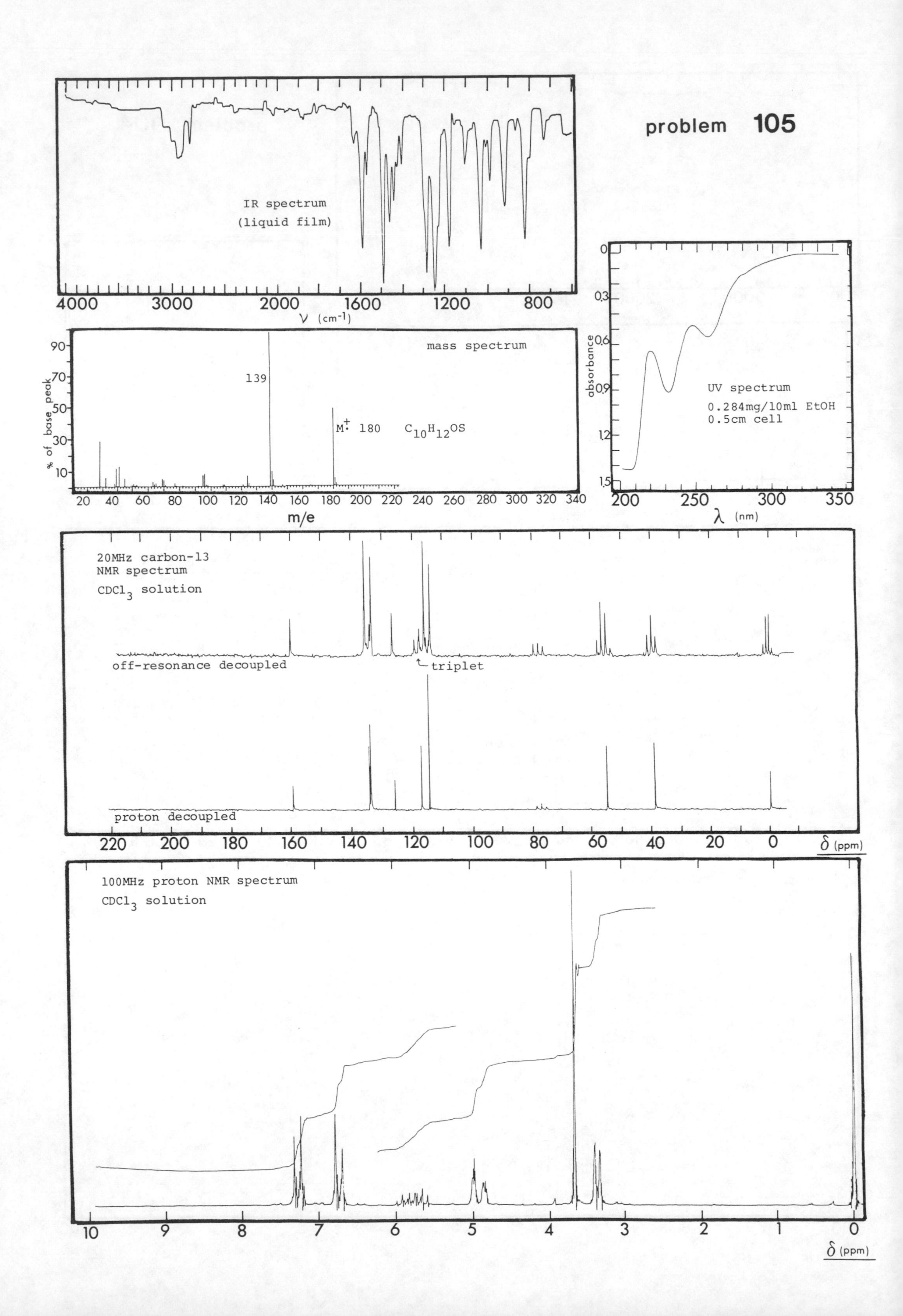

IR spectrum
(liquid film)
4000
3000
2000
1600
1200
800
ν (cm⁻¹)
mass spectrum
139
M⁺· 180
C10H12OS
% of base peak
m/e
UV spectrum
0.284mg/10ml EtOH
0.5cm cell
absorbance
λ (nm)
20MHz carbon-13
NMR spectrum
CDCl3 solution
off-resonance decoupled
triplet
proton decoupled
δ (ppm)
100MHz proton NMR spectrum
CDCl3 solution
δ (ppm)

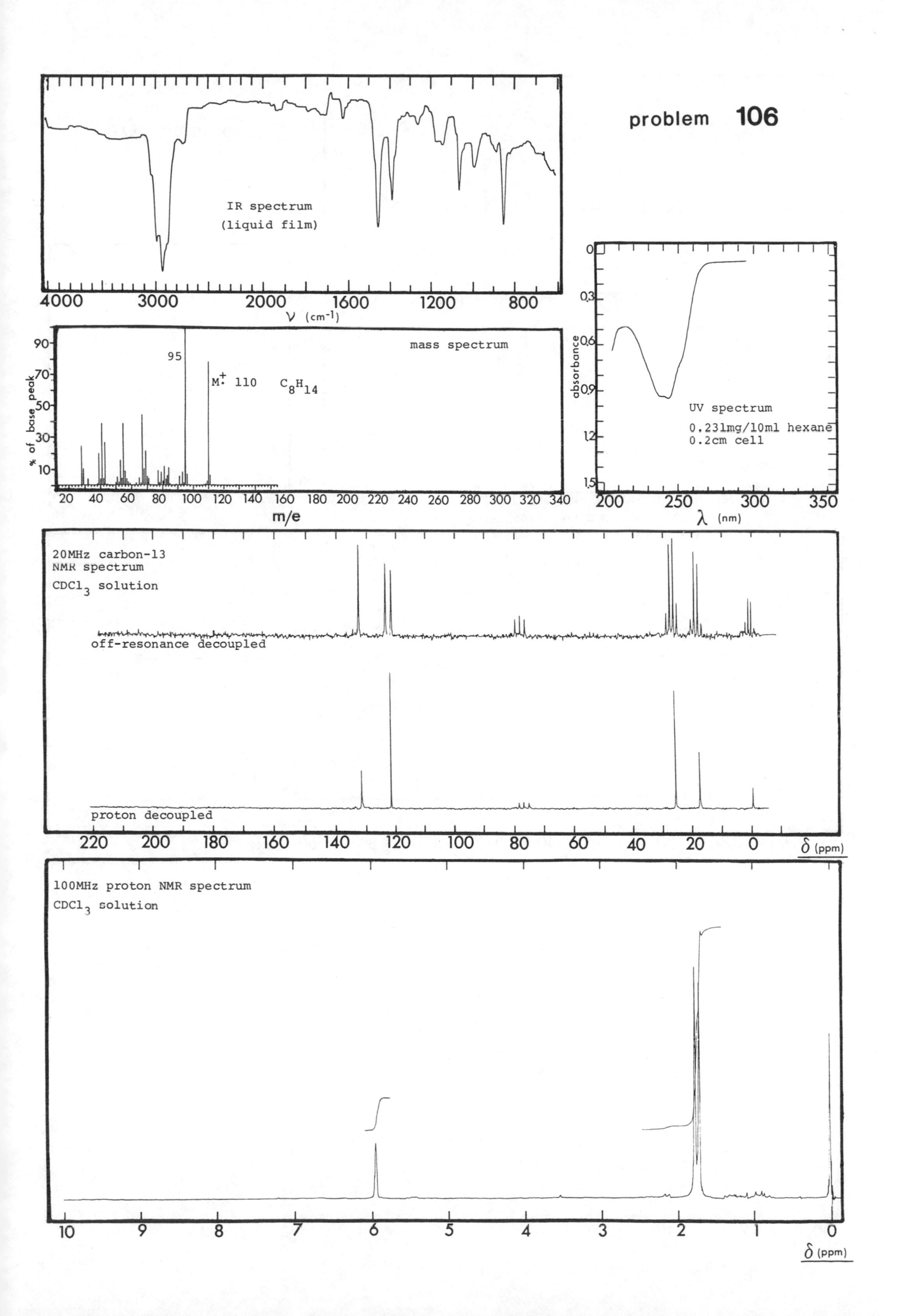

problem 106
IR spectrum
(liquid film)
4000
3000
2000
1600
1200
800
ν (cm-1)
mass spectrum
95
M+ 110
C8H14
% of base peak
m/e
UV spectrum
0.231mg/10ml hexane
0.2cm cell
absorbance
λ (nm)
20MHz carbon-13
NMR spectrum
CDCl3 solution
off-resonance decoupled
proton decoupled
δ (ppm)
100MHz proton NMR spectrum
CDCl3 solution
δ (ppm)

problem 107

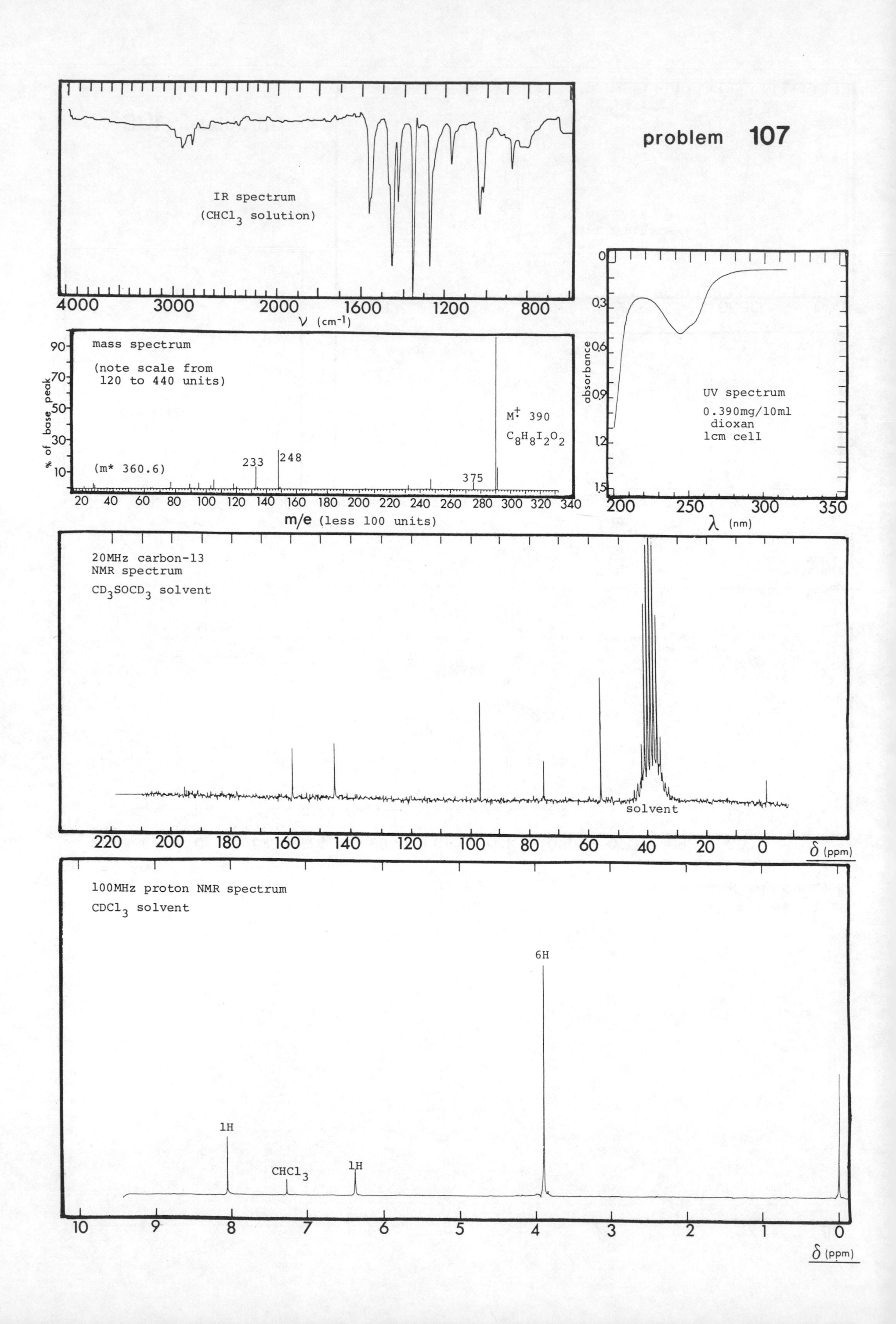

IR spectrum
(CHCl3 solution)
ν (cm-1)
mass spectrum
(note scale from 120 to 440 units)
% of base peak
M+· 390
C8H8I2O2
(m* 360.6)
m/e (less 100 units)
absorbance
UV spectrum
0.390mg/10ml
dioxan
1cm cell
λ (nm)
20MHz carbon-13
NMR spectrum
CD3SOCD3 solvent
solvent
δ (ppm)
100MHz proton NMR spectrum
CDCl3 solvent
6H
1H
CHCl3
1H
δ (ppm)

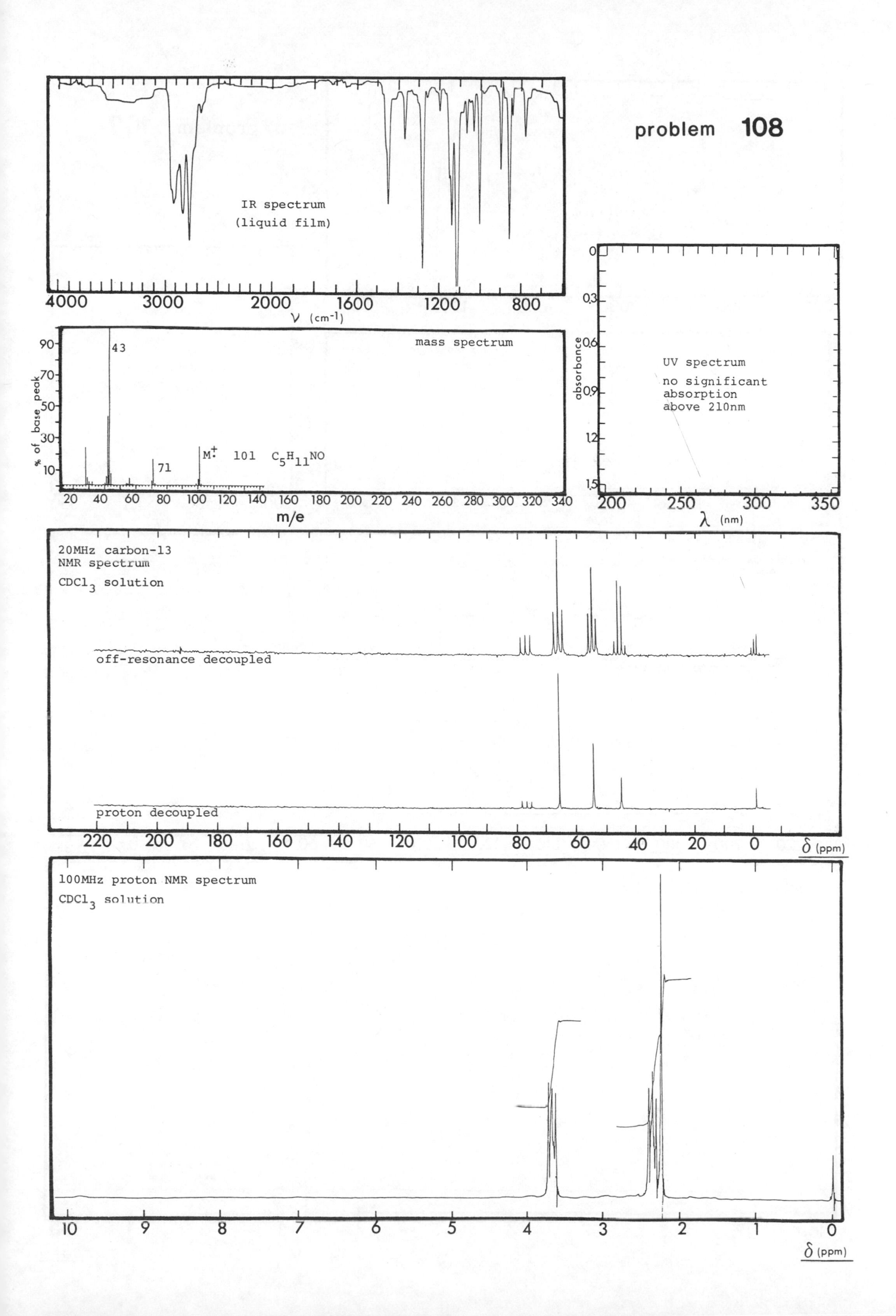

problem 108
IR spectrum
(liquid film)
4000
3000
2000
1600
1200
800
ν (cm-1)
mass spectrum
% of base peak
43
71
M+· 101 C5H11NO
m/e
UV spectrum
no significant absorption above 210nm
absorbance
λ (nm)
20MHz carbon-13 NMR spectrum
CDCl3 solution
off-resonance decoupled
proton decoupled
δ (ppm)
100MHz proton NMR spectrum
CDCl3 solution
δ (ppm)

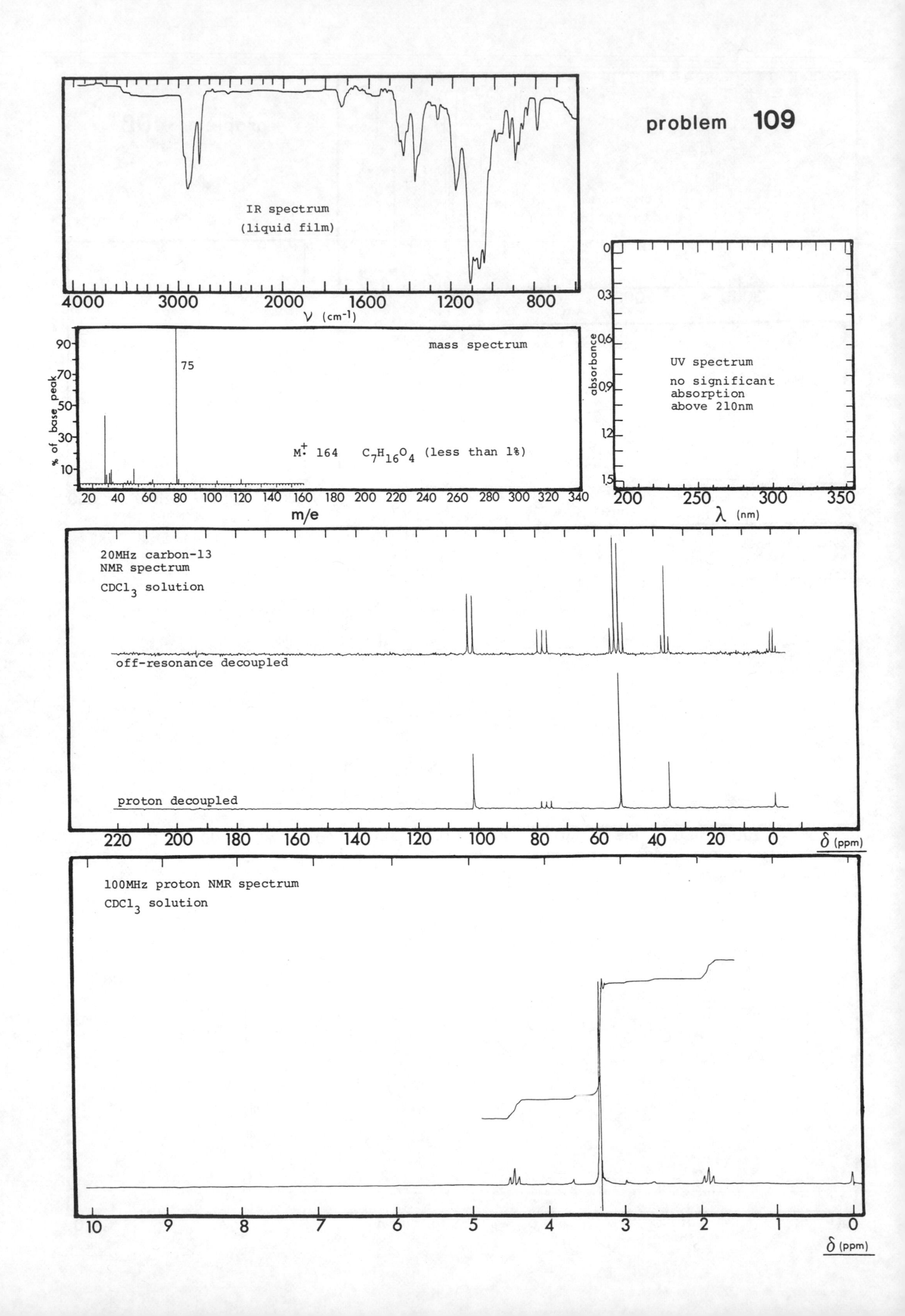
problem 109
IR spectrum
(liquid film)
4000
3000
2000
1600
1200
800
ν (cm-1)
mass spectrum
75
M+· 164 C7H16O4 (less than 1%)
% of base peak
m/e
UV spectrum
no significant
absorption
above 210nm
absorbance
λ (nm)
20MHz carbon-13
NMR spectrum
CDCl3 solution
off-resonance decoupled
proton decoupled
δ (ppm)
100MHz proton NMR spectrum
CDCl3 solution
δ (ppm)

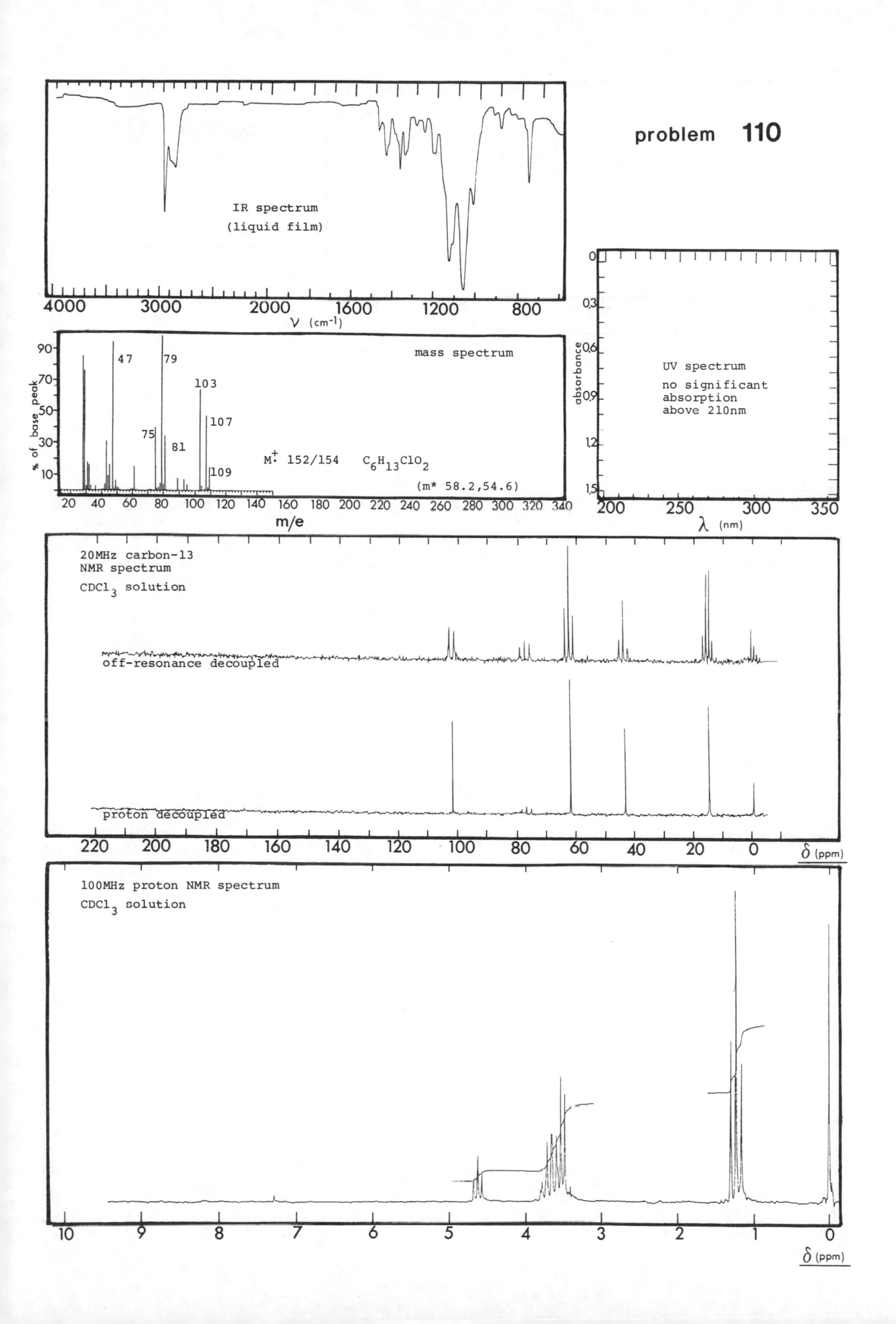
problem 110
IR spectrum
(liquid film)
4000
3000
2000
1600
1200
800
ν (cm-1)
mass spectrum
% of base peak
47
79
103
107
75
81
109
M+ 152/154
C6H13ClO2
(m* 58.2,54.6)
m/e
UV spectrum
no significant
absorption
above 210nm
absorbance
λ (nm)
20MHz carbon-13
NMR spectrum
CDCl3 solution
off-resonance decoupled
proton decoupled
δ (ppm)
100MHz proton NMR spectrum
CDCl3 solution
δ (ppm)

problem **111**

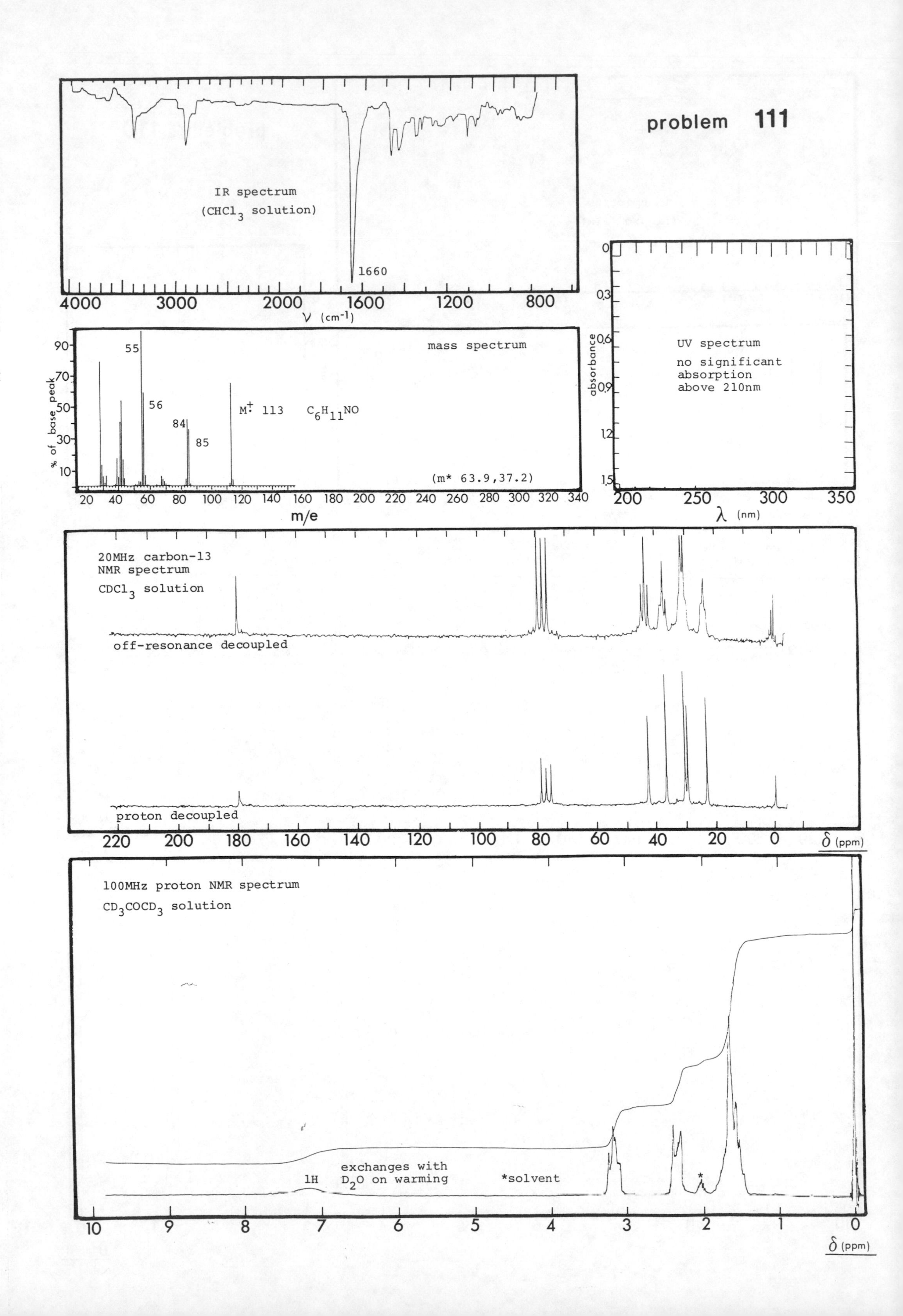

IR spectrum
(CHCl3 solution)
1660
4000
3000
2000
1600
1200
800
ν (cm-1)
mass spectrum
55
56
84
85
M+ 113
C6H11NO
(m* 63.9,37.2)
% of base peak
m/e
UV spectrum
no significant absorption above 210nm
absorbance
λ (nm)
20MHz carbon-13 NMR spectrum
CDCl3 solution
off-resonance decoupled
proton decoupled
δ (ppm)
100MHz proton NMR spectrum
CD3COCD3 solution
1H
exchanges with D2O on warming
*solvent
δ (ppm)

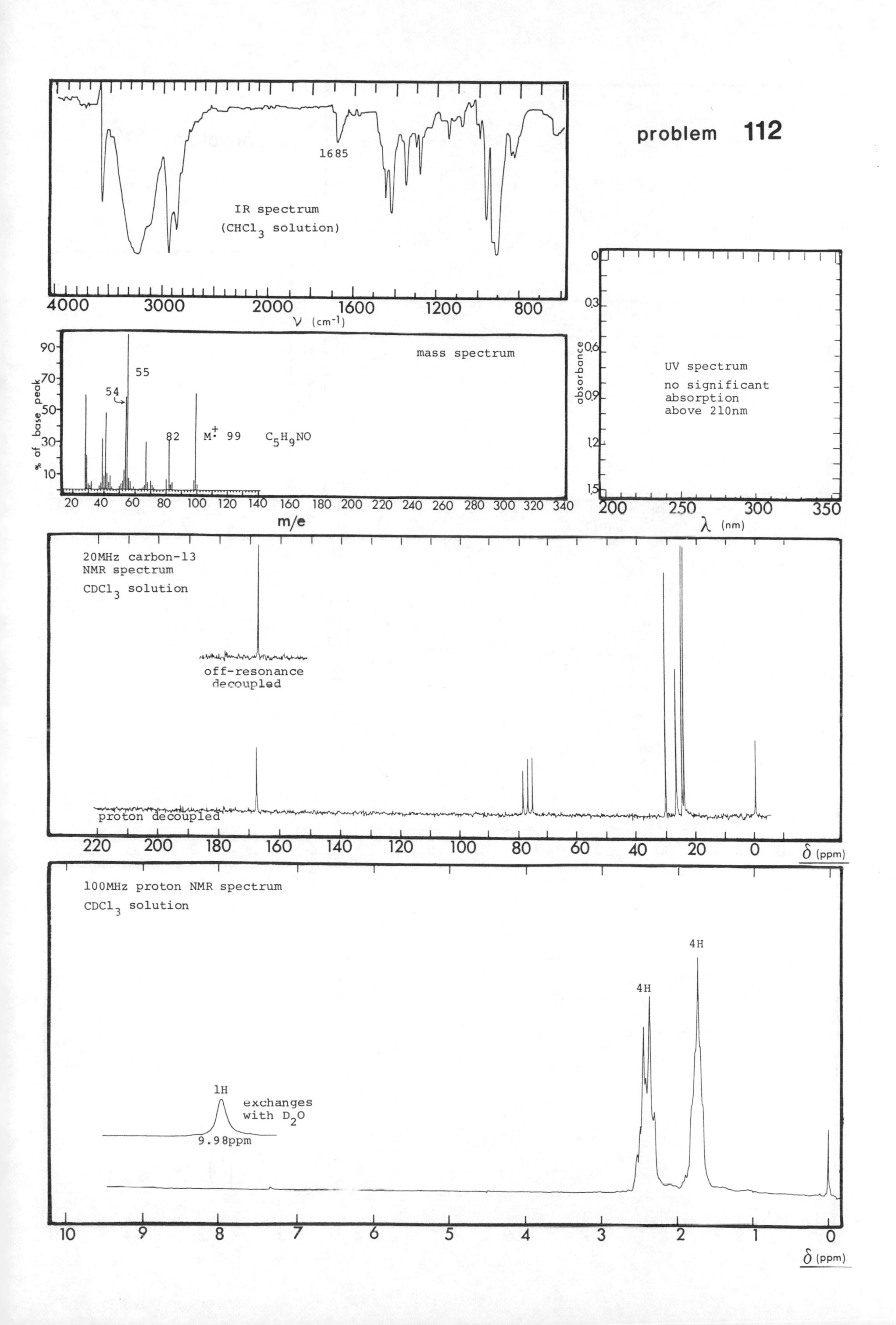

problem 112
IR spectrum
(CHCl3 solution)
1685
4000
3000
2000
1600
1200
800
ν (cm-1)
mass spectrum
55
54
82
M+· 99
C5H9NO
% of base peak
m/e
UV spectrum
no significant
absorption
above 210nm
absorbance
λ (nm)
20MHz carbon-13
NMR spectrum
CDCl3 solution
off-resonance
decoupled
proton decoupled
δ (ppm)
100MHz proton NMR spectrum
CDCl3 solution
4H
4H
1H
exchanges
with D2O
9.98ppm
δ (ppm)

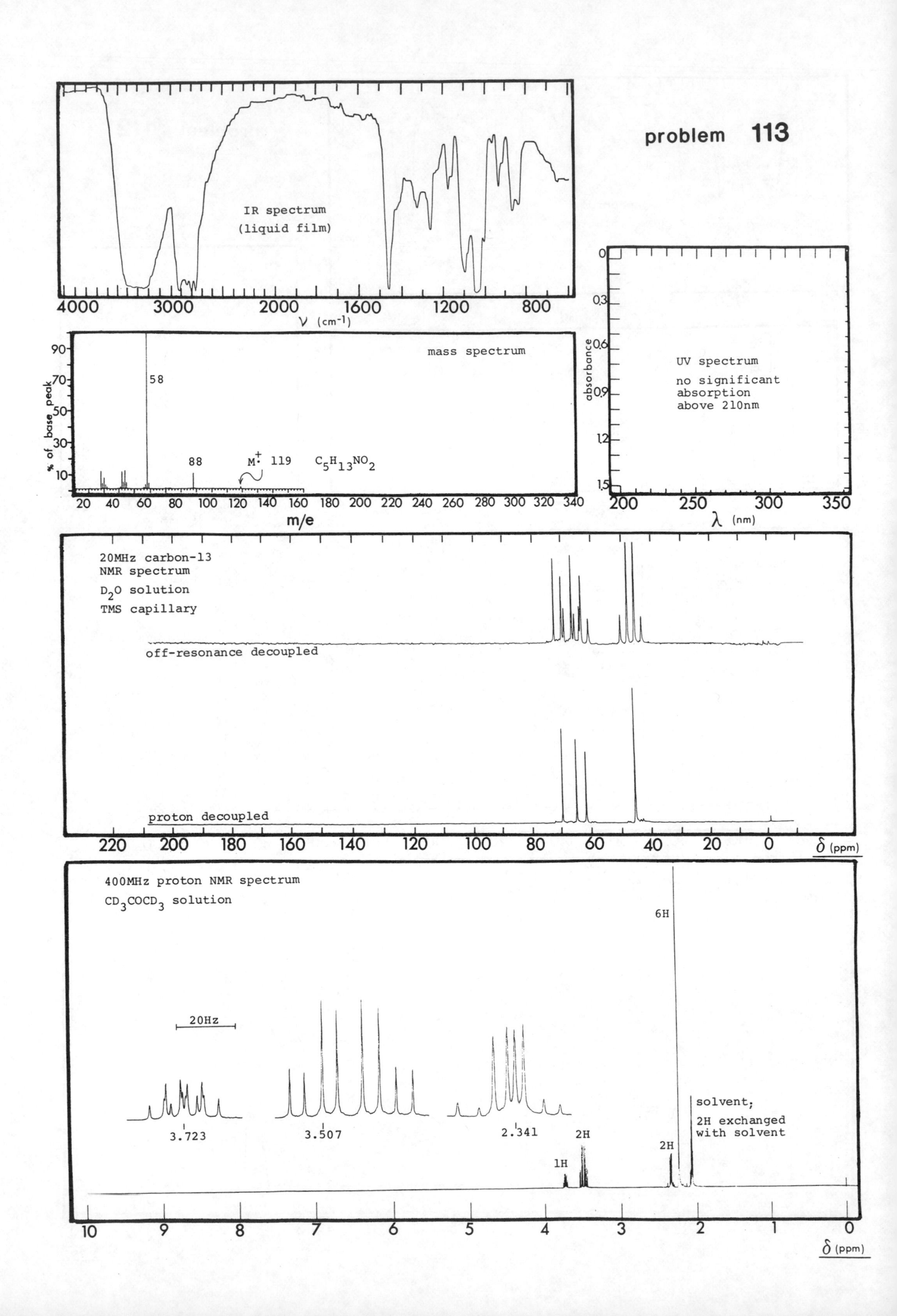

problem 113
IR spectrum
(liquid film)
4000
3000
2000
1600
1200
800
ν (cm-1)
mass spectrum
58
88
M+ 119
C5H13NO2
% of base peak
m/e
UV spectrum
no significant
absorption
above 210nm
absorbance
λ (nm)
20MHz carbon-13
NMR spectrum
D2O solution
TMS capillary
off-resonance decoupled
proton decoupled
δ (ppm)
400MHz proton NMR spectrum
CD3COCD3 solution
6H
20Hz
3.723
3.507
2.341
2H
2H
1H
solvent;
2H exchanged
with solvent
δ (ppm)

problem **114**

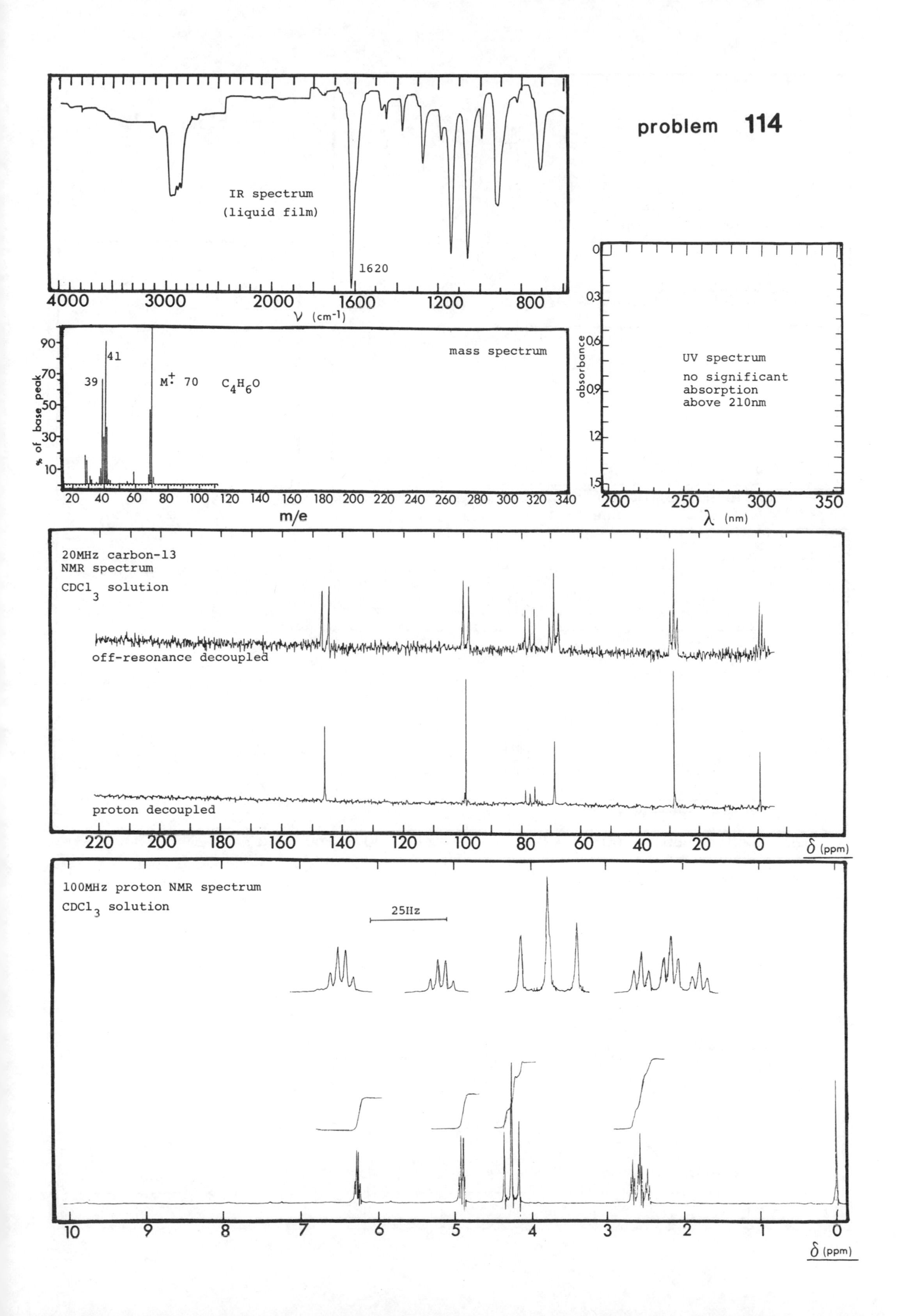

IR spectrum
(liquid film)
1620
4000
3000
2000
1600
1200
800
ν (cm-1)
mass spectrum
% of base peak
41
39
M+· 70
C4H6O
m/e
UV spectrum
no significant absorption above 210nm
absorbance
λ (nm)
20MHz carbon-13 NMR spectrum
CDCl3 solution
off-resonance decoupled
proton decoupled
δ (ppm)
100MHz proton NMR spectrum
CDCl3 solution
25Hz
δ (ppm)

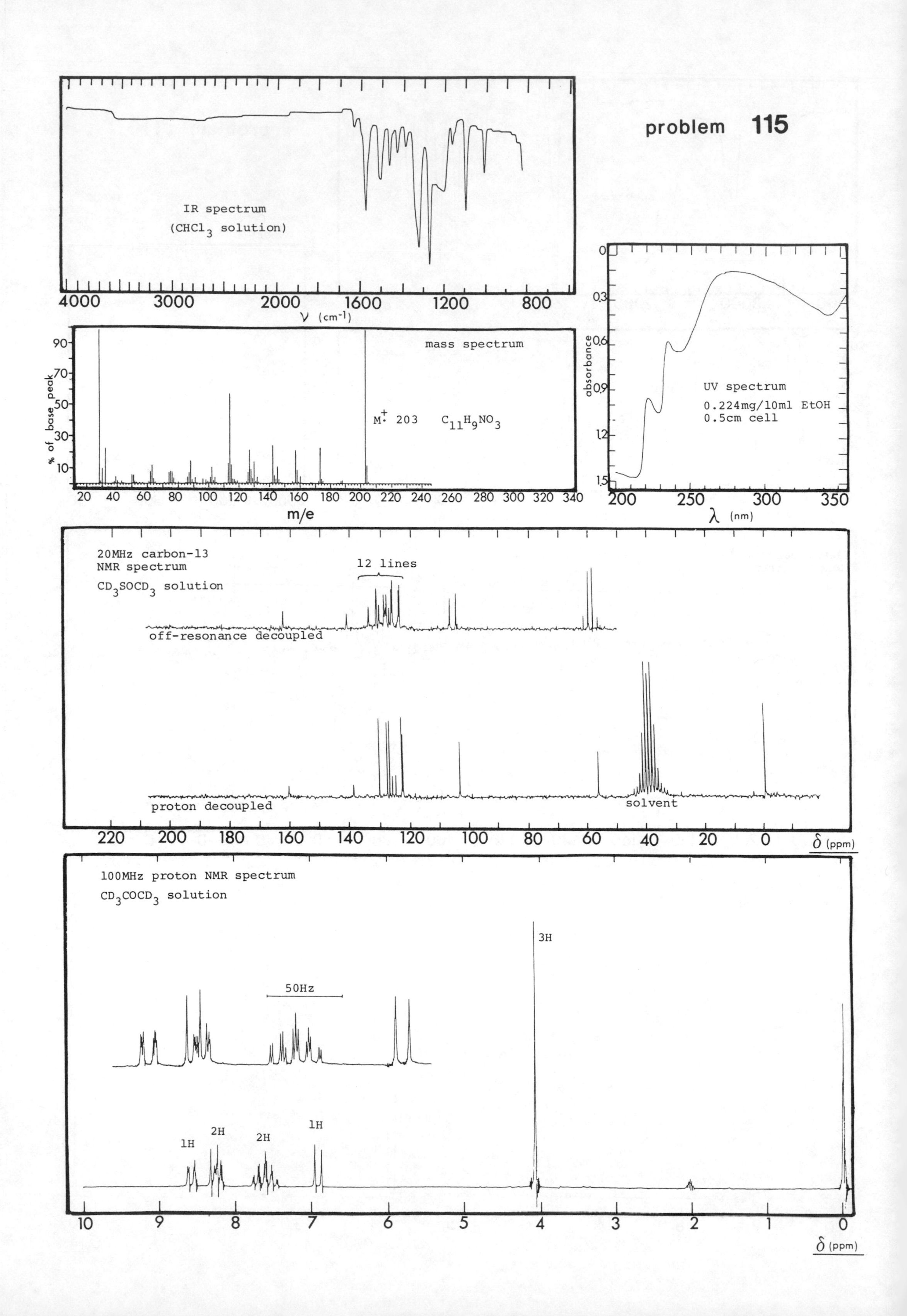
problem 115
IR spectrum
(CHCl3 solution)
4000
3000
2000
1600
1200
800
ν (cm-1)
mass spectrum
M+ 203
C11H9NO3
% of base peak
m/e
UV spectrum
0.224mg/10ml EtOH
0.5cm cell
absorbance
λ (nm)
20MHz carbon-13
NMR spectrum
CD3SOCD3 solution
12 lines
off-resonance decoupled
proton decoupled
solvent
δ (ppm)
100MHz proton NMR spectrum
CD3COCD3 solution
50Hz
3H
1H
2H
2H
1H
δ (ppm)

problem 115

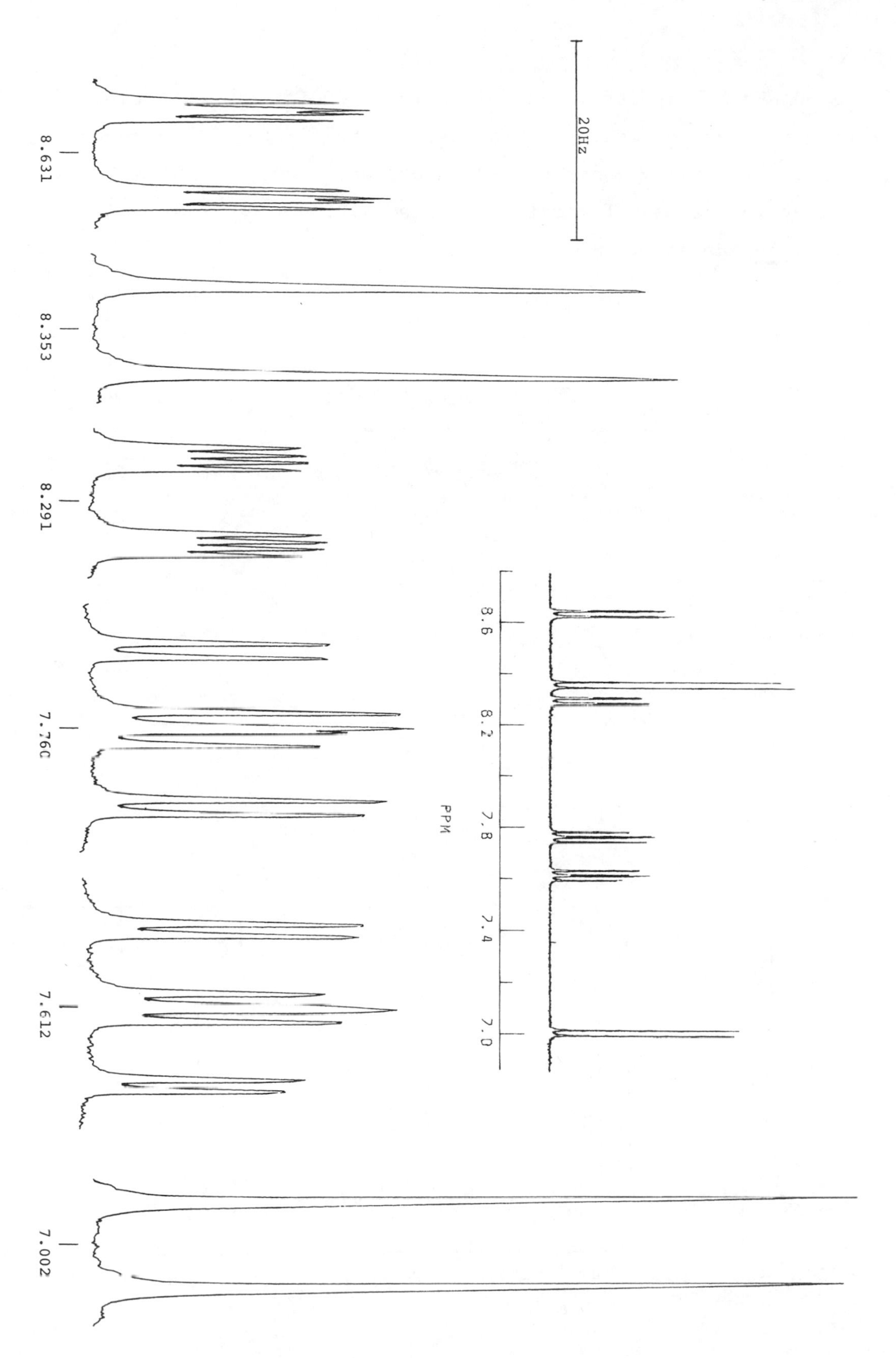

400MHz proton NMR spectrum
20Hz
8.631
8.353
8.291
7.760
7.612
7.002
8.6
8.2
7.8
7.4
7.0
PPM

Portion of the 60 MHz NMR spectrum 2-furoic acid in $CDCl_3$ is shown below. Draw a splitting diagram and analyse this spectrum by first-order methods to give all relevant coupling constants (J in Hz) and chemical shifts (δ in ppm). Justify the use of first-order analysis.

<u>Note</u>: This is a <u>60</u> MHz spectrum

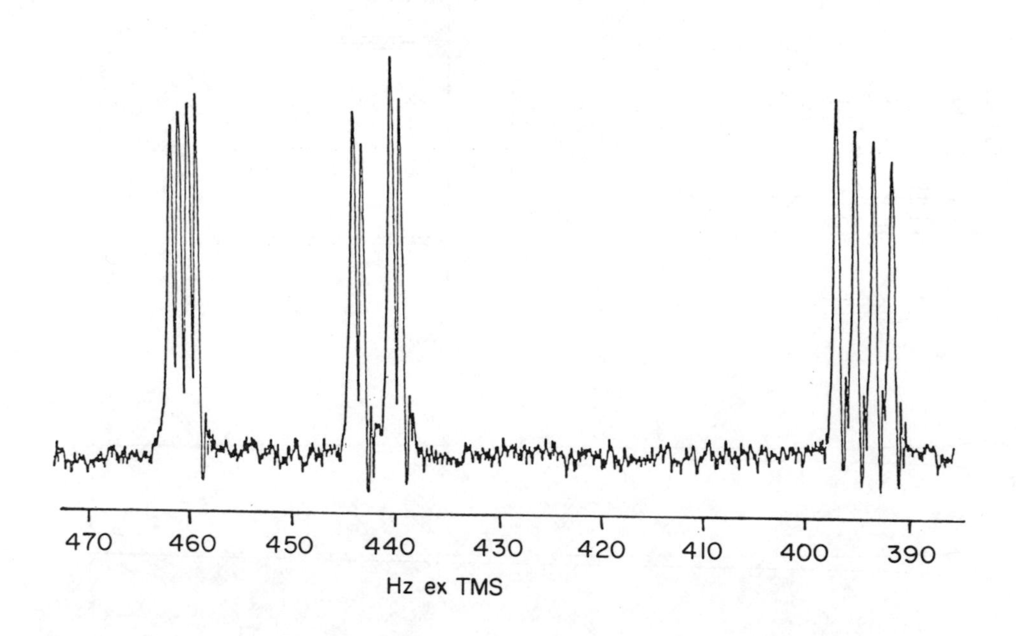

A 100 MHz spectrum of a 3-proton system. Draw a splitting diagram on the Figure and analyse the spectrum by first-order methods to give all relevant coupling constants (J in Hz) and chemical shifts (δ in ppm). Justify the use of first-order analysis.

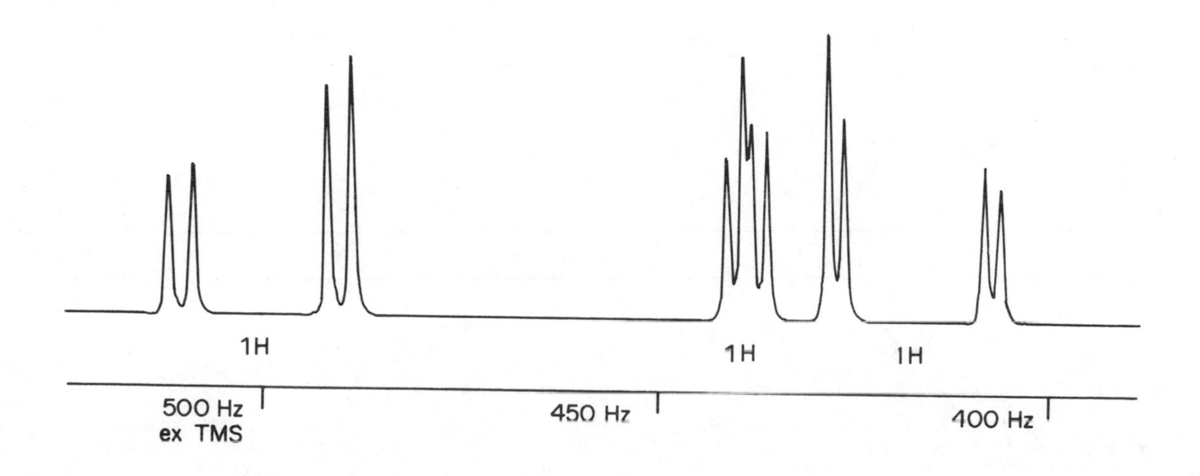

A 100 MHz NMR spectrum of a 4-proton system. Draw a splitting diagram on the Figure and analyse the spectrum by first-order methods to give all the relevant coupling constants (J in Hz) and chemical shifts (δ in ppm). Justify the use of first-order analysis.

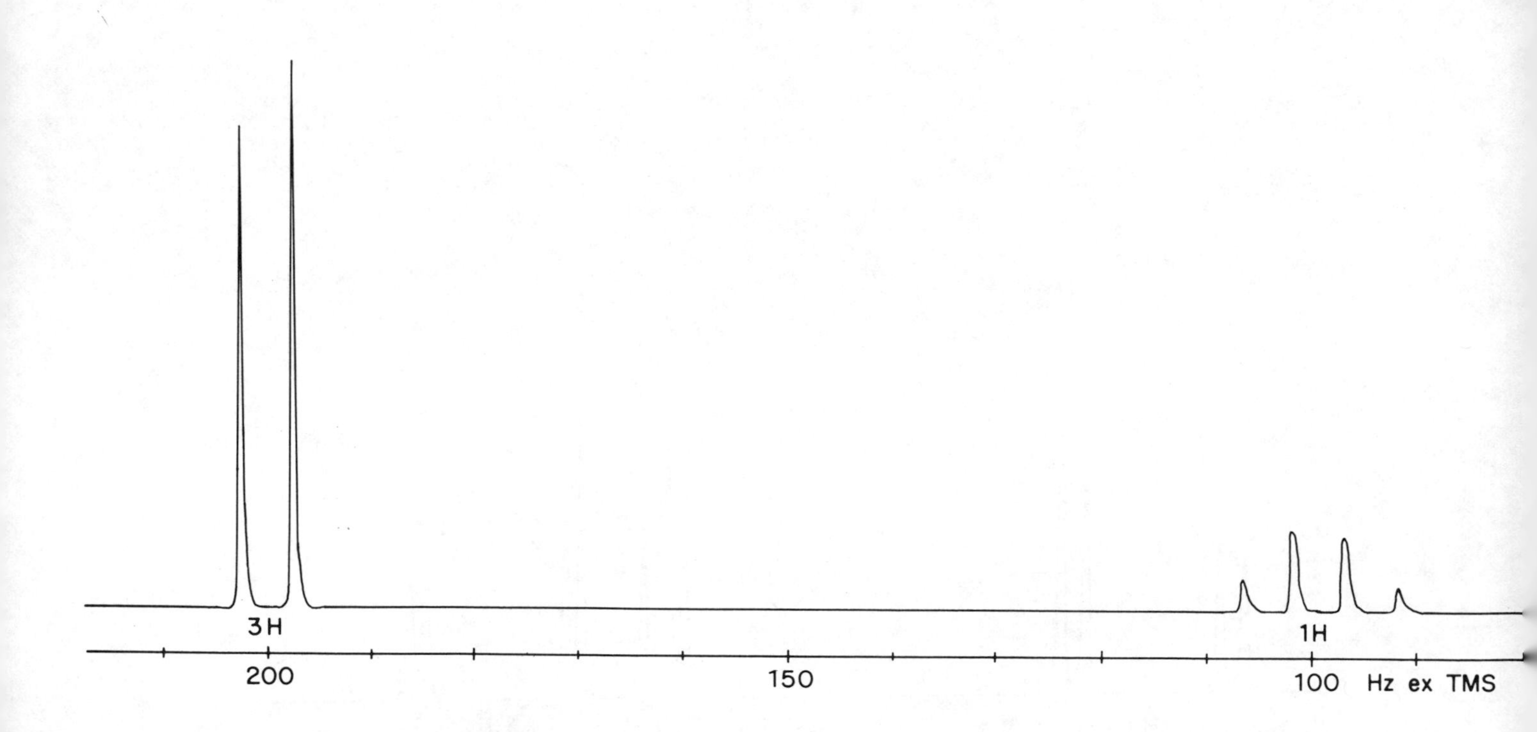

A 100 MHz NMR spectrum of a 4-proton system. Draw a splitting diagram on the Figure and analyse the spectrum by first-order methods to give all relevant coupling constants (J in Hz) and chemical shifts (δ in ppm). Justify the use of first-order analysis.

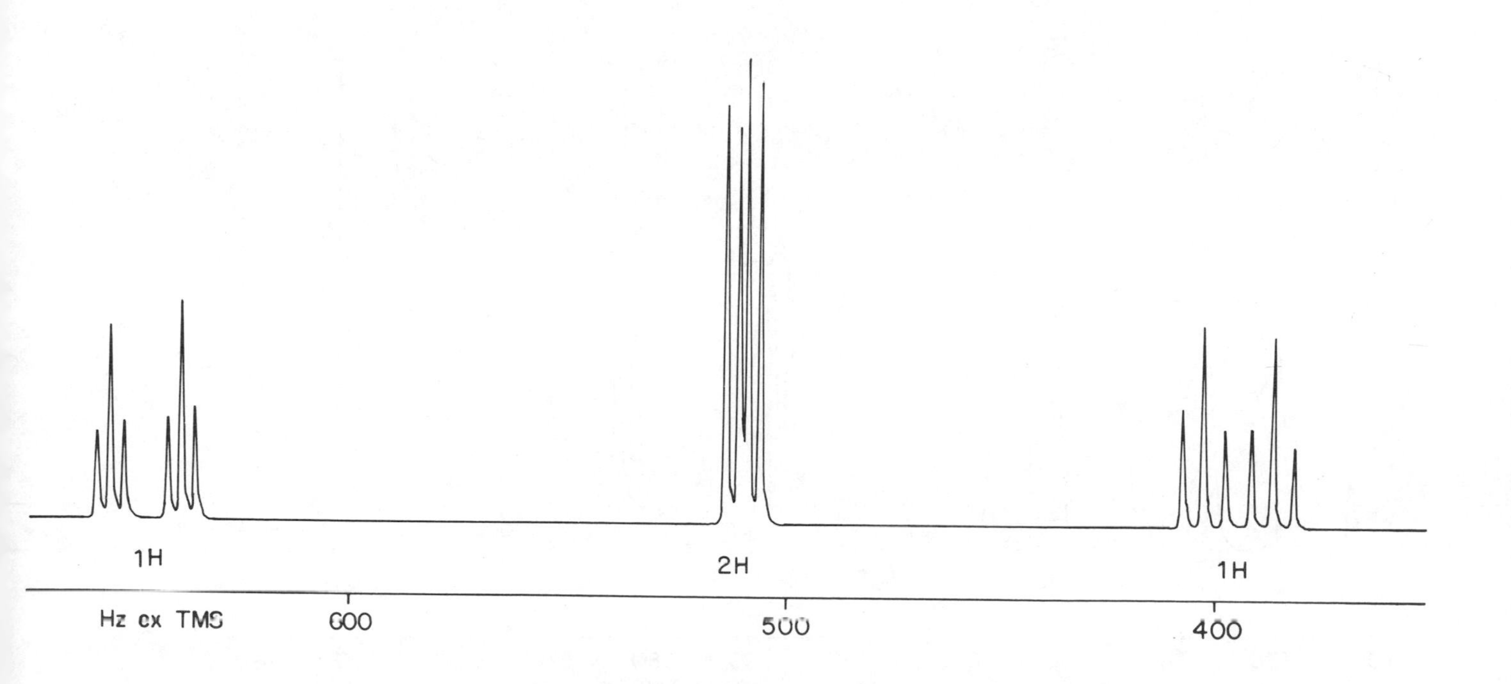

Portion of 100 MHz NMR spectrum of crotonic acid in $CDCl_3$. The upfield part of the spectrum, which is due to the methyl group, is less amplified to fit the page.

(i) Draw a splitting diagram and analyse this spectrum by first-order methods, i.e. extract all relevant coupling constants (J in Hz) and chemical shifts (δ in ppm) by direct measurement. Justify the use of first-order analysis.

(ii) There are certain conventions used for naming spin-systems. Name the spin system responsible for this spectrum.

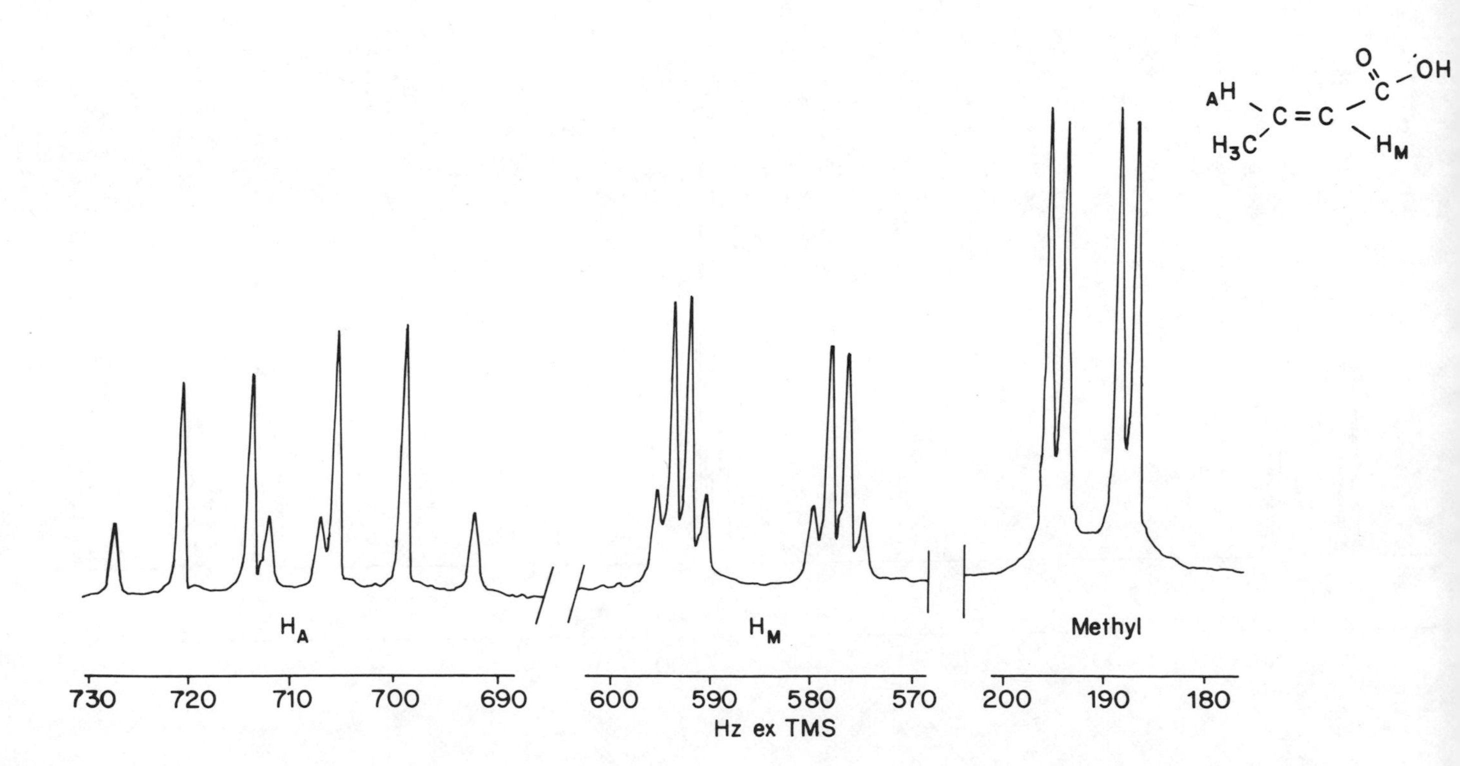

Portion of 100 MHz NMR spectrum of methyl acrylate (5% in C_6D_6) showing resonances due to the olefinic protons H_A, H_B and H_C. (i) Draw a first-order splitting diagram. (ii) Analyse this spectrum by first-order methods, i.e. extract all relevant coupling constants and chemical shifts by direct measurement. (iii) Justify the statement: "This spectrum is really a borderline second-order (strongly coupled) case". Point out the most conspicuous deviation from first-order character in this spectrum. (iv) Assign the three multiplets to H_A, H_B and H_C on the basis of coupling constants only.

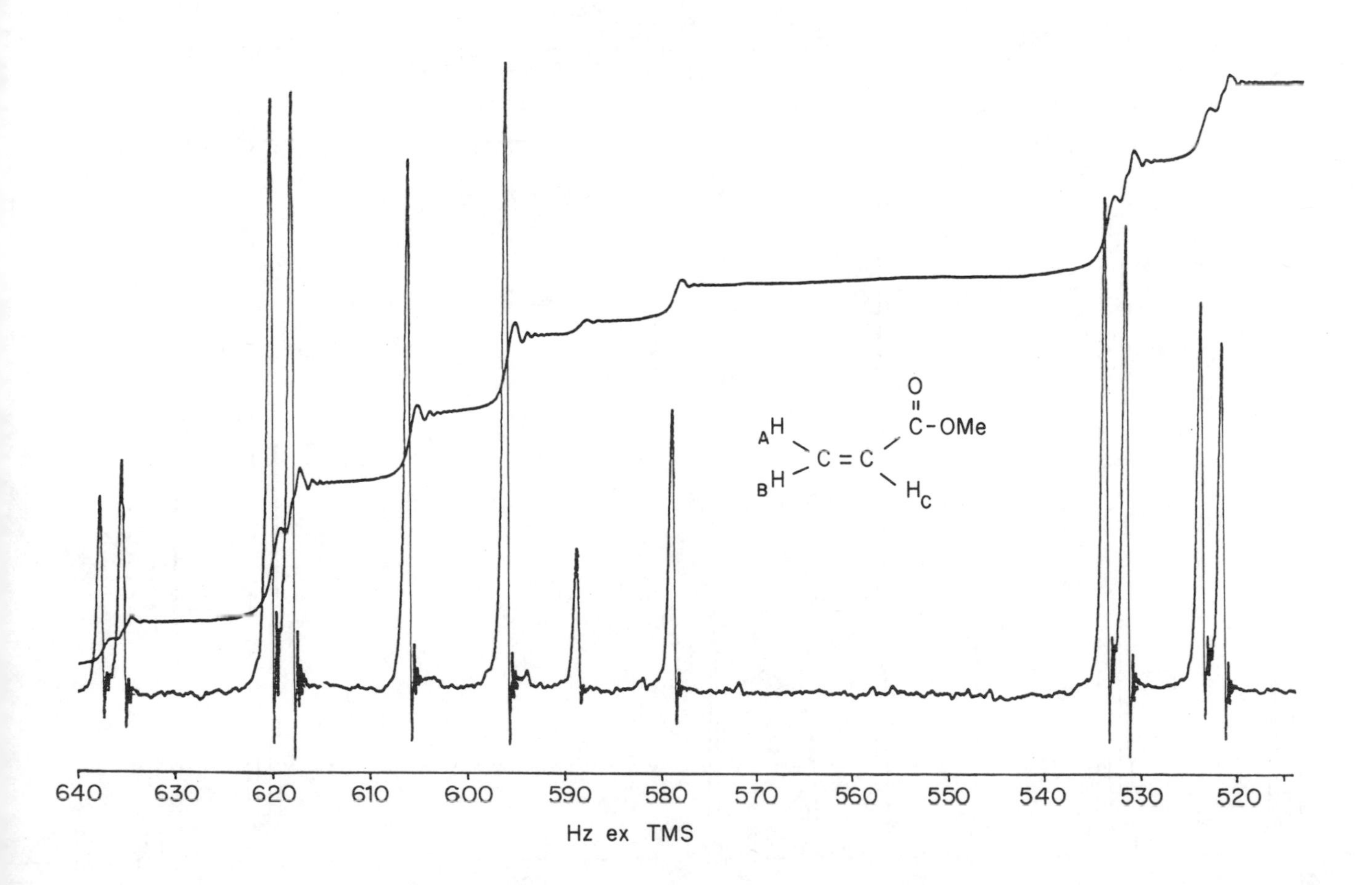

This is the computer-simulated spectrum corresponding to the <u>complete analysis</u> of the spectrum shown in Problem 121, i.e. an analysis in which first-order assumptions were not made. The simulated spectrum fits the experimental spectrum, showing that the analysis was correct. Compare your (first-order) results with the actual solution given here.

```
SPINS   =    3
F(1)    = +    528.500
F(2)    = +    594.531
F(3)    = +    626.093

J(1,2)  =   +    10.539
J(1,3)  =   +     1.589
J(2,3)  =   +    17.278

            PLØT

START   (HZ)   =  +   750.000
FINISH  (HZ)   =  +   500.000
LINE WIDTH     =  +     0.427
SCALE FACT     =  +    80.000
```

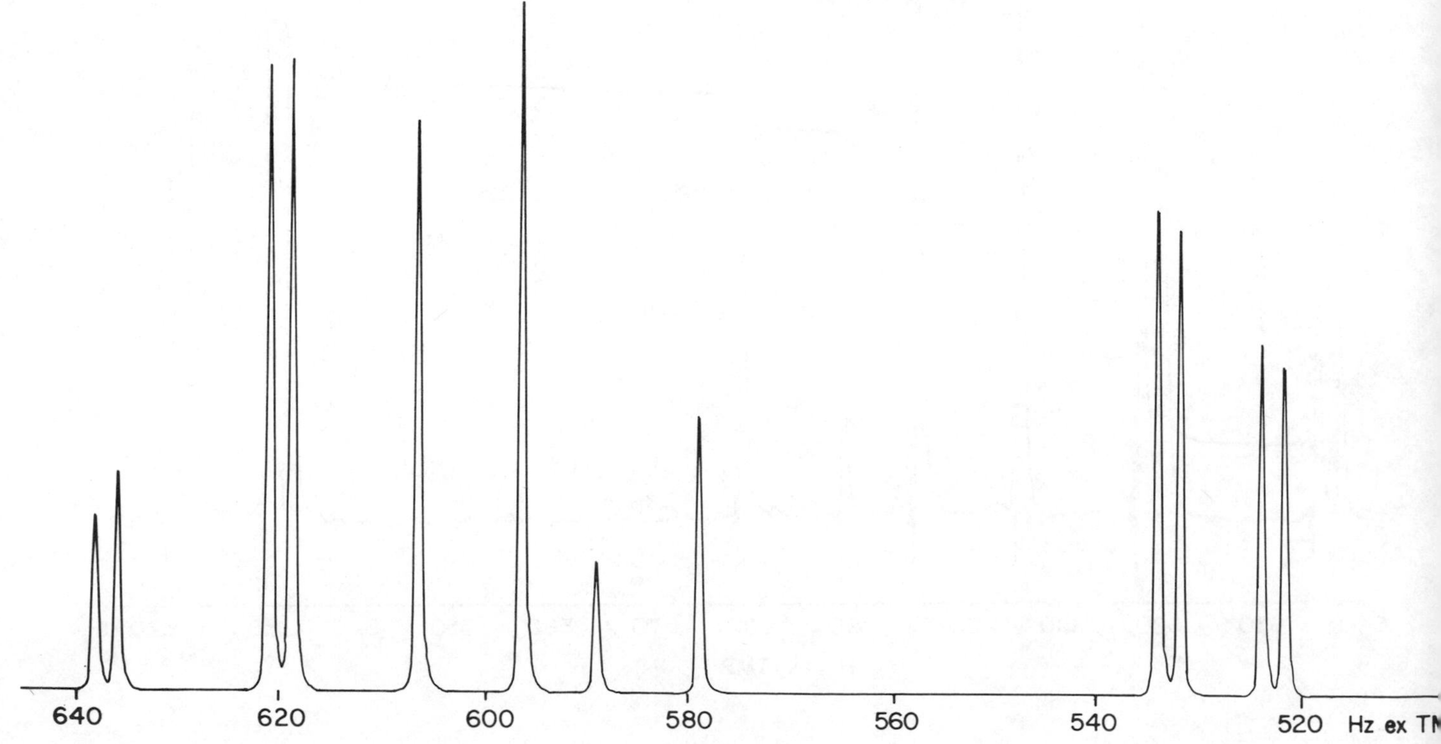

Draw a schematic (line) representation of the pure first-order spectrum (AMX) corresponding to the following parameters: Frequencies (Hz from TMS): $\nu_A = 80$; $\nu_M = 120$; $\nu_X = 200$ Coupling constants (Hz): $J_{AM} = 10$; $J_{AX} = 12$; $J_{MX} = 5$ Ignore small distortions in relative intensities of lines which would, in fact, be apparent in a real spectrum corresponding to the above parameters (not pure first-order). Sketch in "coupling diagrams" above the schematic spectrum to indicate which splittings correspond to which coupling constants. Give the chemical shifts on the δ and τ scales corresponding to the above spectrum obtained with an instrument operating at 60 MHz for protons.

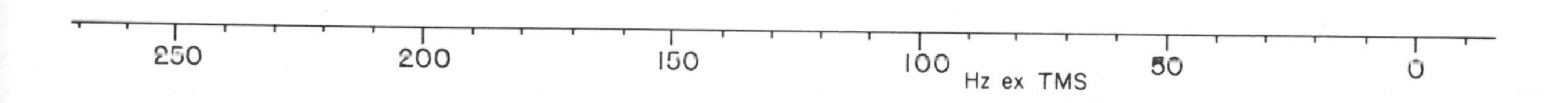

Draw a schematic (line) representation of the pure first-order spectrum (AM_2X) corresponding to the following parameters: Frequencies (Hz from TMS): ν_A=110; ν_M=200; ν_X=290 Coupling constants (Hz): J_{AM}=10; J_{AX}=12; J_{MX}=3. Ignore small distortions in relative intensities of lines which would, in fact, be apparent in a real spectrum corresponding to the above parameters (not pure first-order). Sketch in "splitting diagrams" above the schematic spectrum to indicate which splittings correspond to which coupling constants. Give the chemical shifts on the δ scale corresponding to the above spectrum obtained with an instrument operating at 60 MHz for protons.

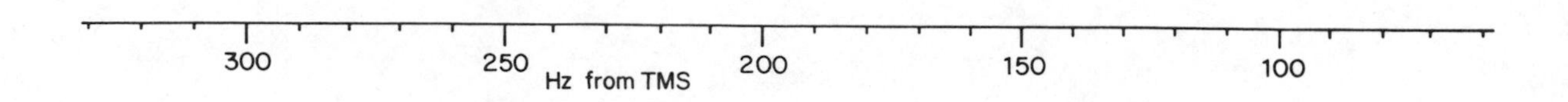

Draw a schematic (line) representation of the pure first-order spectrum (AMX_3) corresponding to the following parameters: Frequencies (Hz from TMS): $\nu_A = 80$; $\nu_M = 220$; $\nu_X = 320$ Coupling constants (Hz): $J_{AM} = 10$; $J_{AX} = 12$; $J_{MX} = 0$ Ignore small distortions in relative intensities of lines which would, in fact be apparent in a real spectrum corresponding to the above parameters (not pure first-order). Sketch in "coupling diagrams" above the schematic spectrum to indicate which splittings correspond to which coupling constants. Give the chemical shifts on the δ and τ scales corresponding to the above spectrum obtained with an instrument operating at 60 MHz for protons.

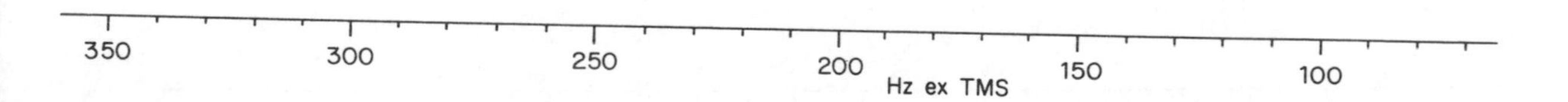

Draw a schematic (line) representation of the pure first-order spectrum (A_2MX) corresponding to the following parameters: Frequencies (Hz from TMS): ν_A=80; ν_M=220; ν_X=320 Coupling constants (Hz): J_{AM}=12; J_{AX}=18; J_{MX}=5. Ignore small distortions in relative intensities of lines which would, in fact be apparent in a real spectrum corresponding to the above parameters (not pure first order). Sketch in "coupling diagrams" above the schematic spectrum to indicate which splittings correspond to which coupling constants. Give the chemical shifts on the δ and τ scales corresponding to the above spectrum. obtained with an instrument operating at 60 MHz for protons.

350 300 250 200 Hz ex TMS 150 100

Draw a schematic (line) representation of the pure first-order spectrum (AMX_2) corresponding to the following parameters: Frequencies (Hz from TMS): ν_A=350; ν_M=240 ν_X=100 Coupling constants (Hz): J_{AM}=10; J_{AX}= 2 ; J_{MX}=6 Ignore small distortions in relative intensities of lines which would, in fact, be apparent in a real spectrum corresponding to the above parameters (not pure first-order). Sketch in "coupling diagrams" above the schematic spectrum to indicate which splittings correspond to which coupling constants. Give the chemical shifts on the δ and τ scales corresponding to the above spectrum obtained with an instrument operating at 60 MHz for protons.

350 300 250 200 Hz ex TMS 150 100

60 MHz NMR spectrum of diethyl ether. Note that the spectrum is calibrated only in parts per million (ppm) ex tetramethylsilane (TMS), i.e., in δ units

(i) Assign the signals due to the $-CH_2-$ and $-CH_3$ groups respectively.

(ii) Obtain the chemical shifts of each group in Hz at 60 MHz, then convert to ppm.

(iii) Obtain the value of the first-order coupling constant J_{CH_2,CH_3} (in Hz).

(iv) Demonstrate that first-order analysis was justified.

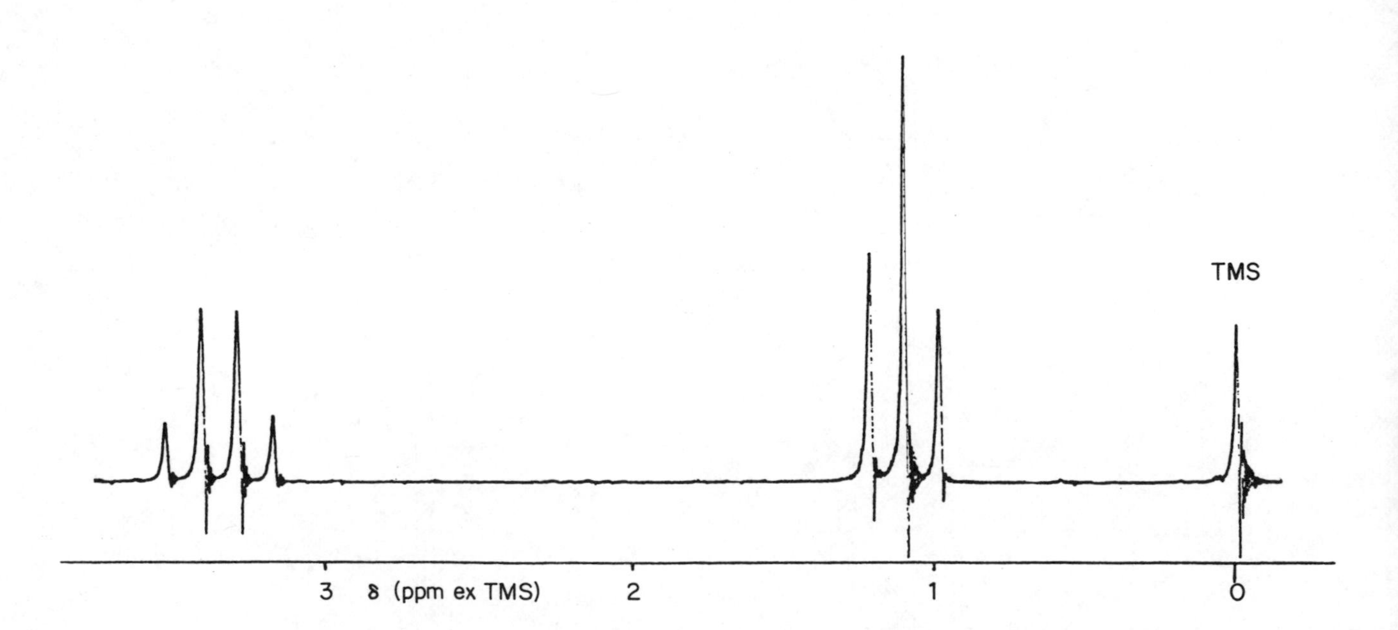

Portion of 100 MHz NMR spectrum of 2-amino-5-chlorobenzoic acid in CD_3OD due to the three aromatic protons.

(i) Draw a splitting diagram and analyse this spectrum by first-order methods, i.e. extract all relevant coupling constants (J in Hz) and chemical shifts (δ in ppm) by direct measurement. Justify the use of first-order analysis.

(ii) Assign the three multiplets to H-3, H-4 and H-6 given: (a) the characteristic ranges for coupling constants between aromatic protons; (b) the fact that H-3 will give rise to the resonance at the highest field due to the strong +R effect of the amino group.

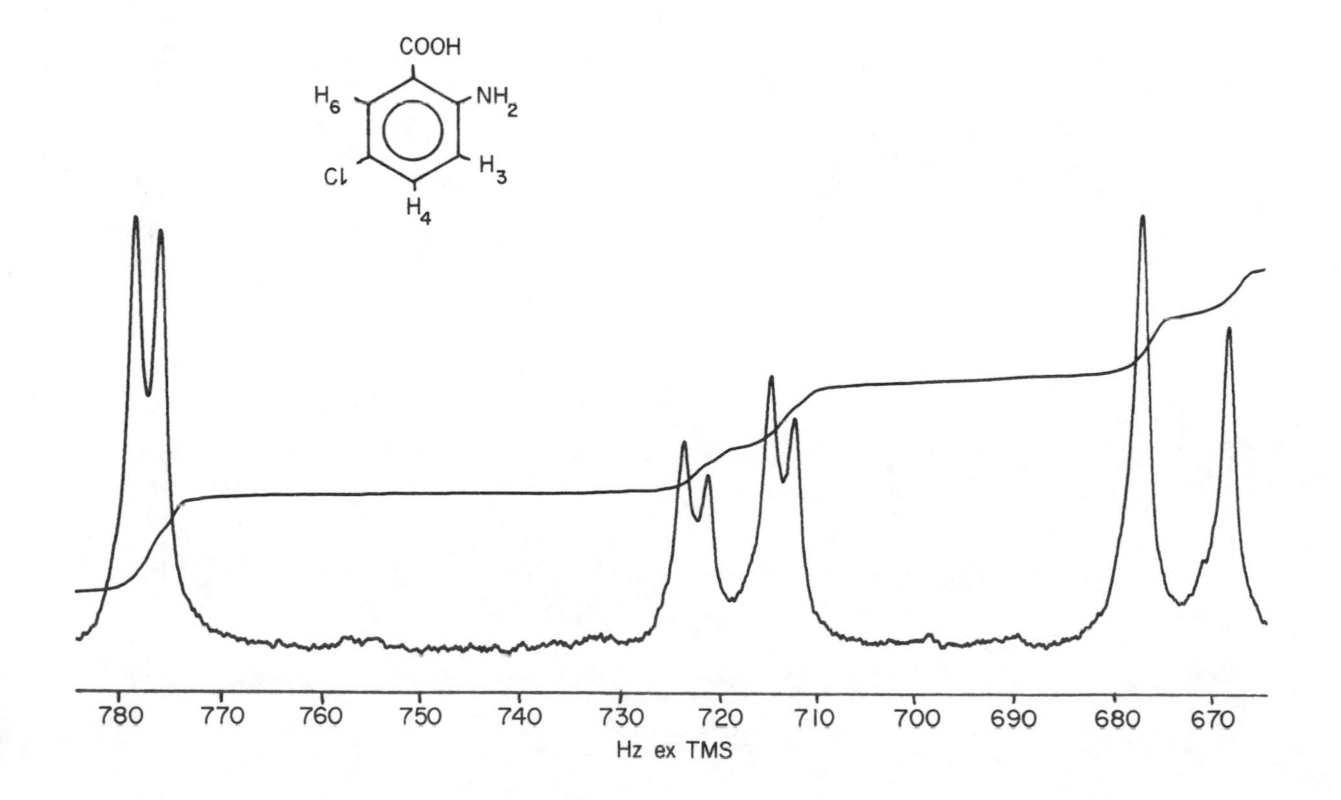

100 MHz NMR spectrum (5% in $CDCl_3$) of an α,β-unsaturated aldehyde C_4H_6O. The relative intensities of the multiplets in regions I, II, III and IV are 1:1:1:3.

(i) Draw a splitting diagram and analyse this spectrum by first-order methods, i.e. extract all relevant coupling constants (J in Hz) and chemical shifts (δ in ppm) by direct measurements. Justify the use of first-order analysis.

(ii) Use these results to obtain the structure (including stereochemistry about the double bond) of the compound.

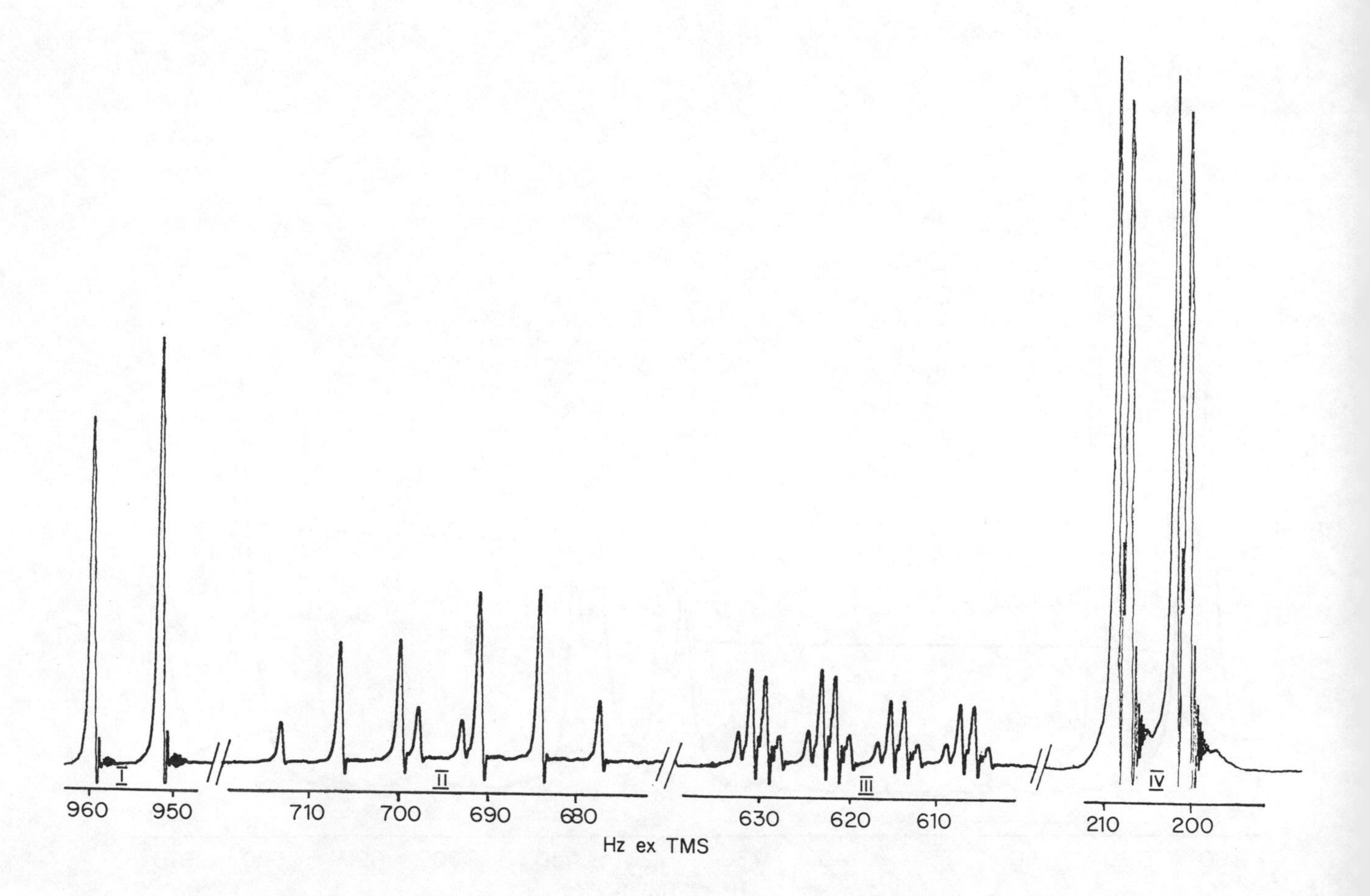

Portion of 90 MHz NMR spectrum of one of the six possible isomeric dibromoanilines (aromatic protons only). Determine the structure of this compound using arguments based on symmetry and magnitudes of spin-spin coupling constants.

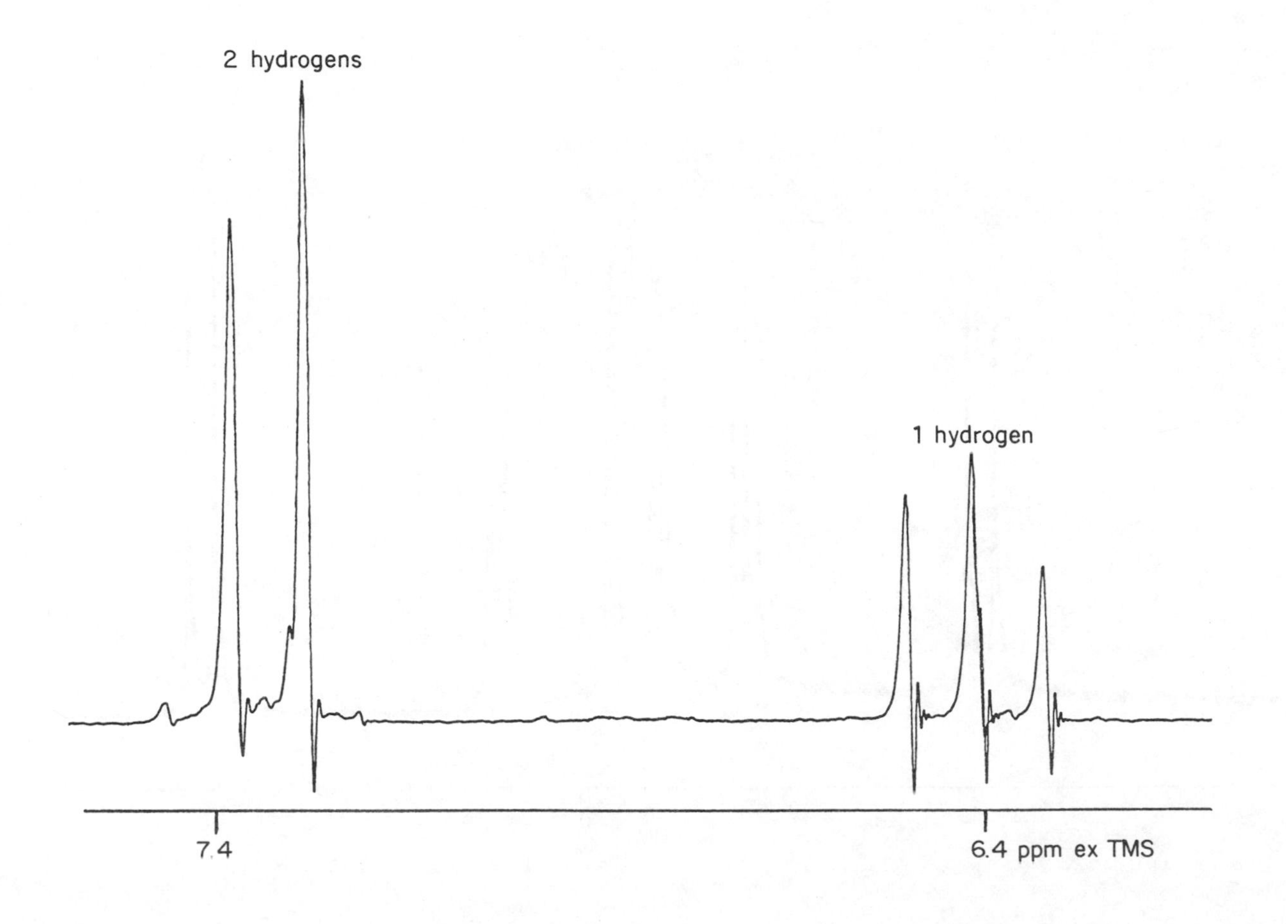

400 MHz ^{1}H NMR spectrum of a hydroxycinnamic acid in $CDCl_3$ after exchange with D_2O. Determine which of the 6 possible isomers is present.

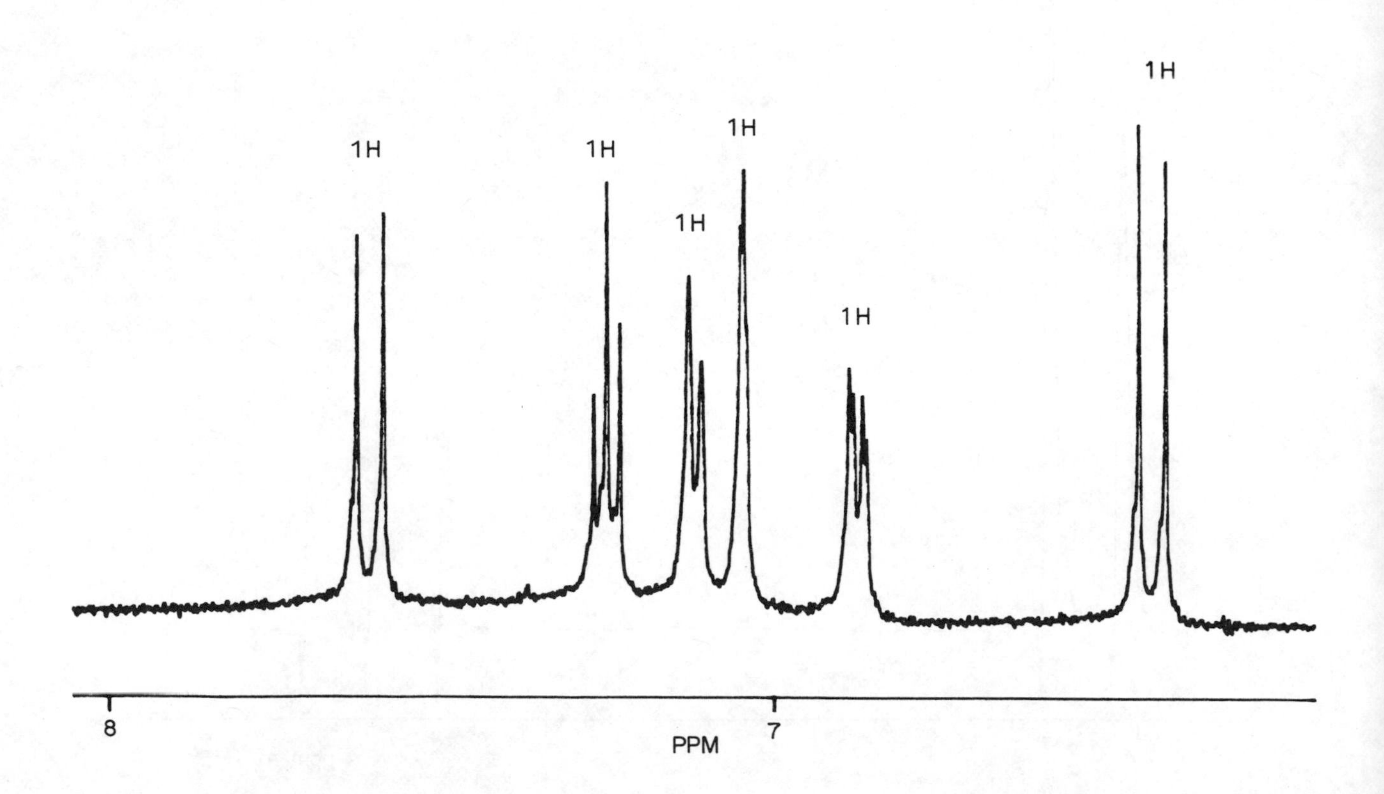

SUBJECT INDEX

Key:

^{13}C NMR = Carbon 13 nuclear magnetic resonance spectrometry

^{1}H NMR = Proton nuclear magnetic resonance spectrometry

IR = Infrared spectrometry

MS = Mass spectrometry

UV = Ultraviolet spectrometry